Biology and Ecology of
Anguillid Eels

Biology and Ecology of Anguillid Eels

Editor

Takaomi Arai
Professor
Environmental and Life Sciences Programme
Faculty of Science
Universiti Brunei Darussalam

CRC Press
Taylor & Francis Group
Boca Raton London New York

CRC Press is an imprint of the
Taylor & Francis Group, an **informa** business

A SCIENCE PUBLISHERS BOOK

Cover Illustration reproduced by kind courtesy of Dr. Don Jellyman.

CRC Press
Taylor & Francis Group
6000 Broken Sound Parkway NW, Suite 300
Boca Raton, FL 33487-2742

First issued in paperback 2021

© 2016 by Taylor & Francis Group, LLC
CRC Press is an imprint of Taylor & Francis Group, an Informa business

No claim to original U.S. Government works

Version Date: 20151207

ISBN 13: 978-0-367-78316-7 (pbk)
ISBN 13: 978-1-4822-5515-7 (hbk)

Library of Congress Cataloging-in-Publication Data

Names: Arai, Takaomi, author.
Title: Biology and ecology of anguillid eels / Takaomi Arai [and others]..
Description: Boca Raton : Taylor & Francis, 2016. | Includes bibliographical references and index.
Identifiers: LCCN 2015040833 | ISBN 9781482255157 (hardcover : alk. paper)
Subjects: LCSH: Anguilla (Fish)
Classification: LCC QL638.A55 A73 2016 | DDC 597/.432--dc23
LC record available at http://lccn.loc.gov/2015040833

Visit the Taylor & Francis Web site at
http://www.taylorandfrancis.com

and the CRC Press Web site at
http://www.crcpress.com

Preface

Freshwater eels of the genus *Anguilla* are exotic animals, and despite a large number of scientific studies conducted on eels, crucial aspects of their biology still remain a mystery. No one has yet observed eels spawning in the natural environment, as spawning areas are located in the open ocean. Since these eels have a unique catadromous life history and are used as food resources, they are one of the most important eel families from a conservation standpoint. Recently, however, the juvenile population has declined dramatically, in all probability due to the use of wild juveniles in cultivation. These eels are captured in estuaries and almost all of the world's total eel supply comes from aquaculture. Therefore, the supply of eel resources for human consumption is completely dependent on wild catch.

The causes of decline in stock and recruitment are not well understood. Overfishing, habitat loss, migration barriers, increased natural predation, parasitism, ocean climate variation, and pollution might all have an impact. Ever since the European eel was listed by CITES under Appendix II and came under protection in March 2009, and since the export/import ban was issued by the EU in 2010, the international trade of juvenile eels has changed drastically. Most recently, the Japanese and American eels were added to the IUCN's list with an endangered classification, suggesting that they too face a high risk of extinction. In the future, even more eel species could be classified as endangered unless there is comprehensive conservation and protection. Therefore, it is highly possible that we may not be able to see such a unique animal on the earth in the near future.

In this book, the latest information regarding the biology and ecology of the freshwater eel genus *Anguilla*, has been compiled. It will be useful not only for scientists and researchers but also for students and members of the public in order to better understand the eel itself. I would like to express sincere thanks to all authors for this book. I hope the book will be a milestone for fish biology and ecology.

May 2015   **Takaomi Arai**

Contents

Taxonomy and Distribution

Takaomi Arai

Introduction

Freshwater eels of the genus *Anguilla* Schrank (1798) consist of 16 species, three of which are further divided into two subspecies (Ege 1939; Castle and Williamson 1974; Watanabe 2003; Watanabe et al. 2004, 2005, 2009a, 2013, 2014a,b). All of these species make oceanic migrations at various scales from a few hundred to thousands of kilometers (Arai 2014). They are globally distributed in temperate, tropical, and subtropical areas and are considered to be prevalent nearly worldwide, except for the land masses adjacent to the South Atlantic and the eastern Pacific oceans (Ege 1939). They all have a catadromous life-history strategy, spawning in remote tropical seas with larvae that are transported back by currents to their nursery grounds in freshwater or coastal areas. Tesch (2003) divided the freshwater eels into tropical and temperate species according to their geographic distributions. The tropical and temperate eels consist of 11 and 5 species, respectively. The basic biology of temperate eels is generally known, but information about the tropical eels that comprise two-thirds of all anguillid eel species is much more limited. A recent molecular genetic study indicated that the genus *Anguilla* originated in the deep ocean (Inoue et al. 2010). Furthermore, studies have revealed that tropical eels are the most basal species originating in the Indonesian region and that freshwater eels radiated out from the tropics to colonise the temperate regions (Minegishi et al. 2005).

Environmental and Life Sciences Programme, Faculty of Science, Universiti Brunei Darussalam, Jalan Tungku Link, Gadong, BE 1410, Brunei Darussalam.
E-mail: takaomi.arai@ubd.edu.bn

The only comprehensive revisions of the genus *Anguilla* were conducted by Kaup (1856), Günther (1870), and Ege (1939) using morphological analyses. In the last revision by Ege (1939), this genus was classified into 16 species, three of which were divided into two subspecies, i.e., *A. celebesensis* Kaup (1856); *A. interioris* Whitley (1938); *A. ancestralis* Ege (1939); *A. megastoma* Kaup (1856); *A. nebulosa nebulosa* McClelland (1844); *A. nebulosa labiata* Peters (1852); *A. marmorata* Quoy and Gaimard (1824); *A. reinhardtii* Steindachner (1867); *A. borneensis* Popta (1924); *A. japonica* Temminck and Schlegel (1846); *A. rostrata* Lesueur (1817); *A. anguilla* Linnaeus (1758); *A. dieffenbachii* Gray (1842); *A. mossambica* Peters (1852); *A. bicolor pacifica* Schmidt (1928); *A. bicolor bicolor* McClelland (1844); *A. obscura* Günther (1872); *A. australis australis* Richardson (1841) and *A. australis schmidtii* Phillips (1925) (Table 1). However, 35 years later, Castle and Williamson (1974) reported that based on morphological

Table 1. Comparison of anguillid eel taxonomies by Ege (1939) and the Latest.

Ege (1939)	Latest*
A. celebesensis	*A. celebesensis*
A. interioris	*A. interioris*
A. ancestralis	*A. megastoma*
A. megastoma	*A. bengalensis bengalensis*
A. nebulosa nebulosa	*A. bengalensis labiata*
A. nebulosa labiata	*A. marmorata*
A. marmorata	*A. reinhardtii*
A. reinhardtii	*A. borneensis*
A. borneensis	*A. japonica*
A. japonica	*A. rostrata*
A. rostrata	*A. anguilla*
A. anguilla	*A. dieffenbachii*
A. dieffenbachii	*A. mossambica*
A. mossambica	*A. luzonensis*
A. bicolor pacifica	*A. bicolor pacifica*
A. bicolor bicolor	*A. bicolor bicolor*
A. obscura	*A. obscura*
A. australis australis	*A. australis australis*
A. australis schmidtii	*A. australis schmidtii*

*: Referred from Castle and Williamson (1974), Watanabe (2003) and Watanabe et al. (2004, 2005, 2009a, 2013, 2014a,b).

analysis, *A. ancestralis*, which was described by Ege (1939) using only glass eels, was a synonym of *A. celebesensis*. Furthermore, *A. nebulosa nebulosa* and *A. nebulosa labiata* are synonyms of *A. bengalensis bengalensis* (Gray 1831) and *A. bengalensis labiata* (Peters 1852) respectively. Although *A. borneensis* was thought to be a synonym of *A. malgumora* (Kaup 1856), it was determined that *A. borneensis* is the valid species after re-examination of morphological characters, and *A. malgumora* is instead suggested to be a junior synonym of *A. borneensis* (Watanabe et al. 2014a).

Ege's (1939) systematics have long been widely accepted by biologists. However, Watanabe et al. (2004) suggested that the morphological characters described by Ege (1939) were not sufficient to classify all species of this genus and that geographic distribution was an indispensable character for Ege's taxonomy of freshwater eels. When only morphological characters were used, the freshwater eels could be classified into only four groups (Watanabe et al. 2004). The geographic distribution of each species is not a suitable taxonomic character of the freshwater eels because it overlaps among species and is highly variable depending on the environment (Shapovalov et al. 1959; Jellyman et al. 1996). In this regard, Ege's (1939) taxonomy is not robust. Furthermore, the only complete morphological key for species identification of the genus *Anguilla* was provided by Ege (1939), but this has been suggested to be unsatisfactory due to critical underestimates of the intraspecific variation among the characters (Aoyama et al. 2000a). Watanabe et al. (2005) examined molecular genetic data in combination with morphological information in order to evaluate Ege's (1939) taxonomy of the genus *Anguilla* from a different perspective. The results suggested that the present morphologically based taxonomy of the freshwater eels proposed by Ege (1939) is basically sound, and 15 taxa were confirmed within the genus *Anguilla* (Watanabe et al. 2005).

Recently, two new species, *Anguilla luzonensis* (Watanabe et al. 2009a) and *A. huangi* (Teng et al. 2009) were discovered almost simultaneously in the Cagayan River system on the northern Luzon Island of the Philippines and from glass eels collected from the Cagayan River estuary that were reared in a culture pond in Taiwan, respectively. However, further study suggested that *A. huangi* is a junior synonym of *A. luzonensis* (Watanabe et al. 2013). Thus, the genus *Anguilla* is currently recognised as being comprised of 16 species (Ege 1939; Castle and Williamson 1974; Watanabe et al. 2004, 2005, 2009a, 2014b), although three of these are further divided into two subspecies (Ege 1939).

This chapter summarizes the latest information about the taxonomy and distribution of the genus *Anguilla*. The information should contribute towards further biological and ecological studies in the genus. Most of the content in this chapter is cited and referenced from the latest taxonomical studies by Watanabe (2003), Watanabe et al. (2004, 2005, 2009a, 2013, 2014b).

Morphological identification

Ege (1939) examined a total of 25,265 specimens, which included 12,793 adults and 12,472 juveniles (glass eels and elvers). Although a great number of specimens were examined, Ege's raw data have never been located (Silfvergrip 2009), and many statistical analyses have therefore not been possible to perform, as they require data sets in a matrix shape. However, recent taxonomic studies by Watanabe et al. (2004, 2005, 2009a, 2013, 2014b) have provided useful morphological characters in combination with molecular analysis.

There are some technical requirements for identification of freshwater eels of unknown origin (Silfvergrip 2009). Morphological identification of glass eels requires a binocular stereoscope ("stereomicroscope") and a sharp-jawed vernier caliper graded to 0.1 mm. Specimens larger than 200 mm in total length also require low-voltage radiograph equipment (Silfvergrip 2009).

There is no comprehensive information available for the identification and comparison of eel eggs using morphology (Castle 1984), but in general, the eggs of *Anguilla* resemble large clupeid eggs, e.g., like those of herring (McGowan and Berry 1984). Identification of anguillid eggs and leptocephali requires molecular analysis (Aoyama et al. 2007).

Identification of eel samples using morphology is often the fastest means of eel identification. Based on the ano-dorsal length (ADL) as a percentage of total length (TL), the glass eels and elvers can first classify as either long-finned or short-finned (Ege 1939; Castle and Williamson 1974; Tabeta et al. 1976a; Arai et al. 1999) without counting the vertebrae. However, further identification for unknown origin samples is difficult, especially in case of the tropical anguillid species that occur sympatrically in a single habitat. Even in known origin samples, it is hard to identify the species using morphological characters only due to the overlap of such characters among different species (Ege 1939; Watanabe et al. 2004). To determine the species composition of the tropical eels, Arai et al. (1999) examined a total of 21,633 glass eels collected in the North Sulawesi of Indonesia. According to Ege (1939) and Castle and Williamson (1974), 1 short-finned eel, *A. bicolor pacifica*, and 3 long-finned eels, *A. celebesensis*, *A. marmorata* and *A. borneensis*, are known in the area. However, *A. celebesensis* and *A. borneensis* are difficult to distinguish by their vertebral characteristics alone (Tabeta et al. 1976a). Thus, Arai et al. (1999) examined the species composition using molecular analysis. Because juvenile eels such as glass eels and elvers develop less colour markings on their skin, and because the representative morphological characters are scarce and less developed when compared to those of adult specimens (Tabeta et al. 1976a; Arai et al. 1999; Silfvergrip 2009), molecular analysis in combination with morphological analysis should be conducted for precise species identification. Silfvergrip (2009) suggested that intact specimens smaller than 200 mm TL require a sharp-legged vernier caliper graded to 0.1 mm and a binocular stereomicroscope for identification.

Regarding species identification of adult specimens, Ege's (1939) systematics have long been widely accepted. However, Watanabe et al. (2004) found that the morphological characters described by Ege (1939) were not sufficient to classify all species of this genus and that geographic distribution was an indispensable character for his taxonomy of these eels. When only morphological characters were used, the freshwater eels could be classified into only four groups (Watanabe et al. 2004). The geographic distribution of each species alone is not a suitable taxonomic character of the freshwater eels because it overlaps among species and is highly variable depending on the environment (Shapovalov et al. 1959; Jellyman et al. 1996). Recently, freshwater eels have been transported around the world for aquaculture. The prevalent trade in glass and young eels has resulted in many reports of accidental or incidental introduction of several exotic species of eels into natural river systems or the sea in some areas (Skinner 1971; Tabeta et al. 1976b, 1977; McCosker 1989; Sasai et al. 2001; Okamura et al. 2008; Arai et al. 2009). In this regard, Ege's (1939) taxonomy is not robust. Furthermore, the only complete morphological key for species identification of the genus *Anguilla* was provided by Ege (1939), but this has been suggested to be unsatisfactory due to critical underestimates of the intraspecific variation among the characters (Aoyama et al. 2000a; Watanabe et al. 2004). Based on the morphological characters such as colour markings, maxillary bands, position of the dorsal fin origin and vertebral counts in combination with genetic clustering, the genus *Anguilla* is recognised as being comprised of 16 species (Ege 1939; Castle and Williamson 1974; Watanabe 2003; Watanabe et al. 2004, 2005, 2009a, 2013).

Molecular identification

Although investigations of morphological characteristics in freshwater eels as well as other taxa are fundamental and important as conventional identification methods, the characteristics are not necessarily useful in identifying the eels. Consequently, this situation requires a critical evaluation of the present classification of the genus *Anguilla* by Ege (1939). It is important to find new defining characteristics for the taxonomy of freshwater eels that are valid regardless of the geographical distribution from which an eel is collected or obtained. Watanabe et al. (2005) found that molecular genetic characteristics are useful in understanding the taxonomy of the genus *Anguilla*. They studied the mitochondrial 16S ribosomal RNA domain (16S rRNA) as a new characteristic for the taxonomy of the freshwater eels. This relatively conservative gene is frequently used for evolutionary studies at the species or genus level in fishes (Meyer 1993). Aoyama et al. (2000b) also suggested that the 16S rRNA was appropriate for identifying eels at the species level.

Recently, the anguillid eels found in the Peninsular Malaysia were identified, using a morphological analysis, as *Anguilla bengalensis bengalensis* and *A. bicolor bicolor* and that identification was further validated by an analysis of the eels' mitochondrial cytochrome oxidase subunit I (COI) and 16S ribosomal RNA (16S rRNA) sequences (Arai and Wong 2016). Previous studies had reported the occurrence of the tropical eel species *A. marmorata* in Peninsular Malaysia without validating the identification genetically. After re-examination of a number of key morphological characteristics of a preserved sample of *A. marmorata*, the sample from Peninsular Malaysia was identified as *A. bengalensis bengalensis*. This was also the first recorded occurrence of *A. bengalensis bengalensis* in Malaysian waters that was confirmed by both morphological and molecular genetic analyses. Although one other sample was identified as *A. borneensis* on the basis of key morphological characteristics, molecular genetic analyses showed that the sample was actually *A. bicolor bicolor*. These results indicate the difficulty of accurately identifying tropical eels solely on the basis of morphological analyses, due to the sympatric distribution of a number of closely related eels.

When using molecular identification of freshwater eels, Watanabe et al. (2005) suggested that because of potentially high levels of variation in population structure, large sample sizes should be used when developing genetic techniques for species identification. They analysed the 16S rRNA gene from 8–66 specimens for each species and found a considerable amount of genetic variation both within and among species (Watanabe et al. 2005). These results suggest that exact species identification may be difficult due to greater sequence variations if a small sample size is utilized. However, once these techniques are refined using larger sample sizes, it should be possible for any eel specimen to be identified using genetic characters, regardless of the stage of growth (Watanabe et al. 2005).

Recent progress in taxonomy studies

Before Watanabe (2003) and Watanabe et al. (2004, 2005, 2009a, 2013, 2014b) examined Ege's work on the taxonomy of the genus *Anguilla*, the comprehensive revisions of the genus *Anguilla* were conducted by Kaup (1856), Günther (1870), and Ege (1939) using morphological analyses only. In the last revision by Ege (1939), the genus was classified into 16 species, three of which were divided into two subspecies. Thereafter, a total of 1736 specimens comprising of 1501 specimens collected worldwide and 235 specimens obtained from museums around the world were examined for both morphological and molecular analyses (Watanabe 2003; Watanabe et al. 2004, 2005, 2009a, 2013, 2014b). Specimen collection and research is ongoing at present.

Watanabe (2003) and Watanabe et al. (2004) have summarized the history of the taxonomy of the genus *Anguilla*, problems with Ege's taxonomy revision and have introduced a new taxonomy. The results suggested that there is not a major difference between the morphologically based taxonomy of the freshwater eels proposed by Ege (1939) and that by Watanabe (2003) and Watanabe et al. (2004). The results of morphological and meristic measurements, counts, and observations suggested that skin with or without variegated marking, wide or narrow maxillary bands of teeth, and short or long dorsal fin were important as valid characteristics for the taxonomy of the genus *Anguilla* (Watanabe 2003; Watanabe et al. 2004). Using those three characteristics, Watanabe (2003) and Watanabe et al. (2004) suggested that the freshwater eels could be classified into four groups as follows (Fig. 1):

Group 1. Variegated body marking, with broad maxillary bands of teeth
Group 2. Variegated body marking, with narrow maxillary bands of teeth
Group 3. Without variegated body markings and with a long dorsal fin
Group 4. Without variegated body markings and with a short dorsal fin

Group 1. Variegated body marking, with broad maxillary bands of teeth

Group 2. Variegated body marking, with narrow maxillary bands of teeth

Group 3. Without variegated body markings and with a long dorsal fin

Group 4. Without variegated body markings and with a short dorsal fin

Fig. 1. Classification of the genus *Anguilla* with no consideration of the geographic distribution by Watanabe (2003) and Watanabe et al. (2004). The eels could be classified into four groups.

In groups 3 and 4, the classifications were same as those of Ege's groups of III and IV as follows. Groups 1 and 2 were partly different from those of Ege's groups of I and II as follows.

I. Variegated species with broad, undivided maxillary and mandibular bands of teeth.
II. Variegated species with a toothless longitudinal groove in the maxillary and mandibular bands of teeth.
III. Species without variegated markings and with a long dorsal fin.
IV. Species without variegated markings and with a short dorsal fin.

The main difference between Ege (1939) and Watanabe (2003) and Watanabe et al. (2004) in their classifications were the characteristics of the teeth. Watanabe (2003) and Watanabe et al. (2004) used the width of the midpart of the maxillary band, divided by the length of the maxillary band and the number of teeth of the midpart of the maxillary band, as the dentition characteristic, instead of groove, as Ege did (1939). Watanabe (2003) suggested that his classification was more accurate than that of Ege (1939).

If we do not take into account the geographic distribution of the genus *Anguilla*, each specimen can be classified into four groups by means of their key morphological characters. Watanabe et al. (2005) studied those four groups using a molecular genetic analysis for further classification. The 16S rRNA region was processed with 10 restriction enzymes; *Alu*I, *Apa*I, *Bsp*1286I, *Eco*OI109I, *Eco*T14I, *Hha*I, *Msp*I, *Mva*I and *Van*91I (Takara Shuzo Co., Ltd.) and *Bbr*PI (Toyobo Co., Ltd.) (Watanabe 2003; Watanabe et al. 2005). Aoyama et al. (2000b) had examined the species identification in the genus using six of these restriction enzymes and suggested that the 16S rRNA was appropriate for identifying eels at the species level. There were 14 clades in the dendrogram, which were distinguished using both molecular genetic markers and the four groups classified by morphological characteristics (Watanabe et al. 2004, 2005). Furthermore, 1 of the 14 clades was divided into 2 clades based on the number of vertebrae (Watanabe 2003). A total of 15 taxa were found in the genus *Anguilla* using their morphological and genetic characteristics (Watanabe et al. 2004, 2005).

Problems of subspecies

Ege (1939) was able to name 16 species (and 3 subspecies) by systematic analyses. Castle and Williamson (1974) suggested that *A. ancestralis* was a synonym of *A. celebesensis*, and therefore reduced the genus to 15 species and 3 subspecies. Thereafter, Watanabe et al. (2004, 2005) identified 15 species, mostly consistent with Ege's (1939) milestone study. However, "18 species, including subspecies" until 2008 or "19 species including subspecies with a recently discovered new species in 2009" is still referred to and cited. In other words, systematic analyses by Ege (1939) have been widely accepted until now.

In a molecular genetic study on the genus *Anguilla,* one specimen of each of the 15 taxa was used to sequence a 1485 base pair segment of the 16S rRNA region (Watanabe et al. 2005). The genetic distance between each species ranged from 0.0115 to 0.0571 (Watanabe 2003). The genetic distance between *A. anguilla* and *A. rostrata* was 0.0115, which was the lowest value between taxa (Watanabe 2003). Watanabe (2003) also examined the differences between specimens of the subspecies of *A. bengalensis, A. bicolor* and *A. australis* classified by Ege (1939) and *A. australis* classified by Dijkstra and Jellyman (1999). The values between the subspecies within *A. bengalensis, A. bicolor* and *A. australis* were 0.0061, 0.0068 and 0.0034, respectively, and these values were much lower than those between the each of the 15 species (Watanabe 2003). Watanabe (2003) suggested that there were 15 definite taxa that presented 15 species, but that the subspecies designations were questionable.

More recently, Watanabe et al. (2014b) argued the subspecies designation details as follows. Recent information about the population structures of anguillid eels also raises questions about how to approach the use of the subspecies or population taxonomic categories for these unique catadromous fishes. There is now clear evidence of both morphological and/or molecular genetic differences between not only each of the subspecies pairs of *A. bicolor,* *A. bengalensis* and *A. australis* (Dijkstra and Jellyman 1999; Watanabe et al. 2006, 2008a; Shen and Tzeng 2007; Minegishi et al. 2009, 2012) but also between the multiple populations of *A. marmorata* (Minegishi et al. 2008; Watanabe et al. 2008b, 2009b). The genetic distances between these subspecies or between the populations of *A. marmorata* are slightly less than those between the most recently diverged species of *A. rostrata* and *A. anguilla* that are present in the North Atlantic, which have the smallest genetic divergence of all species pair comparisons of anguillid eels (Watanabe et al. 2008b; Minegishi et al. 2009). Considering the morphological and molecular genetic differences of these taxonomic groups of eels and the concept of reproductively isolated biological units (populations), there are three possible taxonomic approaches to the subspecies/populations issue of the genus *Anguilla*: (i) because recent molecular genetic (Ishikawa et al. 2004; Minegishi et al. 2008; Gagnaire et al. 2011) and morphological (Watanabe et al. 2008a, 2009b) studies recognised several populations in *A. marmorata,* the two subspecies of *A. bicolor,* *A. bengalensis* and *A. australis* could also be regarded merely as population variations within species. However, the levels of divergences among or between the populations of *A. marmorata, A. bicolor, A. bengalensis* and *A. australis* appear to differ slightly in terms of both morphological characteristics and the degree of genetic divergence (Watanabe et al. 2008b; Minegishi et al. 2009), possibly due to when the divergences occurred or due to how much subsequent gene exchange has occurred. (ii) If the validity of the two subspecies of *A. bicolor, A. bengalensis* and *A. australis* is to be accepted, the populations or regional metapopulations of *A. marmorata* with similar differences could also be regarded as subspecies. (iii) If the focus is on the reproductive isolation of populations of the genus *Anguilla,* without considering the degree of differences in morphological and molecular genetic characters, then each population should be regarded as a species, and the subspecies taxonomic concept would not be used with the genus *Anguilla*.

The morphological and molecular genetic differences that have been found and the allopatric distributions of each of the subspecies pairs of *A. bicolor*, *A. bengalensis* and *A. australis* agree with the subspecies concept defined by Mayr and Ashlock (1991). However, in order to use and expand the use of the subspecies concept to *A. marmorata* as in case (ii), the use of two taxonomic (species and subspecies) and one ecological (population) units are required. Case (i) or (iii) would be simpler concepts than case (ii) because these cases use just species and population designations without any subspecies. Furthermore, some biologists have suggested that the subspecies as a category rank should be abolished (Wilson and Brown 1953; Burt 1954; Hagmeier 1958).

Most recently, a molecular genetic study accurately identified the two subspecies of *A. bicolor* (Tanaka et al. 2014). A single DNA nucleotide substitution in the mitochondrial DNA 16S rRNA gene was employed to identify the two subspecies where *A. bicolor bicolor* and *A. bicolor pacifica* possessed adenine and guanine, respectively (Tanaka et al. 2014). This substitution was highly conserved at 100% in *A. bicolor bicolor* (108/108) and 99.9% in *A. bicolor pacifica* (181/182), and the misidentification rate was estimated to be 0.34% (Tanaka et al. 2014). These results suggest that *A. bicolor* is further divided taxonomically into the two subspecies of *A. bicolor bicolor* and *A. bicolor pacifica*, although they are almost identical in their morphological characteristics except for the mode of their number of vertebrate (Watanabe et al. 2014b).

Based on the discussion, concept and idea proposed by Watanabe et al. (2014b) and recent genetic molecular analysis by Tanaka et al. (2014), the subspecies of *Anguilla bicolor*, *A. bengalensis* and *A. australis* are valid at the moment. However, further molecular genetic identification study is required for each species, in combination with population structure and life history studies.

New species

The genus *Anguilla* has long been recognised as being comprised of 15 species (Ege 1939; Castle and Williamson 1974; Watanabe 2003; Watanabe et al. 2004, 2005). After the first comprehensive taxonomic revision by Ege (1939), three new species, *A. breviceps* Chu and Jin (1984), *A. foochowensis* Chu and Jin (1984), and *A. nigricans* Chu and Wu (1984) were described from China (Chu 1984). However, their identifications were based only on the external morphology and body proportions of single individuals, so a re-examination of these doubtful species was suggested (Tabeta 1994). Watanabe et al. (2009) and Teng et al. (2009) discovered new species of the genus *Anguilla* almost simultaneously, *A. luzonensis* and *A. huangi*, respectively, from the Cagayan River system on the northern Luzon Island of the Philippines and from the glass eels collected from the Cagayan River estuary that were reared in a culture pond in Taiwan, respectively. These discoveries are the first in seventy years since Ege's (1939) revision. Comparisons using morphological and molecular genetic characteristics clearly showed that *A. luzonensis* and *A. huangi* are the same species (Watanabe et al. 2013). Thus, *A. luzonensis* became the valid species name and *A. huangi* became a junior synonym of *A. luzonensis* (Watanabe et al. 2013). The results all lead to the conclusion that the genus *Anguilla* has been recognised as being comprised of 16 species (Ege 1939; Castle and Williamson 1974; Watanabe 2003; Watanabe et al. 2004, 2005, 2009a, 2013, 2014b).

Latest taxonomy

Based on the systematic analyses by Ege (1939) and further analyses by morphological and molecular genetic characters by Castle and Williamson (1974), Watanabe (2003) and Watanabe et al. (2004, 2005, 2009a, 2013, 2014a,b), the genus *Anguilla* is classified as 16 species, three of which were divided into two subspecies (Table 1); *A. celebesensis, A. interioris, A. megastoma, A. luzonensis, A. bengalensis bengalensis, A. bengalensis labiata, A. marmorata, A. reinhardtii, A. borneensis, A. japonica, A. rostrata, A. anguilla, A. dieffenbachii, A. mossambica, A. bicolor bicolor, A. bicolor pacifica, A. obscura, A. australis australis,* and *A. australis schmidtii* (Table 1).

Worldwide distribution

The genus *Anguilla* is currently recognised as being comprised of 16 species. They are widely distributed in most of the tropical and subtropical areas of the world, except for South America and West Africa (Fig. 2). Tesch (2003) divided the freshwater eels into tropical and temperate species according to their geographic distributions. The tropical and temperate eels consist of 11 and 5 species, respectively. The continental distributions of the temperate species appear to be related to the subtropical circulation of the oceans, with most species located along the west sides of the Atlantic, Indian, and Pacific Oceans, except for the European eel, *A. anguilla* (Ege 1939; Watanabe 2003; Fig. 2). However, anguillid eels are absent along the east coast of South America, despite the existence of the warm Brazil Current. Based on this geographic pattern, the Atlantic species (*A. anguilla* and *A. rostrata*) are geographically separated from their other congeners in the Pacific and Indian Oceans (Minegishi et al. 2005). Such a unique geographic distribution has recently attracted the attention of biologists, and numerous molecular phylogenetic studies have been conducted (e.g., Aoyama and Tsukamoto 1997, 2001; Bastrop et al. 2000; Lehmann et al. 2000; Lin et al. 2001; Minegishi et al. 2005; Teng et al. 2009).

Of the eleven species found in tropical areas, six species occur in the western Pacific around Indonesia, i.e., *A. celebesensis, A. interioris, A. bengalensis, A. marmorata, A. borneensis,* and *A. bicolor* (Fig. 3) (Ege 1939; Castle and Williamson 1974; Arai et al. 1999). Molecular phylogenetic researches on

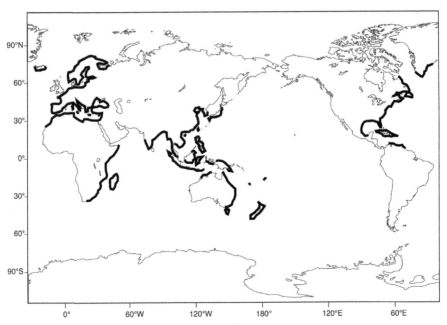

Fig. 2. Global distribution of genus *Anguilla* (thick lines).

freshwater eels have recently revealed that tropical eels are the most basal species originating in the Indonesian region and that freshwater eels radiated out from the tropics to colonise the temperate regions (Minegishi et al. 2005).

Knowledge of the geographic distributions of the genus *Anguilla* has increased dramatically since Ege's (1939) study. Passive transport of leptocephali for long periods of 3 months or more can easily change the range of recruitment of glass eels, thus the geographic distribution of the species can change from one year to another. In fact, there have been many new findings regarding the geographic distributions of the freshwater eels since Ege's (1939) comprehensive studies, e.g., *A. rostrata* in Iceland (Boëtius 1980; Williams and Koehn 1984; Avise et al. 1990), *A. reinhardtii* in New Zealand (Jellyman et al. 1996), *A. marmorata* in the Galápagos Islands (McCosker et al. 2003), and *A. celebesensis's* absence in New Guinea (Aoyama et al. 2000a). Furthermore, there have been several reports which introduced non-native species of freshwater eels that have been caught in several areas of the world (e.g., Shapovalov et al. 1959; Skinner 1971; Tabeta et al. 1976b, 1977; McCosker 1989; Zhang et al. 1999; Sasai et al. 2001; Han et al. 2002; Okamura et al. 2008; Arai et al. 2009) as a result of the international trade in glass eels and young eels for aquaculture.

Representative geographical distribution range for each anguillid species by distinct morphological characteristics classified into four groups by Ege (1939) Watanabe (2003) and Watanabe et al. (2004, 2005, 2009, 2013, 2014a,b) is summarized in Table 2.

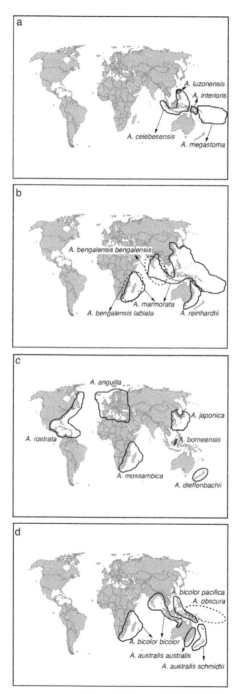

Fig. 3. Geographical distribution of genus *Anguilla* shown by Ege (1939) with the latest taxonomy. Distribution of each species from "a" to "d" is followed by the classification of morphological characters from group 1 to group 4 as shown in Fig. 1.

Table 2. Distinct morphological characters classified into four groups with representative geographical distribution range for each anguillid species.

Group	Species	Distinct morphological characters	Geographical distribution range
1	A. celebesensis	variegated skin	Jawa, Sumatra, Sulawesi Island in Indonesia
		broad maxillary bands of teeth	Philippines, New Guinea
	A. interioris	long dorsal fin	New Guinea
	A. megastoma		Solomon Islands, New Caledonia, Fiji Islands, Cook Islands
	A. luzonensis		Northern Philippines
2	A. bengalensis bengalensis	variegated skin	Sri Lanka, Bangladesh, India, Myanmar, Malaysia
		narrow maxillary bands of teeth	Sumatra Island in Indonesia and Andaman Islands
	A. bengalensis labiata	long dorsal fin	mid-southeastern part of Africa
	A. marmorata		longitudinally from the east coast of Africa to the Marquesas Islands in the southeast Pacific Ocean and as far north as southern Japan
	A. reinhardtii		Eastern Australia, Northern New Zealand
3	A. borneensis	nonvariegated skin	Borneo Island
	A. japonica	long dorsal fin	Japan, China, Korea, Taiwan, Northern Philippines
	A. rostrata		North and South America
	A. anguilla		Europe, North Africa
	A. dieffenbachii		New Zealand
	A. mossambica		mid-southeastern part of Africa, Madagascar
4	A. bicolor pacifica	nonvariegated skin	Philippines, Sulawesi Island in Indonesia, New Guinea
	A. bicolor bicolor	short dorsal fin	Africa, India, Sri Lanka, Bangladesh, Myanmar, Malaysia northwestern Australia, Greater Sunda Islands
	A. obscura		northeastern Australia, New Caledonia, Fiji Islands, Samoa, Tahiti, Cook Islands, Maluku Islands
	A. australis australis		Southeastern Australia, Tasmania
	A. australis schmidtii		New Zealand, New Caledonia, North Norfolk Island

References: Ege (1939), Watanabe (2003), Watanabe et al. (2004, 2005, 2009, 2013, 2014a,b) and Arai and Wong (2016).

Group 1: *A. celebesensis, A. interioris, A. megastoma* and *A. luzonensis* (Fig. 3a). Ege (1939) described that the geographical range of *A. celebesensis* is southward from Luzon Island (Philippines) to Rotti Island (Indonesia) and eastward from Noas Island (Indonesia) to Humboldt Bay on the coast of New Guinea. However, *A. celebesensis* was not found in New Guinea based on a molecular genetic study (Aoyama et al. 2000a). *A. interioris* is an endemic species in New Guinea only (Ege 1939). *A. megastoma* is distributed along a belt of islands south of the equator in the western South Pacific Ocean (WSP) (Ege 1939; Beumer et al. 1981; Allen 1991; Marquet and Galzin 1991). *A. luzonensis* is a recently discovered species just found in the Cagayan River at the northern tip of the Luzon Island, Philippines (Teng et al. 2009; Watanabe et al. 2009, 2013).

Group 2: *A. bengalensis, A. marmorata* and *A. reinhardtii* (Fig. 3b). The distribution range of *A. bengalensis* includes India, from Bombay in the northwest to Calcutta in the northeast, Ceylon, Sandoway, the coast of Burma, Andamans, and Serdang, the west coast of North Sumatra and Peninsular Malaysia, and the east coast of southern Africa from Kenya to South Africa (Fig. 3b, Ege 1939). *A. marmorata* is a unique tropical anguillid that reaches large sizes of almost 2 m in length with a maximum weight of 21 kg (Castle 1984). This species has the widest geographic distribution of the 16 species of the genus *Anguilla* (Ege 1939; Castle and Williamson 1974; Fig. 1) and is found longitudinally from the east coast of Africa to the Marquesas Islands in the southeast Pacific Ocean and as far north as southern Japan (Ege 1939). Recently this species was found at the Palmyra Atoll in the central Pacific (Handler and James 2006) and even farther to the east in the Galápagos Islands (McCosker et al. 2003), which indicates that it has an even wider geographic range than previously thought. *A. reinhardtii* is found mostly in New Caledonia, eastern Australia and New Zealand (Ege 1939; Beumer 1996; Jellyman et al. 1996).

Group 3: *A. borneensis, A. japonica, A. rostrata, A. anguilla, A. dieffenbachii* and *A. mossambica* (Fig. 3c). The freshwater growth habitat of *A. borneensis* is limited to the east-central part of Borneo (Ege 1939). The freshwater distribution of *A. japonica* ranges from Taiwan, through China, Korea, and north to Japan (Tesch 2003). Tabeta et al. (1976c) investigated the anguillid glass eels in the Cagayan River at the northern tip of Luzon Island of the Philippines, which is the southern-limit of the distribution range of the species. The habitat range of *A. rostrata* spans more than 50 degrees of latitude along the Atlantic from the southern tip of Greenland to north-eastern South America (Boëtius 1985; Nilo and Fortin 2001; Tesch 2003). *Anguilla anguilla* is distributed from North Cape in Northern Norway, southwards along the coast of Europe, all coasts of the Mediterranean and along the North African Coast (Schmidt 1909; Dekker 2003). *A. dieffenbachii* is endemic in New Zealand (Ege 1939; Jellyman 2003). *A. mossambica* inhabits rivers and lakes in south-eastern Africa and Madagascar (Jubb 1964; Tesch 2003).

Group 4: *A. bicolor*, *A. obscura* and *A. australis* (Fig. 3d). *A. bicolor* has the second widest geographic distribution in the genus *Anguilla* (Ege 1939), and is distributed from the eastern coast of Africa through the Indonesian Seas to New Guinea adjacent to the Pacific Ocean (Ege 1939). The species was found in Taiwan (Tzeng and Tabeta 1983) and Yagushima Island in southern Japan (Yamamoto et al. 2001). *A. obscura* is distributed along a belt of islands south of the equator in the WSP ranging from western New Guinea to Tahiti (Ege 1939; Jellyman 1991). *A. australis* is widely distributed in southeast Australia, Tasmania, New Caledonia, Norfolk Island, Lord Howe Island and New Zealand (Ege 1939; Beumer 1996; Jellyman et al. 1996).

Conclusion

The recent comprehensive revisions of the genus *Anguilla* have been performed by Watanabe (2003) and Watanabe et al. (2004, 2005, 2014b) with reports on a new species in the genus (Teng et al. 2009; Watanabe et al. 2009a, 2013) using morphological and molecular genetic analyses after the Ege's (1939) revision. In the last revision, this genus is definitively classified into 16 species, three of which are divided into two subspecies, although subspecies problems are still in discussion. Regarding juvenile eels such as glass eels and elvers, specimens can be classified as either long-finned or short-finned without vertebrae counting. Further identification, however, is difficult for samples of unknown origin. Thus, molecular genetic analyses are indispensable in order to precisely identify species, especially in case of tropical species, as a number of key morphological characteristics are overlapping among these species. Regarding adult specimens, Ege's (1939) systematics have long been widely accepted. However, Watanabe et al. (2004) found that the morphological characters described by Ege (1939) were not sufficient to classify all species of this genus and that geographic distribution was an indispensable characteristic for his taxonomy of these eels. The geographic distribution of each species is not a suitable taxonomic characteristic of the freshwater eels by itself because it overlaps among species and is highly variable depending on the environment. Furthermore, the new species discovered recently suggests the possibility of finding other new species in the future. When only morphological characters were used, the freshwater eels could be classified into only four groups (Watanabe et al. 2004). Thus, analyses of morphological characters in combination with molecular genetic characters are indispensable for precise species identification in the genus.

Keywords: *Anguilla*, taxonomy, classification, distribution, morphology, molecular genetics, subspecies

References

Allen, G.R. 1991. Field Guide to the Freshwater Fishes of New Guinea. Christensen Research Institute, Madang.

Aoyama, J. and K. Tsukamoto. 1997. Evolution of the freshwater eels. Naturwissenschaften 84: 17–21.

Aoyama, J., S. Watanabe, S. Ishikawa, M. Nishida and K. Tsukamoto. 2000a. Are morphological characters distinctive enough to discriminate between two species of freshwater eels, *Anguilla celebesensis* and *A. interioris*? Ichthyol. Res. 47: 157–161.

Aoyama, J., S. Watanabe, M. Nishida and K. Tsukamoto. 2000b. Discrimination of catadromous eel species, genus *Anguilla*, using PCR-RFLP analysis of the mitochondrial 16S rRNA domain. Trans. Am. Fish. Soc. 129: 873–878.

Aoyama, J., M. Nishida and K. Tsukamoto. 2001. Molecular phylogeny and evolution of the freshwater eel, genus *Anguilla*. Mol. Phylogenet. Evol. 20: 450–459.

Aoyama, J., S. Wouthuyzen, M.J. Miller, Y. Minegishi, M. Kuroki, S.R. Suharti, T. Kawakami, K.O. Sumardiharga and K. Tsukamoto. 2007. Distribution of leptocephali of the freshwater eels, genus *Anguilla*, in the waters off west Sumatra in the Indian Ocean. Environ. Biol. Fish. 80: 445–452.

Arai, T. 2014. Evidence of local short-distance spawning migration of tropical freshwater eels, and implications for the evolution of freshwater eel migration. Ecol. Evol. 4: 3812–3819.

Arai, T. and L.L. Wong. 2016. Validation of the occurrence of the tropical eels, *Anguilla bengalensis bengalensis* and *A. bicolor bicolor* at Langkawi Island in Peninsular Malaysia, Malaysia. Tropic. Ecol. 57: 23–31.

Arai, T., J. Aoyama, D. Limbong and K. Tsukamoto. 1999. Species composition and inshore migration of the tropical eels *Anguilla* spp. recruiting to the estuary of the Poigar River, Sulawesi Island. Mar. Ecol. Prog. Ser. 188: 299–303.

Arai, T., N. Chino and A. Kotake. 2009. Occurrence of estuarine and sea eels *Anguilla japonica* and a migrating silver eel *Anguilla anguilla* in Tokyo Bay area, Japan. Fish. Sci. 75: 1197–1203.

Avise, J.C., W.S. Nelson, J. Arnold, R.K. Koehn, G.C. Williams and V. Thorsteinsson. 1990. The evolutionary genetic status of Icelandic eels. Evolution 44: 1254–1262.

Beumer, J.P. 1996. Freshwater eels. pp. 39–43. *In*: R. McDowall (ed.). Freshwater Fishes of South Eastern Australia. Reed, Sydney.

Beumer, J.P., R.G. Pearson and L.K. Penridge. 1981. Pacific shortfinned eel, *Anguilla obscura* Günther, 1871 in Australia: recent records of its distribution and maximum size. Proc. R. Soc. Queensland 92: 85–90.

Boëtius, J. 1980. Atlantic *Anguilla*. A presentation of old and new data of total numbers of vertebrae with special reference to the occurrence of *Anguilla rostrata* in Europe. Dana 1: 93–112.

Boëtius, J. 1985. Greenland eels, *Anguilla rostrata* LeSueur. Dana 4: 41–48.

Burt, W.H. 1954. The subspecies category in mammals. Syst. Zool. 3: 99–104.

Castle, P.H.J. 1984. Notacanthiformes and anguilliformes: development. pp. 62–93. *In*: H.G. Moser, W.J. Richards, D.M. Cohen, M.P. Fahay, A.W. Kendall, Jr. and S.L. Richardson (eds.). Ontogeny and Systematics of Fishes: Based on an international symposium dedicated to the memory of Elbert Halvor Ahlstrom. ix + 766pp.

Castle, P.H.J. and G.R. Williamson. 1974. On the validity of the freshwater eel species *Anguilla ancestralis* Ege from Celebes. Copeia 2: 569–570.

Chu, Y.T. 1984. Fishes of Fujian Province. Part I. Fuzhou, Fujian, China: Fujian Science and Technology Press.

Dekker, W. 2003 On the distribution of the European eel (*Anguilla anguilla*) and its fisheries. Can. J. Fish. Aquat. Sci. 60: 787–799.

Dijkstra, L.H. and D.J. Jellyman. 1999. Is the subspecies classification of the freshwater eels *Anguilla australis australis* Richardson and *A. a. schmidtii* Phillipps still valid? Mar. Freshw. Res. 50: 261–263.

Ege, V. 1939. A revision of the genus *Anguilla* Shaw. Dana Rep. 16: 1–256.

Gagnaire, P.A., Y. Minegishi, S. Zenboudji, P. Valade, J. Aoyama and P. Berrebi. 2011. Within-population structure highlighted by differential introgression across semipermeable barriers to gene flow in *Anguilla marmorata*. Evolution 65: 3413–3427.

Günther, A. 1870. Catalogue of the fishes in the British Museum, Vol. VIII. British Museum, London. i–xxv, 549p.

Hagmeier, E.M. 1958. Inapplicability of the subspecies concept to North American marten. Syst. Zool. 7: 1–7.

Han, Y.S., C.H. Yu, H.T. Yu, C.W. Chang, I.C. Liao and W.N. Tzeng. 2002. The exotic American eel in Taiwan: ecological implications. J. Fish Biol. 60: 1608–1612.

Inoue, J.G., M. Miya, M.J. Miller, T. Sado, R. Hanel, K. Hatooka, J. Aoyama, Y. Minegishi, M. Nishida and K. Tsukamoto. 2010. Deep-ocean origin of the freshwater eels. Biol. Lett. 6: 363–366.

Ishikawa, S., K. Tsukamoto and M. Nishida. 2004. Genetic evidence for multiple geographic populations of the giant mottled eel *Anguilla marmorata* in the Pacific and Indian oceans. Ichthyol. Res. 51: 343–353.

Jellyman, D.J. 1991. Factors affecting the activity of two species of eel (*Anguilla* spp.) in a small New Zealand lake. J. Fish Biol. 39: 7–14.

Jellyman, D.J. 2003. The distribution and biology of the South Pacific species of *Anguilla*, 1798. pp. 275–292. *In*: K. Aida, K. Tsukamoto and K. Yamauchi (eds.). Eel Biology. Springer, Tokyo.

Jellyman, D.J., B.L. Chisnall, L.H. Dijkstra and J.A.T. Boubee. 1996. First record of the Australian longfinned eel, *Anguilla reinhardtii*, in New Zealand. Mar. Freshw. Res. 47: 1037–1040.

Jubb, R.A. 1964. The eels of South African rivers and observations on their ecology. pp. 186–206. *In*: D.H.S. David (ed.). Ecological Studies in Southern Africa. Junk, Den Haag, Holland.

Kaup, J.J. 1856. Catalogue of apodal fish in the collection of the British Museum. British Museum. London. 163p.

Lehmann, D., H. Hettwer and H. Taraschewski. 2000. RAPD-PCR investigations of systematic relationships among four species of eels (Teleostei: Anguillidae), particularly *Anguilla anguilla* and *A. rostrata*. Mar. Biol. 137: 195–204.

Lin, Y.S., Y.P. Poh and C.S. Tzeng. 2001. A phylogeny of freshwater eels inferred from mitochondrial genomes. Mol. Phylogenet. Evol. 20: 252–261.

Marquet, G. and R. Galzin. 1991. The eels of French Polynesia: taxonomy, distribution and biomass. Lar. Mer. 29: 8–17.

Mayr, E. and P.D. Ashlock. 1991. Principles of Systematic Zoology. McGraw-Hill, New York. p. 475.

McCosker, J.E. 1989. Freshwater eels (family Anguillidae) in California: current conditions and future scenarios. Calif. Fish Game 75: 4–10.

McCosker, J.E., R.H. Bustamante and G.M. Wellington. 2003. The Freshwater Eel, *Anguilla marmorata*, discovered at Galápagos. Noticias de Galápagos 62: 2–6.

Meyer, A. 1993. Evolution of mitochondrial DNA in fishes. pp. 1–38. *In*: P.W. Hochachka and T.P. Mommsen (eds.). Biochemistry and Molecular Biology of Fishes. Elsevier, Amsterdam.

Minegishi, Y., J. Aoyama, J.G. Inoue, M. Miya, M. Nishida and K. Tsukamoto. 2005. Molecular phylogeny and evolution of the freshwater eels genus *Anguilla* based on the whole mitochondrial genome sequences. Mol. Phyl. Evol. 34: 134–146.

Minegishi, Y., J. Aoyama and K. Tsukamoto. 2008. Multiple population structure of the giant mottled eel *Anguilla marmorata*. Mol. Ecol. 17: 3109–3122.

Minegishi, Y., J. Aoyama, J.G. Inoue, R.V. Azanza and K. Tsukamoto. 2009. Inter-specific and subspecific genetic divergences of freshwater eels, genus *Anguilla* including a recently described species, *A. luzonensis*, based on whole mitochondrial genome sequences. Coast. Mar. Sci. 33: 64–77.

Minegishi, Y., P.-A. Gagnaire, J. Aoyama, P. Bosc, E. Feunteun, K. Tsukamoto and P. Berrebi. 2012. Present and past genetic connectivity of the Indo-Pacific tropical eel *Anguilla bicolor*. J. Biogeogr. 39: 408–420.

Nilo, P. and R. Fortin. 2001. Synthèse des connaissances et établissement d'une programmation de recherche sur l'anguille d'Amérique (*Anguilla rostrata*). Société de la faune et des parcs du Québec Direction de la recherche sur la faune.

Okamura, A., H. Zhang, N. Mikawa, A. Kotake, Y. Yamada, N. Horie, T. Utoh, S. Tanaka, H.P. Oka and K. Tsukamoto. 2008. Decline in non-native freshwater eels in Japan: ecology and future perspectives. Env. Biol. Fish. 81: 347–358.

Sasai, S., J. Aoyama, S. Watanabe, T. Kaneko, M.J. Miller and K. Tsukamoto. 2001. Occurrence of migrating silver eels, *Anguilla japonica*, in the East China Sea. Mar. Ecol. Prog. Ser. 212: 305–310.

Schmidt, J. 1909. On the distribution of the freshwater eels (*Anguilla*) throughout the world. I. Atlantic Ocean adjacent region. Meddelelser fra Kommissionen for Havundersogelser. Serie Fiskeri 3: 1–45.

Shapovalov, L., A.D. William and A.J. Cordone. 1959. A revised checklist of the freshwater and anadromous fishes of California. Calif. Fish Game 45: 159–180.

Shen, K.N. and W.N. Tzeng. 2007. Genetic differentiation among populations of the shortfinned eel *Anguilla australis* from East Australia and New Zealand. J. Fish Biol. 70: 177–190.

Silfvergrip, A.M.C. 2009. CITES Identification guide to freshwater eels (Anguillidae), with focus on the European eel *Anguilla anguilla*. Swedish Environmental Protection Agency, Stockholm.

Skinner, J.E. 1971. *Anguilla* recorded from California. Calif. Fish Game 57: 76–79.

Tabeta, O. 1994. Eel research in the world. Kaiyo Month. 287: 270–273.

Tabeta, O., T. Takai and I. Matsui. 1976a. The sectional counts of vertebrae in the anguillid elvers. Japan. J. Ichthyol. 22: 195–200.

Tabeta, O., T. Takai and I. Matsui. 1976b. Record of short finned eel from Nagata River, Shimonoseki, Japan. Bull. Jap. Soc. Sci. Fish. 42: 1333–1338.

Tabeta, O., T. Tanimoto, T. Takai, I. Matsui and T. Imamura. 1976c. Seasonal occurrence of anguillid elvers in Cagayan River, Luzon Island, the Philippines. Bull. Japan. Soc. Sci. Fish. 42: 424–426.

Tabeta, O., T. Takai and I. Matsui. 1977. On the foreign eel species in Japan. Suisan Zoushoku 24: 116–122.

Tanaka, C., F. Shirotori, M. Sato, M. Ishikawa, A. Shinoda, J. Aoyama and T. Yoshinaga. 2014. Genetic identification method for two subspecies of the Indonesian short-finned eel, *Anguilla bicolor*, using an allelic discrimination technique. Zool. Stud. 53: 57.

Teng, H.Y., Y.S. Lin and C.S. Tzeng. 2009. A new *Anguilla* species and reanalysis of the phylogeny of freshwater eels. Zool. Stud. 48: 808–822.

Tesch, F.W. 2003. The Eel. Biology and Management of Anguillid Eels. Chapman and Hall, London.

Tzeng, W.N. and O. Tabeta. 1983. First record of the short-finned eel *Anguilla bicolor pacifica* elvers from Taiwan. Bull. Jpn. Soc. Sci. Fish. 49: 27–32.

Watanabe, S. 2003. Taxonomy of the freshwater eels, genus *Anguilla* Schrank, 1798. pp. 3–18. *In*: K. Aida, K. Tsukamoto and K. Yamauchi (eds.). Eel Biology. Springer, Tokyo.

Watanabe, S., J. Aoyama and K. Tsukamoto. 2004. Reexamination of Ege's (1939) use of taxonomic characters of the genus *Anguilla*. Bull. Mar. Sci. 74: 337–351.

Watanabe, S., J. Aoyama, M. Nishida and K. Tsukamoto. 2005. A molecular genetic evaluation of the taxonomy of eels of the genus *Anguilla* (Pisces: Anguilliformes). Bull. Mar. Sci. 76: 675–690.

Watanabe, S., J. Aoyama, M. Nishida and K. Tsukamoto. 2006. Confirmation of morphological differences between *Anguilla australis australis* and *A. australis schmidtii*. N.Z. J. Mar. Freshw. Res. 40: 325–332.

Watanabe, S., J. Aoyama and K. Tsukamoto. 2008a. The use of morphological and molecular genetic variations to evaluate subspecies issues in the genus *Anguilla*. Coast. Mar. Sci. 32: 19–29.

Watanabe, S., J. Aoyama, M.J. Miller, S. Ishikawa, E. Feunteun and K. Tsukamoto. 2008b. Evidence of population structure in the giant mottled eel, *Anguilla marmorata*, using total number of vertebrae. Copeia 2008: 680–688.

Watanabe, S., J. Aoyama and K. Tsukamoto. 2009a. A new species of freshwater eel, *Anguilla luzonensis* (Teleostei: Anguillidae) from Luzon Island of the Philippines. Fish. Sci. 75: 387–392.

Watanabe, S., M.J. Miller, J. Aoyama and K. Tsukamoto. 2009b. Morphological and meristic evaluation of the population structure of *Anguilla marmorata* across its range. J. Fish Biol. 74: 2069–2093.

Watanabe, S., J. Aoyama, S. Hagihara, B. Ai, R.V. Azanza and K. Tsukamoto. 2013. *Anguilla huangi* Teng, Lin, and Tzeng, 2009, is a junior synonym of *Anguilla luzonensis* Watanabe, Aoyama, and Tsukamoto, 2009. Fish. Sci. 79: 375–383.

Watanabe, S., J. Aoyama and K. Tsukamoto. 2014a. On the Identities of *Anguilla borneensis*, *A. malgumora*, and *Muraena malgumora*. Copeia 2014: 568–576.

Watanabe, S., M.J. Miller, J. Aoyama and K. Tsukamoto. 2014b. Evaluation of the population structure of *Anguilla bicolor* and *A. bengalensis* using total number of vertebrae and consideration of the subspecies concept for the genus *Anguilla*. Ecol. Freshw. Fish 23: 77–85.

Williams, G.C. and R.K. Koehn. 1984. Icelandic eels: evidence for a single species of *Anguilla* in the North Atlantic. Copeia 1984: 221–223.

Wilson, E.O. and W.L. Brown. 1953. The subspecies concept and its taxonomic application. Syst. Zool. 3: 97–111.

Yamamoto, T., N. Mochioka and A. Nakazono. 2001. Seasonal occurrence of anguillid glass eels at Yakushima Island, Japan. Fish. Sci. 67: 530–532.

Zhang, H., N. Mikawa, Y. Yamada, N. Horie, A. Okamura, T. Uto, S. Tanaka and T. Motonobu. 1999. Foreign eel species in the Natural waters of Japan detected by polymerase chain reaction of mitochondrial cytochrome *b* region. Fish. Sci. 65: 684–686.

2

Overview and Current Trends in Studies on the Evolution and Phylogeny of *Anguilla*

Mei-Chen Tseng

Introduction

Anguilliforms are a distinctive group of teleosts that comprise of 820 species which are further classified into 20 families and 147 genera and share a unique leptocephalus larva (Nelson 2006; Inoue et al. 2010; Johnson et al. 2011). Anguilliforms first appeared as fossils in the Aptian age, about 113–119 million years ago (Mya) (Patterson 1993; Belouze et al. 2003). Johnson et al. (2011) discovered the living fossil eel *Protoanguilla palau* at Palau. Phylogenetic analysis and divergence time estimation based on whole mitogenome sequences from various actinopterygians indicated that *P. palau* is one of the most primitive and independent lineages of true eels. Its evolutionary history dates back to the Triassic Age (251–199.6 Mya) and suggests that the origin of Anguilliformes should be about 200–250 Mya (Fig. 1).

Fossil records of the genus *Anguilla* have been found in Europe from the Eocene epoch (Ypresian, about 50–55 Mya) (Patterson 1993). Nevertheless, according to mitochondrial analyses using the molecular clock for bony fishes calibrated by Kumazawa et al. (1999), the estimated divergence time between *Anguilla* and *Serrivomer* is 52 Mya and the speciation of extant anguillid

Department of Aquaculture, National Pingtung University of Science and Technology, Pingtung 912, Taiwan.
E-mail: mctseng@mail.npust.edu.tw

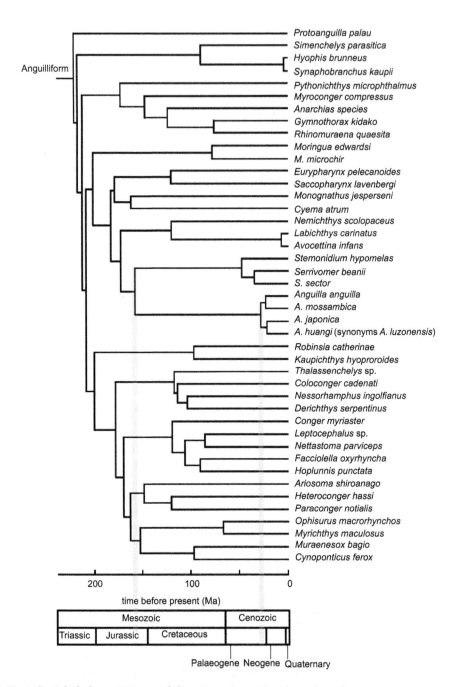

Fig. 1. Partial phylogenetic tree and divergence time estimation using mitogenome sequences from Anguilliform fishes. The figure was modified from Johnson et al. (2011).

species began approximately 20 Mya (Minegishi et al. 2005; Johnson et al. 2011). However, the fossil evidence is slightly inconsistent with molecular evidence, as the latter infers a more recent speciation pattern in *Anguilla* eels. Minegishi et al. (2005) indicated that the divergence time deduced from molecular data could be underestimated due to the group's peculiar ecological characteristics associated with slow metabolism. On the other hand, Teng et al. (2009) described the existence of *Anguilla* species 55 Mya based on fossil records, and several radiation events that probably occurred around 20–55 Mya. The last common ancestor of all living *Anguilla* began speciation about 20 Mya. However, these inferences need to be explicitly proven by more substantial evidence.

In all relative studies of *Anguilla* eel evolution, several questions have always mystified us. How many species of *Anguilla* are there in the world? Where did the *Anguilla* genus originate? Are all temperate freshwater eels closely related sister species? Which species of *Anguilla* is the most ancestral? How did *A. anguilla* and *A. rostrata* come to occupy the Atlantic Ocean? Fortunately, there have been major advances in molecular technologies during the past two decades that have enabled researchers to solve some of the foregoing problems using molecular genetic markers.

How many *Anguilla* species have occurred over time?

There is no information on how many *Anguilla* species have existed on earth during the last 50 million years. More than 30 species of *Anguilla* eels have been recorded, but some of them are synonyms (http://www.fishbase.org/ Nomenclature/Scientific NameSearchList.php?). Nineteen species/subspecies are unequivocally identified via morphological characters and DNA barcoding, including 2 Atlantic eels, 3 Oceanian eels, 1 Western Pacific eel, and 13 Indo-Pacific eels (Minegishi et al. 2005; Teng et al. 2009). Even so, it is possible that some cryptic species or undetected species of *Anguilla* eels exist. More precisely, the names of uncommon species still need to be verified, e.g., the species *A. breviceps* Chu and Jin 1984, seems to have never been examined. Cryptic species are worth examining more thoroughly because there are instances that prove their existence; for example, a new species was recently discovered from the Cagayan River estuary, northern Luzon, Philippines, by Lin et al. (2002). In order to identify this species, they spent five years rearing elvers into adulthood and measured their morphological characters, naming the species *A. huangi* (Teng et al. 2009). This species was coincidentally described in the same year under another name, *A. luzonensis*, by Watanabe et al. (2009). Cryptic *Anguilla* species must be found and identified if the genus' phylogeny is to be accurately determined.

Where is the original center of *Anguilla* eel evolution?

Almost all eel researchers believe that the Indo-Pacific waters are the original location of *Anguilla* eels (Table 1). The distributions of inshore marine fishes can be divided into four regions based on the order of decreasing biodiversity: Indo-West Pacific, western Atlantic, eastern Pacific, and eastern Atlantic. The Indo-West Pacific region ranges from South Africa and the Red Sea, east through Indonesia and Australia to Hawaii and through the South Pacific islands to Easter Island, harboring approximately one-third of all shallow marine fishes (*ca.* 3000 species). This region is also a significant reservoir of biodiversity for other marine taxa (Briggs 1974; Helfman et al. 1997), and is considered a "center of origin" from where many species evolved and then dispersed to wherever they are found today. In addition, Carpenter and Springer (2005) indicate that the global maxima of marine biodiversity is found in the Indo-Malay-Philippines Archipelago (IMPA), which is part of the Indo-Pacific Ocean. This has long been considered to be the area of highest marine biodiversity, with decreasing latitudinal and longitudinal gradients in species richness radiating from this center (Mora et al. 2003). Apparently, almost half of the *Anguilla* eels also reside in the IMPA. This high level of biodiversity is inferred to have possibly resulted from numerous vicariant and island integration events (Carpenter and Springer 2005). Present distributions can be compared to phylogenetic relationships in order to assess the relative movements of biota. In all pertinent phylogenetic studies of *Anguilla* eels, *A. borneensis* and *A. mossambica* are deduced as possible ancestors (Aoyama et al. 2001; Minegishi et al. 2005) of the other species distributed around the Indo-Pacific Ocean. It is firmly believed that this ocean is the original center of *Anguilla* speciation.

Are all temperate freshwater eels close sister species?

All temperate eels are like their sister group except *A. japonica*. Ege (1939) subdivided *Anguilla* eels into four distinct groups based on morphological characters. Four temperate *Anguilla* eels (*A. rostrata*, *A. anguilla*, *A. japonica*, and *A. dieffenbachii*) and two tropical species (*A. mossambica* and *A. malgumora*) having similar long dorsal fins and uniform coloration are in the same group. In contrast, temperate *A. australis* and tropical *A. bicolor* and *A. obscura*, which have shorter dorsal fins and uniform coloration, were placed in the same group (Fig. 2a). Although *A. anguilla*, *A. rostrata* and *A. japonica* have similar adult-phase morphological characteristics and rest in temperate water, their mitochondrial DNA (Minegishi et al. 2005) (Fig. 2b) and microsatellite flanking regions (Tseng 2012) suggest that *A. japonica* is not clustered with other temperate eels in the phylogenetic tree. It is interesting that inconsistent morphological and molecular phylogenetic trees cover *A. australis* and *A. japonica*. In brief, *A. rostrata*, *A. anguilla*, *A. australis*, and *A. dieffenbachii*

Table 1. Hypotheses of *Anguilla* eels dispersion patterns have been proposed below.

Published	Species number	Molecular Marker	Description
Tethys Sea Route			
Aoyama and Tsukamoto (1997) Tsukamoto and Aoyama (1998)	8 species	Cytochrome b (*Cyt b*) (410 bp)	*A. celebesensis* was the possible ancestral species of the Anguillidae. In the Eocene epoch (57–36 Mya) the ancestral eel originated in the western Pacific in and around Indonesia. The ancestors of Atlantic eels may thus have dispersed through the Tethys Sea to enter the paleo-Atlantic Ocean from the western Pacific occurred over 25–30 Mya. The split between two Atlantic eels is estimated at approximately 10 Mya.
Aoyama et al. (2001)	18 species/ subspecies	16S rRNA+ *Cyt b* (1,427 bp + 1,140 bp)	*Anguilla* was a monophyletic group. *A. borneensis* was the earliest derived species. Species disperse from Indonesia and move to Atlantic through the Tethys Corridor. The ancestral species of Atlantic eels moved into the Atlantic Ocean at least before the closure of the Tethys Sea at around 30 Mya. *Anguilla* originated about 50–60 Mya.
Bastrop et al. (2000)	7 species	16S rRNA (558 bp)	*A. marmorata* has the most ancestral lineage. *Anguilla* eels had originated in the Indo-Malaysian region. Atlantic eels were separated from congeners by circumglobal Tethys Seaway about 18 Mya.
Multiple radiation events + Central American Isthmus route Isthmus			
Lin et al. (2001)	12 species/ subspecies	12S rRNA + *Cyt b* (950 bp + 1140 bp)	No ancient species was inferred. *Anguilla* first radiated about 20 million years ago. The ancestors of Atlantic eels trekked across the Central American Isthmus to the Sargasso Sea for spawning. Multiple radial speciation events have occurred in the Indo-West Pacific region during *Anguilla* evolution.
Teng et al. (2009)	19 species/ subspecies	13 mt protein coding sequences (11,394 bp)	A new species *Anguilla huangi* was discovered, description and nomenclature. The phylogeny of freshwater eels is polytomies which are likely derived from multiple radiation events. Atlantic eels dispersed by the Central American Isthmus.
Multidirectional dispersion			
Minegishi et al. (2005)	18 species/ subspecies	protein-coding genes and RNA genes of Whole mitochondrial (mt) genome (15,187 bp)	*A. mossambica* was the most basal species of *Anguilla* eels. The unique geographic distribution of *Anguilla* eels has resulted from multiple dispersal events. The beginning of speciation was estimated as 20 Mya but it may be underestimated.

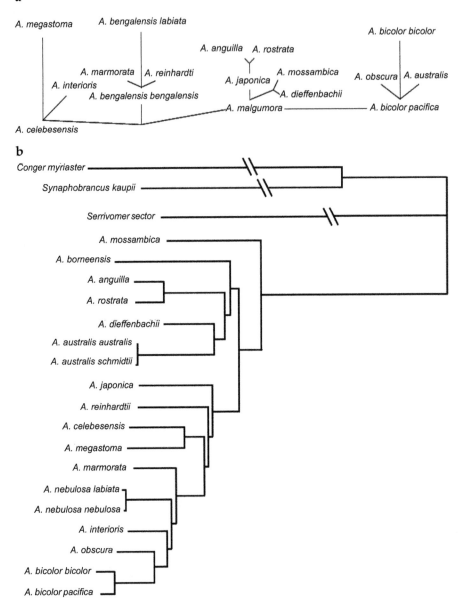

Fig. 2. contd....

Fig. 2. contd....

c

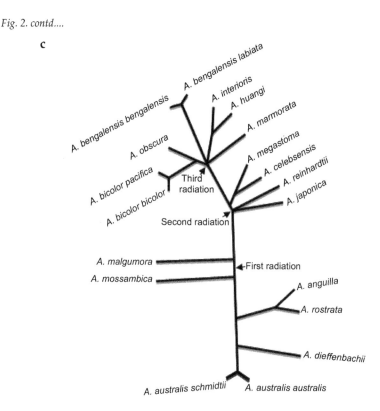

Fig. 2. (a) The phylogenetic synopsis of *Anguilla* eels from Ege's (1939). The figure was modified according to Lin et al. (2001). (b) The consensus tree of 18 *Anguilla* eels using the Bayesian analysis. The tree was modified according to Minegishi et al. (2005). (c) Inferred molecular phylogeny of 19 *Anguilla* species using 13 mitochondrial protein coding sequences. The figure was modified according to Teng et al. (2009).

have a recent common ancestor, while *A. japonica* separated early from other temperate eels, indicating that parallel evolution occurred subsequent to the common ancestor.

Fish speciation mechanisms include allopatric speciation, parapatric speciation, sympatric speciation, hybrid speciation, and polyploidy (Coyne and Allen Orr 2004). The 19 *Anguilla* species are extremely extensive distributions across the Atlantic, Pacific, and Indian Oceans. Unlike tropical eels, each of the six temperate eels has a unique ambit (Fig. 3). For example, *A. rostrata* and *A. anguilla* dwell in the northeastern and northwestern Atlantic Ocean, respectively; *A. japonica* resides in the northwestern Pacific Ocean, and *A. australis* and *A. dieffenbachii* are distributed from southeastern Australia to New Zealand waters. According to their phylogenetic status and distribution, the most likely explanation is that allopatric speciation occurred first, followed by sympatric speciation. Allopatric speciation requires complete geographic

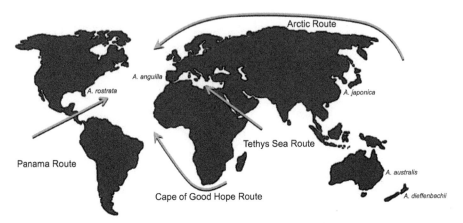

Fig. 3. The distributions of five temperate *Anguilla* eels. Four arrows were the possible dispersal routes of ancestral eels into Atlantic Ocean.

isolation by a physical barrier extrinsic to the organism or by the biological characters of the organisms themselves. The theory of vicariant allopatric speciation describes reproductive isolation evolving after the geographic range of a species is split into two or more large isolated populations. Many scholars believe that the geographic separation of previously interbreeding populations can occur through diverse climatic or geological events including continental drift, climatic change, or extirpation of intermediate populations (Coyne and Allen Orr 2004). Under these circumstances, genetic differentiation is nearly inevitable: populations usually occupy ecologically different habitats that impose different forms of selection, different mutations arise in different populations, genetic drift fixes different genes in different isolates, and initial evolutionary divergence can be amplified by later selection (Coyne and Allen Orr 2004). Apparently, allopatrically divergent taxa are likely to have non-overlapping distributions. Temperate eels, the Japanese eel, Atlantic eels, and Oceania eels reside in various oceans, so dispersal abilities probably play only modest roles in allopatric speciation in the marine realm, as with other groups of organisms (Mayr 1954; Hellberg 1998). Allopatric speciation, as observed within *Anguilla* eels, can also be found in numerous genetic patterns in marine fish (Lacson and Clark 1995; Leray et al. 2010; McMillan and Palumbi 1995). However, *Anguilla* eels can undergo different speciation modes during their evolution. We can interpret *A. australis* and *A. dieffenbachii* to be sister species that share overlapping habitat, suggesting sympatric speciation because they also have more closely related genetic components than allopatric temperate eels. *Anguilla* spawns in the sea and grows to adulthood either in rivers, estuaries or the sea. Arai et al. (2001) examined the larval duration of anguillid eels using the otolith microstructure of temperate and tropical eels and found that tropical species have a much shorter larval duration than temperate species. Recently proposed phylogenetic relationships plus the present knowledge about the spawning migrations of anguillid eels strongly suggest that the long spawning migrations of temperate species are not phylogenetically genealogical, but are an adaptation for species that migrated to temperate regions. This seems to explain some of the complications of *Anguilla* evolution.

How did *A. anguilla* and *A. rostrata* colonize the Atlantic Ocean?

Six hypothesized *Anguilla* eel dispersal routes are proposed: (1) Tethys Sea route: the ancestor of Atlantic eels passed through the Tethys Sea to current locations; (2) Cape of Good Hope route: the ancestor invaded the Atlantic Ocean via the southern tip of Africa; (3) Central American Isthmus route: the ancestor moved eastward in the Pacific Ocean and across the Central American gateway to the north Atlantic Ocean; (4) Arctic route: the ancestor migrated through the Arctic Ocean to the north Atlantic Ocean (Fig. 3); (5) multidirectional dispersion; and (6) multiple radiation events (Aoyama and Tsukamoto 1997; Tsukamoto and Aoyama 1998; Aoyama et al. 2001; Bastrop et al. 2000; Lin et al. 2001; Minegishi et al. 2005; Teng et al. 2009). However, there are two more important reports, from Minegishi et al. (2005) and Teng et al. (2009). Minegishi et al. (2005) analyzed protein coding sequences, tRNA, and RNA genes (15,187 nucleotides) of whole mitogenomes to construct phylogeny and estimate the divergence times of 18 *Anguilla* species/subspecies and three anguilliform fishes as outgroups (Fig. 2b). Their results suggest that the preceding four hypotheses are not supported by mtDNA data, implying that the unique geographic distribution of present-day species of *Anguilla* is a consequence of multiple dispersal events. The initiation of *Anguilla* speciation is estimated to have begun more than 20 Mya. Teng et al. (2009) infer the molecular phylogeny of *Anguilla* eels based on 13 mitochondrial protein coding sequences (excluded mitochondrial rRNA, tRNA, and control region), and imply that three possible radiation events accompanied extensive expansion (Fig. 2c). They also indicated that the Central American Isthmus hypothesis for the dispersal route of the Atlantic eels was supported by molecular evidence. In both of the foregoing studies, however, dispersal modes based on phylogenetic inferences seem to be mere conjectures. Dissimilarities in these inferences result from differences in sampling completeness, genetic markers, methods for restructuring phylogeny, and the authors' favorite hypotheses.

  Anguilla eels have been found in Europe from the Eocene epoch (Ypresian, about 50–55 Mya) according to fossil records (Patterson 1993). The evolutionary history of *Anguilla* can be traced back to 50 Mya. Johnson et al. (2011) analyzed the molecular phylogeny of anguilliforms via whole mitogenome sequencing. The divergence time between *Stemonidium* and *Anguilla* was approximately 150 Mya (Fig. 1). The ancestor of *Anguilla* appeared roughly 50–150 Mya, a period that can be narrowed down only after more research is completed in the future.

It follows from what has been said that the geologic history of earth is worth mentioning in passing. The supercontinent, Pangaea, existed from approximately 510 to 180 Mya (Condie 1989). About 200–180 Mya, during the mid-Mesozoic era, Pangaea began to split into a southern portion, Gondwana, and a northern portion, Laurasia. Around 60 Mya, Laurasia split to form North America and Eurasia (Pielou 1979; Smith et al. 1994). The Panthalassic Ocean was the great global ancestral Pacific Ocean surrounding Pangaea (about 300–200 Mya) (Taira 2001). The Tethys Ocean formed gradually between Cimmeria and Gondwana during the Jurassic period (150 Mya). Cimmeria finally collided with Laurasia. Water levels rose and the western Tethys came to shallowly cover significant portions of Europe, forming the Tethys Sea. Between the Jurassic and Late Cretaceous periods (which started about 100 Mya), even Gondwana began breaking up, pushing Africa and India north across the Tethys and opening up the Indian Ocean. As these land masses pushed in on the Tethys ocean from all sides, the final closing of the Tethys occurred near the Oligocene/Miocene boundary (20 Mya) (Steininger and Rögl 1984; Dercourt et al. 1986) (Fig. 4). Although the North Atlantic opened during the final breakup of Pangaea, a land bridge appears to have remained between North America and Europe (Tiffney 1985). Assuming that the ancestor of *Anguilla* appeared in the Tethys Ocean 50–100 Mya, it may have colonized this sea and its coastal freshwater regions, but its life cycle then is unknown. When North America drifted away from Eurasia, the ancestor of Atlantic eels may have remained on both continents, allowing vicariant allopatric speciation to occur. The closing of the Tethys severed

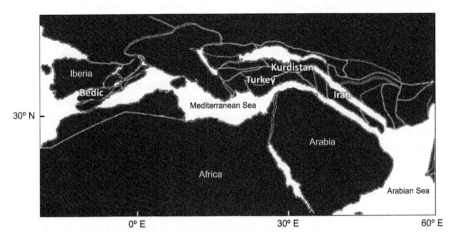

Fig. 4. A schematic map of the Mediterranean region 25 Mya modified according to Hrbek and Meyer (2003). Remnant of the Tethys Sea is the region between the Mediterranean and the Arabian Seas. Closing of the Tethys proceeded by northward movement of the African and Arabian plates.

the marine seaway and thus terminated the exchange of tropical marine elements between the Indian and Atlantic Oceans (Hrbek and Meyer 2003). The formation and isolation of many new geological units likely resulted in the vicariant speciation of *Anguilla* eels.

Gondwana broke up about 125 Mya, with Africa splitting off from Antarctica and India separating from Antarctica/Australia. Africa separated from South America about 100–120 Mya (Smith et al. 1994; Hay et al. 1999), India collided with Asia and formed the Himalayas about 50 Mya, and Australia finally separated from Antarctica about 45 Mya (White 2006).

The continents continued to drift towards their present positions during the Eocene epoch. The system of oceanic currents was radically different from that of today due to continental drift. The *Anguilla* eel origin occurred in what is today known as the Indo-Pacific region. These geological events, the former oceanic current system, and climate change had profound effects on *Anguilla* distribution and speciation. In addition, the extinction events of species include the Eocene–Oligocene extinction event (about 33.9 Mya), Middle Miocene disruption (about 14.5 Mya), Pliocene-Pleistocene boundary marine extinction event (about 2 Mya), Quaternary extinction events (0.64, 0.074, 0.013 Mya), and the Holocene extinction event. The latter is ongoing and has unpredictable consequences. However, species extinction may cause perplexed phylogenetic inference.

The current trend of study

In order to explore the evolution of *Anguilla* eels, the question which we must consider next is whether cryptic or undetected species exist in the world. The integrity of *Anguilla* species sampling can influence evolutionary inference. Recently, Teng et al. (2009) discovered a new species from northern Luzon, Philippines, supporting the existence of cryptic species. However, the morphological characters of glass eels are too similar to determine species clearly by eye, and adult color patterns are often variable in variegated eels. Thus, there is room for further investigation of *Anguilla* species. One further point that needs to be clarified is that all phylogenetic investigations of *Anguilla* eels to date rely on the molecular markers within mitochondrial DNA. No one considered any nuclear molecular markers, which can provide genetic information from both parents and infer evolutionary processes across distinct spatio-temporal scales (Sequeira et al. 2008). A final point should be made that cytogenetic studies in *Anguilla* fishes are quite uncommon. Chromosome studies focusing on five *Anguilla* species (*A. anguilla*, *A. australis*, *A. japonica*, *A. rostrata*, and *A. marmorata*) (Table 2) and other eel species have not been done. Whether hybrid speciation or polyploidy occurred in *Anguilla* eels is still a mystery. Cytogenetic studies provide a starting point for such research in the future.

Table 2. Cytogenetic studies within five *Anguilla* eels. All data referred from Arai (2011).

Species	Karyotype	Locality	Reference
Anguilla anguilla 2n = 38			
	12M + 10SM + 16A	Poland	Park and Grimm (1981) Passakas (1976) Passakas (1981)
	22M/SM + 16A	Italy	Hinegardner and Rosen (1972) Sola et al. (1980)
	20M/SM + 18A	Europe	Kobayasi and Kimura (1975)
	21M/SM + 17A	Europe	Kobayasi and Kimura (1975)
A. australis 2n = 38			
	22M/SM + 16A	New Zealand	Sola et al. (1980)
	20M/SM + 18A	New Zealand	Nishikawa et al. (1971)
A. japonica 2n = 38			
	20M/SM + 18A	Japan	Kobayasi and Kimura (1975) Nishikawa et al. (1971)
	10M + 10SM + 18A	China	Yu et al. (1989)
	12M + 8SM + 18A	Korea	Park and Kang (1976) Park and Kang (1979)
A. marmorata 2n = 38			
	20M/SM + 18A	Japan	Kobayasi (1977)
A. rostrata 2n = 38			
	12M + 10M + 16A	USA	Park and Grimm (1981)
	22M/SM + 16A	USA	Hinegardner and Rosen (1972) Sola et al. (1980)
	14M + 6SM + 18A	USA	Ohno et al. (1973)
	10M + 10SM + 18A	USA	Hardie and Hebert (2003) Passakas (1981)

Acknowledgements

I am extremely grateful to YH Hung for her help with make drawings.

Keywords: Cryptic species, Indo-Pacific Ocean, sister species, dispersal route, Cytogenetic, Fossil records

References

Aoyama, J. and K. Tsukamoto. 1997. Evolution of the freshwater eels. Naturwissenschaften 84: 17–21.
Aoyama, J., M. Nishida and K. Tsukamoto. 2001. Molecular phylogeny and evolution of the freshwater eel, genus *Anguilla*. Mol. Phylogenet. Evol. 20: 450–459.
Arai, R. 2011. Fish Karyotypes. A Check List. Springer Press, New York.

Arai, T., D. Limbong, T. Otake and K. Tsukamoto. 2001. Recruitment mechanisms of tropical eels, *Anguilla* spp. and implications for the evolution of oceanic migration in the genus *Anguilla*. Mar. Ecol. Prog. Ser. 216: 253–264.

Bastrop, R., B. Strehlow, K. Jürss and C. Sturmbauer. 2000. A new molecular phylogenetic hypothesis for the evolution of freshwater eels. Mol. Phylogenet. Evol. 14: 250–258.

Belouze, A., M. Gayet and C. Atallah. 2003. Les premiers Anguilliformes. I. Re´vision des genres ce´nomaniens Anguillavus Hay, 1903 et Luenchelys nov. gen. Geobios 36: 241–273. (doi:10.1016/S0016-6995(03)00029-9)

Briggs, J.C. 1974. Marine Biogeography, McGraw-Hill Press, New York.

Carpenter, K.E. and V.G. Springer. 2005. The center of the center of marine shorefish biodiversity: the Philippine Islands. Environ. Biol. Fish. 72: 467–480.

Condie, K.C. 1989. Plate Tectonics and Crustal Evolution, Third Ed. Pergamon Press, Oxford.

Coyne, J.A. and H. Allen Orr. 2004. Speciation. Sinauer Associates Inc. Press, Sunderland, MA.

Dercourt, J., L.P. Zonenshain, L.E. Ricou, V.G. Kazmin, X. Le Pichon, A.L. Knipper, C. Grandjacquet, I.M. Sbortshikov, J. Geyssant, C. Lepvrier, D.H. Pechersky, J. Boulin, J.C. Sibuet, L.A. Savostin, O. Sorokhtin, M. Westphal, M.L. Bazhenov, J.P. Lauer, B. Biju-Duval, X. Le Pichon and A.S. Monin. 1986. Geological evolution of the Tethys belt from the Atlantic to the Pamirs since the Lias. Tectonophysics 123: 241–315.

Ege, V. 1939. A revision of the genus *Anguilla* Shaw: a systematic, phylogenetic and geographical study. Dana Rep. 16: 1–256.

Hardie, D.C. and P.D.N. Hebert. 2003. The nucleotypic effects of cellular DNA content in cartilaginous and ray-finned fishes. Genome 46: 683–706.

Hay, W.W., R.M. DeConto, C.N. Wold, K.M. Wilson, S. Voigt, M. Schulz, A.R. Wold, W.C. Dullo, A.B. Ronov, A.N. Balukhovsky and E. Söding. 1999. An alternative global Cretaceous paleogeography. *In*: E. Barrera and C. Johnson (eds.). Evolution of the Cretaceous Ocean-Climate System. Spec. Pap. Geol. Soc. Am. 332: 1–48.

Helfman, G.S., B.B. Collette and D.E. Facey. 1997. The Diversity of Fishes. Blackwell Science, Malden, MA.

Hellberg, M.E. 1998. Sympatric seashells along the sea's shore: the geography of speciation in the marine gastropod Tegula. Evolution 52: 1311–1324.

Hinegardner, R. and D.E. Rosen. 1972. Cellular DNA content and the evolution of teleostean fishes. Amer. Natur. 106: 621–644.

Hrbek, T. and A. Meyer. 2003. Closing of the Tethys Sea and the phylogeny of *Eurasian killifishes* (Cyprinodontiformes: Cyprinodontidae). J. Evol. Biol. 16: 17–36.

Inoue, J.G., M. Miya, M.J. Miller, T. Sado, R. Hanel, K. Hatooka, J. Aoyama, Y. Minegishi, M. Nishida and K. Tsukamoto. 2010. Deep-ocean origin of the freshwater eels. Biol. Lett. 6: 363–366. (doi:10.1098/rsbl.2009.0989)

Johnson, G.D., H. Ida, J. Sakaue, T. Sado, T. Asahida and M. Miya. 2011. A 'living fossil' eel (Anguilliformes: Protoanguillidae, fam. nov.) from an undersea cave in Palau. Proc. R. Soc. B doi:10.1098/rspb.2011.1289.

Kobayasi, H. 1977. A chromosome study on *Anguilla marmorata* Quoy & Gaimard. Japan Women's Univ. J. 24: 135–137. (In Japanese with English summary.)

Kobayasi, H. and Y. Kimura. 1975. On the chromosomes of leucocytes from Japanese and European eels (*Anguilla*). Japan Women's Univ. J. 22: 169–174. (In Japanese with English summary.)

Kumazawa, Y., M. Yamaguchi and M. Nishida. 1999. Mitochondrial molecular clocks and the origin of euteleostean biodiversity: familial radiation of perciforms may have predated the Cretaceous/Tertiary boundary. pp. 35–52. *In*: M. Kato (ed.). The Biology of Biodiversity. Springer-Verlag, Tokyo, Japan.

Lacson, J.M. and S. Clark. 1995. Genetic divergence of Maldivian and Micronesian demes of the Damselfishes *Stegastes nigricans, Chrysiptera biocellata, C. glauca* and *C. leucopoma* (Pomacentridae). Mar. Biol. 121: 585–590.

Leray, M., R. Beldade, S.J. Holbrook, R.J. Schmitt, S. Planes and G. Bernardi. 2010. Allopatric divergence and speciation in coral reef fish: the three-spot dascyllus, *Dascyllus trimaculatus,* species complex. Evolution 64: 1218–1230.

Lin, Y.S., Y.P. Poh and C.S. Tzeng. 2001. A phylogeny of freshwater eels inferred from mitochondrial genomes. Mol. Phylogenet. Evol. 20: 252–261.

Lin, Y.S., Y.P. Poh, S.M. Lin and C.S. Tzeng. 2002. Molecular techniques to identify freshwater eels: RFLP analyses of PCR-amplified DNA fragments and allele-specific PCR from mitochondrial DNA. Zool. Stud. 41: 421–430.

Mayr, E. 1954. Change of genetic environment and evolution. pp. 157–180. *In*: E.B. Ford (ed.). Evolution as a Process. Allen & Unwin, London, UK.

McMillan, W.O. and S.R. Palumbi. 1995. Concordant evolutionary patterns among Indo-West Pacific Butterfly fishes. Proc. R. Soc. Lond. B 260: 229–236.

Minegishi, Y., J. Aoyama, J.G. Inoue, M. Miya, M. Nishida and K. Tsukamoto. 2005. Molecular phylogeny and evolution of the freshwater eels genus *Anguilla* based on the whole mitochondrial genome sequences. Mol. Phylogenet. Evol. 34: 134–146.

Mora, C., P.M. Chittaro, P.F. Sale, J.P. Kritzer and S.A. Ludsin. 2003. Patterns and processes in reef fish diversity. Nature 421: 933–936.

Nelson, J.S. 2006. Fishes of the World. John Wiley and Sons, New York.

Nishikawa, S., K. Amaoka and T. Karasawa. 1971. On the chromosomes of two species of eels (*Anguilla*). Chrom. Inform. Serv. 12: 27–28.

Ohno, S., L. Christian, M. Romero, R. Dofuku and C. Ivey. 1973. On the question of American eels, *Anguilla rostrata* versus European eels, *Anguilla anguilla*. Experientia 29: 891.

Park, E.H. and Y.S. Kang. 1976. Karyotype conservation and difference in DNA amount in anguilloid fishes. Science 193: 64–66.

Park, E.H. and Y.S. Kang. 1979. Karyological confirmation of conspicuous ZW sex chromosomes in two species of Pacific anguilloid fishes (Anguilliformes: Teleostomi). Cytogenet. Cell Genet. 23: 33–38.

Park, E.H. and H. Grimm. 1981. Distribution of C-band heterochromatin in the ZW sex chromosomes of European and American eels (Anguillidae, Teleostomi). Cytogenet. Cell Genet. 31: 167–174.

Passakas, T. 1976. Further investigations on the chromosomes of *Anguilla anguilla*. Folia Biol. (Krakow) 24: 239–244.

Passakas, T. 1981. Comparative studies on the chromosomes of the European eel (*Anguilla anguilla* L.) and the American eel (*Anguilla rostrata* Le Sueur). Folia Biol. (Krakow) 29: 41–57.

Patterson, C. 1993. Osteichthyes: Teleostei. pp. 621–656. *In*: M.J. Benton (ed.). The Fossil Record 2. Chapman & Hall Press, London, UK.

Pielou, E.C. 1979. Biogeography. John Wiley and Sons, New York.

Sequeira, A.S., M. Sijapati, A.A. Lanteri and L. Roque Albelo. 2008. Nuclear and mitochondrial sequences confirm complex colonization patterns and clear species boundaries for flightless weevils in the Galápagos archipelago Philos. Trans. R. Soc. Lond. B Biol. Sci. 363: 3439–3451.

Smith, A.G., D.G. Smith and B.M. Funnell. 1994. Atlas of Mesozoic and Cenozoic Coastlines. Cambridge University Press, New York.

Sola, L., G. Gentili and S. Cataudella. 1980. Eel chromosomes: cytotaxonomical interrelationships and sex chromosomes. Copeia 1980: 911–913.

Steininger, F.F. and F. Rögl. 1984. Paleogeography and palinspatic reconstruction of the Neogene of the Mediterranean and Paratethys. pp. 659–668. *In*: J.E. Dixon and A.H.F. Robertson (eds.). The Geological Evolution of the Eastern Mediterranean, Geological Society Special Publication no. 17. Blackwell Scientific, Oxford.

Taira, A. 2001. Tectonic evolution of the Japanese Island Arc System. Annu. Rev. Earth Planet. Sci. 29: 109–134.

Teng, H.Y., Y.S. Lin and C.S. Tzeng. 2009. A new *Anguilla* species and a reanalysis of the phylogeny of freshwater eels. Zool. Stu. 48: 808–822.

Tiffney, B.H. 1985. The Eocene North Atlantic land bridge: its importance in Tertiary and modern phytogeography of the Northern Hemisphere. J. Arnold Arbor. Harv. Univ. 66: 243–273.

Tseng, M.C. 2012. Evolution of microsatellite loci of tropical and temperate *Anguilla* eels. Int. J. Mol. Sci. 13: 4281–4294.

Tsukamoto, K. and J. Aoyama. 1998. Evolution of freshwater eels of the genus *Anguilla*: a probable scenario. Environ. Biol. Fish. 52: 139–148.

Watanabe, S., J. Aoyama and K. Tsukamoto. 2009. A new species of freshwater eel *Anguilla luzonensis* (Teleostei: Anguillidae) from Luzon Island of the Philippines. Fish. Sci. 75: 387–392.

White, M.E. 2006. Environments of the geological past. *In*: J.R. Merrick, M. Archer, G.M. Hickey and M.S.Y. Lee (eds.). Evolution and Biogeography of Australian Vertebrates. Auscipub, Oatlands.

Yu, X.J., T. Zhou, Y.C. Li, K. Li and M. Zhou. 1989. Chromosomes of Chinese freshwater fishes. Science Press, Beijing. (In Chinese.)

Evolutionary Genomics of North Atlantic Eels
Current Status and Perspectives

José Martin Pujolar[1], and Gregory E. Maes[2,3]*

Introduction

Recent genetic studies have greatly advanced our knowledge on the genetic diversity of eels, population genetic structure, effective population size and possible evolutionary responses to anthropogenic environmental stress, among others. Traditionally, these topics had been addressed in the past by studying a very limited number of molecular and transcriptional markers, mainly due to the shortage of genomic sequence resources available for eels. For instance, only five years ago, no genome sequencing had been conducted for any eel species and the number of sequences available in Genbank for the European eel consisted of only 121 expressed sequence tags (ESTs), 404 nucleotide sequences and 232 proteins. All this has changed recently due to the advances in the speed, cost and accuracy of next-generation sequencing (NGS) techniques that allow

[1] Department of Bioscience, Aarhus University, Ny Munkegade 114, Bldg. 1540, DK-8000 Aarhus C, Denmark.
[2] Centre for Sustainable Tropical Fisheries and Aquaculture, Comparative Genomics Centre, College of Marine and Environmental Sciences, James Cook University, Townsville, 4811 QLD, Australia.
E-mail: gregory.maes@jcu.edu.au
[3] Laboratory of Biodiversity and Evolutionary Genomics, University of Leuven (KU Leuven), B-3000 Leuven, Belgium.
* Corresponding author: jmartin@biology.au.dk

for the production of extremely large collections of data and the discovery of genome/transcriptome-wide resources at relatively low and decreasing costs. The advent of genotyping-by-sequencing methods such as RAD (Restriction Associated DNA marker) sequencing (Baird et al. 2008; Hohenlohe et al. 2010; Davey et al. 2011), i.e., sequencing of a reduced representation of the genome followed by the discovery of single nucleotide polymorphisms (SNPs), are revolutionizing the field of population genetics and can provide data on hundreds of thousands of markers distributed across the genome of non-model organisms in a short period of time. In the case of the European eel, the RAD approach has been used to generate a resource of over 350,000 SNPs (Pujolar et al. 2013a), which has been used to provide conclusive evidence for genomic panmixia in the European eel and to test for signatures of selection in the genome (Pujolar et al. 2014a), to study hybridization between the European and the American eel (Pujolar et al. 2014b) and, to study speciation between North Atlantic eels (Jacobsen et al. 2014b). The revolution in NGS methods has also allowed the sequencing and annotation of the first eel transcriptome (Coppe et al. 2010) as well as the complete genome sequencing of the European (Henkel et al. 2012a) and the Japanese eel (Henkel et al. 2012b). These studies provide a rich source of data to better characterize gene expression variation in natural or experimental populations, besides providing more power to discover and identify new genes involved in important pathways that might help our fundamental understanding of eels and ultimately result in better measures for the management and conservation of the species.

Population genetics

Based on the occurrence of a single spawning area (Sargasso Sea) and an extensive migration loop with great opportunity for mixing of individuals, North Atlantic eels have been regarded as classic textbook examples of panmixia, with both the American and the European eel being constituted by a single randomly mating population.

In case of the American eel, the effort involved in the study of its population structure was limited to a few markers and/or samples (Avise et al. 1986; Mank and Avise 2003; Wirth and Bernatchez 2003). However, the recent paper by Côté et al. (2013) showed that the American eel is a panmictic species, in the most extensive study to date on American eel population genetics. Over 2,000 individuals representing 12 cohorts were genotyped with 18 microsatellite loci over a large geographical scale in North America. Results showed a total lack of genetic differentiation, hence providing decisive evidence for panmixia in the American eel. Global genetic differentiation among all samples was virtually zero, with no significant differences among sampling sites or between life stages.

In the case of the European eel, the panmixia hypothesis was challenged by evidence for an isolation-by-distance pattern (Wirth and Bernatchez

2001). However, this pattern proved to be unstable over time when temporal replicates were included in the analysis (Dannewitz et al. 2005), leading to the suggestion of a leading role of temporal processes shaping population differentiation in the European eel (Maes et al. 2006, 2009; Pujolar et al. 2006, 2011a). Recently, the study of Als et al. (2011) also provided conclusive evidence for panmixia in the European eel after conducting the most extensive study to date in terms of samples and markers, also including Sargasso Sea larvae. After genotyping over 1,000 individuals at 21 microsatellite loci, the study showed a very low and non-significant genetic differentiation between locations across Europe ($F_{ST} = 0.00024$ among a total of 21 geographical and temporal samples). Similarly, no genetic differences were found when comparing larvae collected *in situ* in the Sargasso Sea ($F_{ST} = 0.00076$), which provides very strong support for panmixia in the European eel. These results clearly demonstrated the challenges faced by the eel genetics and management community in reliably assessing the neutral pattern of genetic differentiation in such vagile species using available markers, but more importantly the need for additional markers in order to better estimate genome-wide levels of diversity and differentiation (van Ginneken and Maes 2005; Maes and Volckaert 2007).

With the advent of the genomics revolution and more cost-efficient marker development in non-model organisms (Davey et al. 2011), further evidence of panmixia has also been demonstrated at the genome-wide level using a RAD-sequencing approach (Pujolar et al. 2014a). A total of 259 glass eels from eight distinct European locations (from Morocco to Iceland) were RAD-sequenced and the patterns of genome-wide genetic diversity across these locations were examined. Samples were compared using 453,062 single nucleotide polymorphisms (SNPs). No differences were found in values of genetic diversity and overall genetic differentiation was very low ($F_{ST} = 0.0007$) and not significant. Visualization of pairwise genetic distances showed all samples clustering together with no apparent geographic pattern (Fig. 1), while a Mantel test showed no correlation between genetic and oceanic distances, suggesting a lack of isolation by distance (IBD). Similarly, no genetic differentiation was found when considering only mitochondrial markers (41 SNPs), $F_{ST} = 0.0002$. This indicates that most of the genome is homogenized by gene flow, providing further evidence for genomic panmixia in the European eel.

Surprisingly, a recent paper by Baltazar-Soares et al. (2014) reported a significant genetic structure using a portion of a mitochondrial gene (ND5) and 17 microsatellite loci. However, re-examination of the data suggests that the differences are due to low sample sizes and biased sampling. In this sense, most samples consisted of 10 individuals or less, and differences were only found among samples from Ireland, which represented nine out of the 12 samples included in the study.

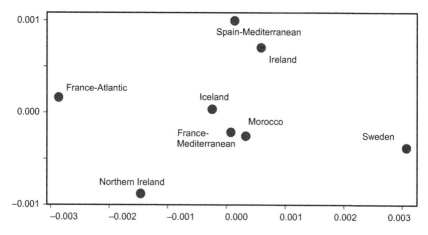

Fig. 1. Multi-dimensional Scaling (MDS) analysis based on RAD-tag SNPs pairwise F_{ST} between European glass eel samples. No spatial clustering was observed and samples did not group according to geographic location, i.e., Iceland clustered with the samples from Morocco and France-Mediterranean. Modified from Pujolar et al. 2014a.

Detection of selection footprints

One of the fields advancing the most in evolutionary biology thanks to the developments in next-generation sequencing is the study of selection and adaptation, since now scientists can interrogate the full genome in search of signatures of selection.

As mentioned above, Gagnaire and collaborators studied selection focusing on its role in hybridization between North Atlantic eels. Gagnaire et al. (2009) suggested an important role for selection in the hybridization process. Gagnaire et al. (2012a) focused more on the detection of signatures of selection and compared sequences obtained from transcriptome data of 58 American and European eel individuals. Signatures of selection were found in 94 nuclear genes. Surprisingly, signatures of selection were also found in a single mitochondrial gene, ATP6. This was also confirmed in a recent study comparing amino acid changes in ATP6 among 18 eel species with the sequence of the putative ancestral species (Jacobsen et al. 2015). The authors found that species with longer migratory loops and consequently higher metabolic requirements showed a higher number of fixed non-synonymous changes and also more changes of stronger physico-chemical and structural importance in comparison with species with shorter migratory loops. This seems to support a link between genotype and phenotype driven by positive selection in ATP6.

Gagnaire et al. (2012b) also tested for spatially varying selection in the American eel using a 100 coding-gene SNP array. A total of 13 genes were candidates under selection. Most of those were genes related to metabolism and showed significant positive correlations with temperature. Building on these results, Ulrik et al. (2014) tested for parallel signatures of selection between North Atlantic eels, by genotyping a total of 321 European eel individuals (glass eels) from 8 geographic locations using the panel of 100 coding-gene SNPs developed by Gagnaire et al. (2012b), thus allowing for comparison with the results observed in the American eel. The study found 11 candidate genes under selection in the European eel and mainly constituted genes with a role in metabolism as well as defense genes. When comparing results across studies, none of the genes under selection in the European eel was also under selection in the American eel. The lack of parallel patterns was somewhat unexpected due to the sister species status of North American eels and the fact that they are found in similar environments and habitats. The different signatures of selection found could be due to the distinct selective pressures encountered, especially since the European eel shows a much larger migration loop relative to the American eel. Alternatively, given that most phenotypic traits are polygenic, the likelihood of selection acting on exactly the same genes in both species is reduced.

A test of selection at the full genomic level was recently conducted by Pujolar et al. (2014a) on the European eel, testing for footprints of spatially varying selection on a total of 259 RAD-sequenced glass eels obtained across the entire geographic distribution of the species (from Morocco to Iceland). Despite panmixia, a small set of SNPs showed high genetic differentiation consistent with single-generation signatures of spatially varying selection acting on glass eels. Candidate genes under putative selection constituted a wide array of functions, including signaling, receptors and circadian rhythm (i.e., period). The authors also investigated the distribution of genes under selection, finding them spread all over the genome showing just a few single narrow peaks (Fig. 2), unlike other species in which genes under selection have been located in so-called genomic islands of divergence (large regions of increased differentiation). This supports the conclusion that in the case of eels, selection occurs within a single generation due to panmixia and all effects of diversifying selection are removed by gene flow.

Speciation and demographic history

The speciation mechanism in eels, and in particular, the estimation of the divergence time between the two North Atlantic eel species remains unsolved. Studies attempting to estimate the split between the American and the European eel have yielded contradictory results in the past: from 1.5 Mya based on Restriction Fragment Length Polymorphism (RFLP) analysis of total mtDNA (Avise et al. 1986) to 10.2 Mya based on the mitochondrial cytochrome b gene (Tsukamoto and Aoyama 1998), while Minegishi et al. (2005) reported an intermediate value (5.8 Mya) using the entire mitogenome.

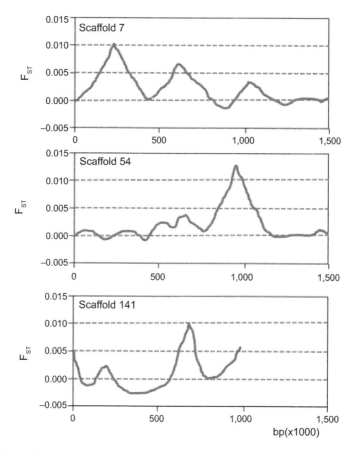

Fig. 2. Plots of average F_{ST} calculated using a 50-kb sliding window for three scaffolds (7, 54 and 141) in the European eel genome. F_{ST} is low across the scaffolds, with just a few narrow peaks. No regions with pronounced divergence peaks (genomic islands) are observed. Modified from Pujolar et al. 2014a.

Two recent studies deal with the speciation and demographic history of North Atlantic eels. Using complete mitogenomes, Jacobsen et al. (2014a) analyzed a total of 104 mitogenomes of the two species along with mitogenomes of other eel species and out-groups. Divergence time was calibrated taking into account recent knowledge of out-group relationships and fossil information in order to avoid misleading calibrations based on standard substitution rates used in previous studies. Divergence between North Atlantic eels was estimated to circa 3.38 Mya, which coincides with the closure of the Panama Gateway. The authors hypothesized that this event could have caused a reinforcement of the Gulf Stream and the advection of larvae towards Europe. Bayesian analyses suggested past population declines during glaciations and important fluctuations in effective population sizes related to climate.

In a second study, Jacobsen et al. (2014b) addressed the issue of speciation in North Atlantic eels using a RAD sequencing approach. The advantage of this method is that a large number of markers are used. In this case, analysis was conducted using over 300,000 SNPs. Comparison of the American vs. the European eel samples showed an overall F_{ST} of 0.041. Although there were some highly differentiated SNPs (a total of 3,757 SNPs showed statistically significant differences), the majority of SNPs showed low values close to zero. The relatively low level of background differentiation between North Atlantic eels suggests a scenario of speciation with gene flow (Feder et al. 2012), resulting in incomplete reproductive isolation and the ongoing interbreeding between species, which is in agreement with the observation of hybrids, although in small numbers (see hybridization section).

Other previous studies on the demography of the European eel include the study of Pujolar et al. (2011a) using microsatellite data to investigate recent population bottlenecks. The main result of the study is that despite the large population decline of the European eel (99% in some locations relative to 1970s and 1980s data), the species has not experienced a genetic decline of the same magnitude. No evidence for a genetic bottleneck in the European eel was found: genetic diversity was moderate to high and allele range showed a continuous distribution including many alleles. Moreover, data suggested a stationary population with growth values not different from zero and a historical effective population size of around N_e = 5,000–10,000. However, genetic analysis of historical samples and temporal comparison of pre-decline vs. present samples over the last century could still provide novel insights into the effect of the decline at the genetic and genomic level over the last century of exploitation (Schaerlaekens et al. 2011).

In regard to effective population size, the recent paper of Pujolar et al. (2013a) estimated a much larger effective population size using a RAD-sequencing approach. After RAD-sequencing 30 individuals, a genome-wide set of over 80,000 loci and over 375,000 associated SNPs was generated. The number of SNPs per locus was surprisingly high, ranging from 1 to 22, with two SNPs being the most frequent (Fig. 3). Nucleotide diversity was high (0.00529 on average). Based on this average, long-term effective population size was estimated to range between 132,000 and 1,320,000, depending on the mutation rate used. The N_e obtained from the RAD data (> 100,000) is a more realistic value than the one obtained by the previous studies on microsatellites.

Hybridization

The first evidence of hybrids between the European and the American eel in Iceland came from the study of Avise et al. (1990) using a combination of allozymes, mitochondrial DNA markers and vertebrae counts. Later studies confirmed the occurrence of hybrids in Iceland using microsatellites (Mank and Avise 2003) and amplified fragment length polymorphism analysis (Albert

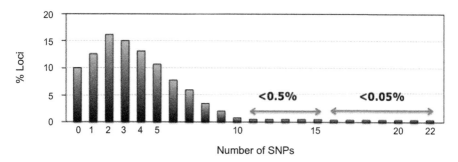

Fig. 3. Distribution of the number of SNPs per locus in RAD sequenced European eels. Modified from Pujolar et al. 2013a.

et al. 2006). However, in all cases the lack of discriminatory power of the markers did not allow determining the status of the hybrids further than F1s.

One of the latest studies by Pujolar et al. (2014b) took a new look at the question of hybridization between the European and the American eel and used for the first time high-resolution markers allowing them to classify later-generation (post F1) hybrids. In order to do so, the authors used a RAD-sequencing approach to identify fixed ($F_{ST} = 1$) species-specific diagnostic markers. Using an array of 96 diagnostics SNPs, they then genotyped a large number of individuals from Iceland. Hybrids constituted 10.7% of the Icelandic individuals (Fig. 4). Most hybrids were identified as F1 hybrids between European eel female and American eel male crosses. Backcrosses were also detected, including: (i) first-generation backcrosses between pure European eel and F1 hybrid and (ii) second-generation backcrosses between first-generation backcrosses (American eel x F1 hybrid) and pure European eel. The authors hypothesized that hybrids might show an intermediate migratory larval behaviour compared to pure individuals, which might explain why hybrids

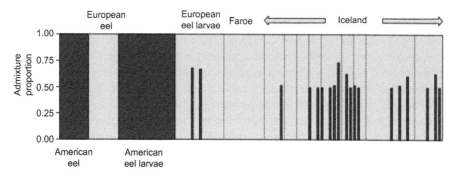

Fig. 4. Admixture analysis assuming the presence of two groups (K = 2: European and American eels). Each vertical line represents one individual partitioned into segments according to admixture proportion of the European (light) and the American eel (dark). Samples include larvae from the American and European eel larvae collected in the Sargasso Sea and the European eel samples collected in the Faroe Islands and Iceland. Modified from Pujolar et al. 2014b.

are found almost exclusively in Iceland, which is somehow situated halfway between North America and Europe. The study also included the analysis of larvae from the Sargasso Sea, with a very low proportion of admixed individuals in the larvae (Fig. 4), namely around 2%, with all hybrids being identified as backcrosses. Surprisingly, no F1 hybrids were identified out of 98 larvae.

Hybridization was also briefly discussed in the Pujolar et al. (2014a) paper on panmixia. Within Europe, very few hybrids were observed. One individual from France (Gironde) showed a 5% admixture proportion plus three further individuals from Ireland and Sweden presented a 3% admixture proportion. Those would correspond to old generation hybrids (over third generation) and would confirm that hybrids are mostly localized in Iceland and that their presence in mainland Europe is rare. Interestingly, some individuals showed a European nuclear genotype but an American mitogenome, suggestive of an old hybridization even (at least more than six generations ago).

In previous studies, which did not look directly at the nature of hybrids in Iceland, Gagnaire et al. (2009) re-examined previous AFLP data from the study of Albert et al. (2006). They concluded that hybridization was likely shaped by selection, both at the intra- and inter-specific level. Later, Gagnaire et al. (2012a), although focusing more on selection (see later), suggested a possible cyto-nuclear incompatibility between North Atlantic eels. Signatures of positive selection were found in both the mitochondrial ATP6 gene and its nuclear receptor. The authors pointed to a possible incompatibility in hybrids showing the American mitochondrial ATP6 gene and the European receptor or vice-versa, which could explain the partial reproductive isolation between North Atlantic eels.

Spatio-temporal modelling

Due to its peculiar life-history and extreme reproductive and larval migrations, one of the main challenges in eel genetics has always been to build realistic test-hypotheses to test empirical data. Andrello et al. (2010) developed the first genetic-demographic model specific for the European eel, including different levels of mixing between adults and also between larvae. The study showed that the very low levels of mixing occurring during larval dispersal or adult migration are sufficient to prevent genetic differentiation. Any genetic differences between putative subpopulations would be erased if those were connected by very small migration rates. The model also showed how small-scale temporal differences can arise if the spawning stock is subdivided into distinct reproductive groups. Finally, the model reinforced the importance of including temporal replicates in population genetic studies as emphasized before by Dannewitz et al. (2005) and Maes et al. (2006), since the use of a limited number of temporal recruits might lead to overestimated differences among geographic samples.

Using a similar approach, Pujolar et al. (2011b) combined population genetics and modelling, based on a panel of 22 EST-derived microsatellites. Microsatellite data showed a pattern of genetic patchiness, e.g., highly significant genetic differentiation with no apparent geographic or temporal pattern. Since genetic patchiness likely results from a limited parental contribution to each spawning event, the authors used a modelling approach to unravel the demographic mechanisms that can produce such patterns. The value of genetic differentiation found was predicted in the model when spawning takes place in a limited number of isolated events in the Sargasso Sea with an average of 130–375 breeders per event.

Environmental transcriptomics

In recent years, small and large-scale studies on gene expression profiling (microarrays) have been conducted for the first time on eels. Microarrays allow for the measurement of the activity of thousands of genes at once, providing a snapshot of which genes are being actively expressed at a particular moment in time. Transcriptomic approaches have the potential to identify and study the genetic basis of those traits affecting fitness and to further identify the genes whose expression changes in response to environmental perturbations (e.g., pollution, parasite infestation) and these genes therefore become candidate genes for being involved in the response.

Using the first eel array consisting of 678 probes, Bernatchez et al. (2011) investigated the different timing in gene expression in larvae of the American and the European eel collected *in situ* in the Sargasso Sea after spawning. When comparing gene expression profiles between species, no gene showed significant differences. However, a total of 146 genes showed different timing of expression between species, some being expressed later in one species relative to the other, involving a large array of biological functions.

Two later transcriptomic studies have focused on the effects of pollution on the European eel. Pujolar et al. (2012) first updated the European eel transcriptome available at that time and also developed a microarray specific for eels consisting of 15,000 annotated genes. Then, the array was applied to detect differentially expressed genes between polluted sites in Italy. As expected, the genes related to detoxification were upregulated in the polluted sites relative to the clean sites, including CYP3A, which takes part in phase I of the xenobiotic metabolism and glutathione-5-transferase that takes part in phase II, or glutathione peroxidase that is involved in oxidative stress. Additionally, several metabolic related genes were down-regulated. A similar pattern was found when comparing polluted and clean sites in Belgium (Maes et al. 2013; Pujolar et al. 2013b). Again, energy-related genes were down-regulated in animals from polluted sites, which points to a poor energetic condition of eels experiencing a high pollution burden.

Using the same 15,000-probe microarray used to study pollution; Churcher et al. (2014) investigated the changes in the gene expression profiles of the brains of male European eels during sexual maturation by comparing which genes were differentially expressed when immature and hormonally-induced mature eels were compared. A total of 1,497 differentially expressed genes were identified, 991 over-expressed and 506 under-expressed. Overall, the study showed that the brains of eels undergo major changes at the molecular level during sexual maturity. Upregulated genes were involved in cell signaling, including receptor and channels, development and differentiation. Several neuroendocrine genes were also expressed at higher levels in the brains of mature males in comparison to the brains of immature males. These genes are likely involved in gonadal maturation and development and act through the brain-pituitary-gonad axis. Genes involved in immune system function were down-regulated, suggesting that a re-allocation of resources occurs at this stage. Interestingly, the expression levels of several receptors and channels changed at sexual maturity, suggesting that some neuronal rewiring occurs in the forebrains of males at this stage. Eyes and olfactory epithelium also changed during sexual maturity, which might indicate that there is a change in the use of sensory system with reproduction.

The same array has also been applied to test the hypothesis that different genes are expressed as a response to upstream eel migration, in particular as a consequence of the passage of dams (Podgorniak et al. 2015). Brain, liver and muscle profiles were compared in individuals caught before and after surpassing two closely-located dams at Canal des Etangs, a freshwater corridor in southern France. Overall, the differences between samples were small, with no differences found in muscle or liver and only 26 out of 14,913 probes significantly differentiated in the brain. Differentially expressed genes included calmodulin, cofilin, keratins, cytokeratins, calcium binding proteins (S100 family), claudin and thy-1 membrane glycoprotein. Main functions of these proteins include cellular signaling, neural development and differentiation. Surprisingly, genes related to thyroid activity, such as iodothyronine deiodinase type I and III were not differentially expressed, despite other studies in fish showing that thyroid hormones might have a role in upstream migration and obstacle passing (Edeline et al. 2004).

Recently, a gene profiling study aimed at investigating the effect of pollution on eels using RNA-seq (Baillon et al. 2015), a more advanced method as compared to microarrays. While microarrays are based on hybridizing probes, RNA-seq utilizes the capabilities of next-generation sequencing methods to generate millions of DNA sequence reads in parallel at relative low costs derived from the entire RNA molecule. Using this approach, several genes over-expressed in the livers of eels showed a positive correlation with individual contaminants including arsenic and cadmium. Enriched pathways included p53 signaling pathway and lipid and protein metabolism in the case of cadmium and protein metabolism, protein folding and vasculotoxicity in the case of arsenium.

Finally, there have been some studies in the field of proteomics, in which the focus is on studying differences in protein abundance. While proteomic studies are important in confirming results from transcriptomics studies, few studies have been conducted to date in eels. Only recently, Roland et al. (2013, 2014) used the 2D-DIGE approach to study the effect of one particular group of pollutants, perfluorooctane sulfonates (PFOS). A total of 48 proteins showed significant differences in abundance between polluted and control animals. Functions of those proteins included cytoskeleton, protein folding, cell signalling, proteolytic pathway and carbonate and energy metabolism.

Eel genomes

Advances in sequencing technology, specifically the advent of relatively affordable massively parallel short read sequencing on the Illumina platforms, made it possible to conduct first the European eel transcriptome and later the European and the Japanese eel genome projects.

The first European eel transcriptome (Coppe et al. 2010) was obtained after sequencing a normalized cDNA library from a pool of 18 glass eels. Over 310,000 reads were obtained and later assembled into 19,631 transcribed contigs. At the time, this transcriptome was the only genomic resource available for eels and was extremely useful as it was the basis for the development of a microarray of 15,000 annotated probes (see above).

The European eel draft genome was recently published (Henkel et al. 2012), making it one of the first available genomes for a non-model fish species. However, the first version of the eel genome (v.1) should be considered a "draft genome" as it shows lower accuracy than finished sequences and some segments are missing or still incorrectly orientated, in the sense, that the genes are not ordered into chromosomes but rather scaffolds of different lengths instead, with the largest scaffolds being 2 million bp. However, it has become a very useful tool for population genetics studies, in particular those using RAD-sequencing and detection of SNPs by comparison to the eel genome; the improved version 2 (v.2) European eel genome version is expected to be released soon.

The Japanese eel draft genome was also recently made available (Henkel et al. 2013). Again, it is a draft genome divided into over 300,000 scaffolds with a maximum scaffold size of 1.14 million bp. As in the case of the European eel genome, its fragmentation is an inevitable result of the technology used. At the time, Illumina sequencing was limited to fragments smaller than 300 bp. Hence, it is understandable that the reconstruction of the entire eel genome (*ca.* 1 Gb or 10^9 bp) and a high-quality annotation were going to be difficult to achieve.

Conclusion and perspectives

Over the past five years of evolutionary genetics research on eels, the genomics revolution has clearly deepened our insights into the fascinating life cycle of eels and the genetic consequences of one of the largest migration loops in fishes.

The rapid development of new genomic resources such as the first transcriptome, genome-wide polymorphisms and the full genome of the European eel, has allowed for subsequent studies to focus on the remaining questions regarding the speciation with gene flow mechanism and the demographic history of North Atlantic eels, hybridization between North Atlantic eels, their respective population structure and the detection of latitudinal footprints of varying selection. Due to the remaining complexities of their life cycle and challenging sampling logistics in the Sargasso Sea, various studies performed advanced spatio-temporal modelling exercises in order to test and validate earlier evolutionary studies and predict population structure.

Following the dramatic decline in European eel stocks, the establishment of strict management plans has resulted in a slight recovery of the population. In this sense, integrating advanced tools such as transcriptomics and genomics into population and environmental management will hopefully improve our understanding of the long term evolutionary effects of the decline over the last decades. On this topic, genetic analysis of historical samples and comparison of pre-decline vs. present samples could provide new insights into the effects of the decline at the genetic and genomic level over the last century of exploitation, although recent studies suggest a lack of long-term genetic decline despite the observed demographic decline. Future research should also focus on a more in-depth analysis of adaptive genetic polymorphism at the full genome level. This could facilitate the discovery of new biomarkers for the quality of spawners that would allow using knowledge on the metabolic status of healthy populations before they start their spawning migrations. Finally, evolutionary genomics research on eels has mainly focused on the temperate European, American and Japanese eels. However, with the costs of sequencing ever decreasing, we are now at a stage where we can start investigating the evolutionary history and population structure of various tropical eel species that remain highly understudied to this day.

Acknowledgements

JMP and GEM would like to thank all their eel collaborators that helped them with sampling, genomics analyses and writing of the many articles throughout the last 10 years. Special thanks go to Filip Volckaert, Lorenzo Zane, Michael Hansen and Louis Bernatchez for their continuous support and enthusiasm in eel evolutionary genetics research.

Keywords: Adaptation, Demographic history, Ecotoxicogenomics, Effective population size, Evolution, Evolutionary modelling, Genome sequencing, Hybridisation, Mitogenome, Panmixia, Population genomics, RAD-tag sequencing, Selection, SNPs, Speciation, Transcriptomics/gene expression

References

Albert, V., B. Jónsson and L. Bernatchez. 2006. Natural hybrids in Atlantic eels (*Anguilla anguilla*, *A. rostrata*): evidence for successful reproduction and fluctuating abundance in space and time. Mol. Ecol. 15: 1903–1916.

Als, T.D., M.M. Hansen, G.E. Maes, M. Castonguay, L. Riemann, K. Aerestrup, P. Munk, T. Sparholt, R. Hanel and L. Bernatchez. 2011. All roads lead to home: panmixia of European eel in the Sargasso Sea. Mol. Ecol. 20: 1333–1346.

Andrello, M., D. Bevacqua, G.E. Maes and G. De Leo. 2010. An integrated genetic-demographic model to unravel the origin of genetic structure in European eel (*Anguilla anguilla*). Evol. Appl. 4: 517–533.

Avise, J.C., G.S. Helfman, N.C. Saunders and L.S. Hales. 1986. Mitochondrial DNA differentiation in North Atlantic eels: population genetic consequences of an unusual life history pattern. Proc. Natl. Acad. Sci. USA 83: 4350–4354.

Avise, J.C., W.S. Nelson, J. Arnold, R.K. Koehn, G.C. Williams and V. Thorsteinsson. 1990. The evolutionary genetic status of Icelandic eels. Evolution 44: 1254–1262.

Baillon, L., F. Pierron, R. Coudret, E. Normandeau, A. Caron, L. Peluhet, P. Labadie, H. Budzinski, G. Durrieu, J. Sarraco, P. Elie, P. Couture, M. Baudrimont and L. Bernatchez. 2015. Transcriptome profile analysis reveals specific signatures of pollutants in Atlantic eels. Ecotoxicology 24: 71–84.

Baird, N.A., P.D. Etter, T.S. Atwood, M.C. Currey, A.L. Shiver, Z.A. Lewis, E.U. Seiker, W.A. Cresko and E.A. Johnson. 2008. Rapid SNP discovery and genetic mapping using sequenced RAD markers. PLoS One 3: e3376.

Baltazar-Soares, M., A. Biastoch, C. Harrod, R. Hanel, L. Marohn, E. Prigge, D. Evans, K. Bodles, E. Behrens, C.W. Böning and C. Eizaguirre. 2014. Recruitment collapse and population structure of the European eel shaped by local ocean current dynamics. Curr. Biol. 24: 1–5.

Bernatchez, L., J. St-Cyr, E. Normandeau, G.E. Maes, T.D. Als, S. Kalujnaia, G. Cramb, M. Castonguay and M.M. Hansen. 2011. Different timing in gene expression regulation between leptocephali of the two *Anguilla* species in the Sargasso Sea. Ecol. Evol. 1: 459–467.

Churcher, A.M., J.M. Pujolar, M. Milan, P.C. Hubbard, R.S.T. Martins, J.L. Saraiva, M. Huertas, L. Bargelloni, T. Patarnello, I.A.M. Marino, L. Zane and A.V.M. Canario. 2014. Changes in the gene expression profiles of the brains of male European eels (*Anguilla anguilla*) during sexual maturation. BMC Genomics 15: 799.

Coppe, A., J.M. Pujolar, G.E. Maes, P.F. Larsen, M.M. Hansen, L. Bernatchez, L. Zane and S. Bortoluzzi. 2010. Sequencing, *de novo* annotation and analysis of the first *Anguilla anguilla* transcriptome: EeelBase opens new perspectives for the study of the critically endangered European eel. BMC Genomics 11: 635.

Côté, C., P.A. Gagnaire, V. Bourret, G. Verrault, M. Castonguay and L. Bernatchez. 2013. Population genetics of the American eel (*Anguilla rostrata*). $F_{ST} = 0$ and North Atlantic Oscillation effects on demographic fluctuations of a panmictic species. Mol. Ecol. 22: 1763–1776.

Dannewitz, J., G.E. Maes, L. Johansson, H. Wickström, F.A.M. Volckaert and T. Jarvi. 2005. Panmixia in the European eel: a matter of time. Proc. R. Soc. Lond. B Biol. Sci. 272: 1129–1137.

Davey, J.W., P.A. Hohenlohe, P.D. Etter, J.Q. Boone, J.M. Catchen and J.L. Blaxter. 2011. Genome-wide genetic marker discovery and genotyping using next-generation sequencing. Nat. Rev. Genet. 12: 499–510.

Edeline, E., S. Dufour, C. Briand, D. Fatin and P. Elie. 2004. Thyroid status is related to migratory behaviour in *Anguilla anguilla* glass eels. Mar. Ecol. Progr. Ser. 282: 261–270.

Feder, J.L., S.P. Egan and P. Nosil. 2012. The genomics of speciation. Trends Genet. 28: 342–350.

Gagnaire, P.A., V. Albert, B. Jónsson and L. Bernatchez. 2009. Natural selection influences AFLP intraspecific genetic variability and introgression patterns in Atlantic eels. Mol. Ecol. 18: 1678–1691.

Gagnaire, P.A., E. Normandeau and L. Bernatchez. 2012a. Comparative genomics reveals adaptive protein evolution and a possible cytonuclear incompatibility between European and American eels. Mol. Biol. Evol. 29: 2909–2919.

Gagnaire, P.A., E. Normandeau, C. Côté, M.M. Hansen and L. Bernatchez. 2012b. The genomic consequences of spatially varying selection in the panmictic American eel (*Anguilla rostrata*). Genetics 106: 404–420.

Henkel, C.V., E. Burgerhout, D.L. de Wijze, R.P. Dirks, Y. Minegishi, H.J. Jansen, H.P. Spaink, S. Dufour, F.-A. Weltzien, K. Tsukamoto and G.E.E.J.M. van den Thillart. 2012a. Primitive duplicate HOX clusters in the European eel's genome. PLoS One 7: e32231.

Henkel, C.V., R.P. Dirks, D.L. de Wijze, Y. Minegishi, J. Aoyama, H.J. Jansen, B. Turner, B. Knudsen, M. Bundgaard, K.L. Hvam, M. Boetzer, W. Pirovano, F.-A. Weltzien, S. Dufour, K. Tsukamoto, H.P. Spaink and G.E.E.J.M. van den Thillart. 2012b. First draft genome sequence of the Japanese eel, *Anguilla japonica*. Gene. 511: 195–201.

Hohenlohe, P.A., S. Basshan, P.D. Etter, N. Stiffler, E.A. Johnson and W.A. Cresko. 2010. Population genomics of parallel adaptation in threespine stickleback using sequenced RAD tags. PLoS Genetics 6: e1000862.

Jacobsen, M.W., J.M. Pujolar, M.T. Gilbert, V. Moreno-Mayar, L. Bernatchez, T. Als, J. Lobón-Cervia and M.M. Hansen. 2014a. Speciation and demographic history of Atlantic eels (*Anguilla anguilla* and *A. rostrata*) revealed by mitogenome sequencing. Heredity 113: 432–442.

Jacobsen, M.W., J.M. Pujolar, L. Bernatchez, K. Munch, J. Jian, Y. Niu and M.M. Hansen. 2014b. Genomic footprints of speciation in Atlantic eels *Anguilla anguilla* and *A. rostrata*. Mol. Ecol. 23: 4785–4798.

Jacobsen, M.W., J.M. Pujolar and M.M. Hansen. 2015. Relationship between amino acid changes in mitochondrial ATP6 and life-history variation in anguillid eels. Biol. Lett. 11: 20150014.

Maes, G.E. and F.A.M. Volckaert. 2007. Challenges for genetic research in European eel management. ICES J. Mar. Sci. 64: 1463-1471.

Maes, G.E., J.M. Pujolar, B. Hellemans and F.A.M. Volckaert. 2006. Evidence for isolation by time in the European eel. Mol. Ecol. 15: 2095–2107.

Maes, G.E., van Vo Binh, A. Crivelli and F.A.M. Volckaert. 2009. Morphological and genetic dynamics of European eel (*Anguilla Anguilla* L.) recruitment waves in Southern France. J. Fish Biol. 74: 2047–2068.

Maes, G.E., J.A.M. Raeymaekers, B. Hellemans, C. Geeraerts, K. Parmentier, L. De Temmerman, F.A.M. Volckaert and C. Belpaire. 2013. Gene transcription reflects poor health status of resident European eel chronically exposed to environmental pollutants. Aquatic Toxicology 126: 242–255.

Mank, J.E. and J.C. Avise. 2003. Microsatellite variation and differentiation in North Atlantic eels. J. Hered. 94: 310–314.

Minegishi, Y., J. Aoyama, J.G. Inoue, M. Miya, M. Nishida and K. Tsukamoto. 2005. Molecular phylogeny and evolution of the freshwater eels genus *Anguilla* based on the whole mitochondrial genome sequences. Mol. Phylogenet. Evol. 34: 134–146.

Podgorniak, T., M. Milan, J.M. Pujolar, G.E. Maes, L. Bargelloni, F. De Oliveira, F. Pierron and F. Daverat. 2015. Differences in brain gene transcription profiles advocate for an important role of cognitive function in upstream migration and water obstacles crossing in European eel. BMC Genomics 16: 378.

Pujolar, J.M., G.E. Maes and F.A.M. Volckaert. 2006. Genetic patchiness among recruits of the European eel, *Anguilla anguilla*. Mar. Ecol. Progr. Ser. 307: 209–217.

Pujolar, J.M., D. Bevacqua, M. Andrello, F. Capoccioni, E. Ciccotti, G.A. De Leo and L. Zane. 2011a. Genetic patchiness in European eel adults evidenced by molecular genetics and population dynamics modelling. Mol. Phylogenet. Evol. 58: 198–206.

Pujolar, J.M., D. Bevacqua, F. Capoccioni, E. Ciccotti, G.A. De Leo and L. Zane. 2011b. No apparent genetic bottleneck in the demographically declining European eel using molecular genetics and forward-time simulations. Conserv. Genet. 12: 813–825.

Pujolar, J.M., I.A.M. Marino, M. Milan, G.E. Maes, F. Capoccioni, E. Ciccotti, C. Belpaire, L. Bervoets, A. Covaci, G. Cramb, T. Patarnello, L. Bargelloni, S. Bortoluzzi and L. Zane. 2012. Surviving in a toxic-world: transcriptomics and gene expression profiling in response to pollution in the critically endangered European eel. BMC Genomics 13: 507.
Pujolar, J.M., M.W. Jacobsen, J. Frydenberg, T.D. Als, P.F. Larsen, G.E. Maes, L. Zane, J.B. Jian, L. Cheng and M.M. Hansen. 2013a. A resource of genome-wide single nucleotide polymorphisms generated by RAD tag sequencing in the critically endangered European eel. Mol. Ecol. Res. 13: 706–714.
Pujolar, J.M., M. Milan, I.A.M. Marino, F. Capoccioni, E. Ciccotti, C. Belpaire, A. Covaci, G. Malarvannan, T. Patarnello, L. Bargelloni, L. Zane and G.E. Maes. 2013b. Detecting genome-wide gene transcription profiles associated with high pollution burden in the critically endangered European eel. Aquat. Toxicol. 132-133: 157–164.
Pujolar, J.M., M.W. Jacobsen, T.D. Als, J. Frydenberg, K. Munch, B. Jónsson, J.B. Jian, L. Cheng, G.E. Maes, L. Bernatchez and M.M. Hansen. 2014a. Genome-wide signatures of within-generation local selection in the panmictic European eel. Mol. Ecol. 23: 2514–2528.
Pujolar, J.M., M.W. Jacobsen, T.D. Als, J. Frydenberg, E. Magnussen, B. Jónsson, X. Jiang, L. Cheng, D. Bekkevold, G.E. Maes, L. Bernatchez and M.M. Hansen. 2014b. Assessing patterns of hybridization between North Atlantic eels using diagnostic single nucleotide polymorphisms. Heredity 112: 627–637.
Roland, K., P. Kestemont, L. Hénuset, M.A. Pierrard, M. Raes, M. Dieu and F. Silvestre. 2013. Proteomic responses of peripheral blood mononuclear cells in the European eel (*Anguilla anguilla*) after perfluorooctane sulfonate exposure. Aquat. Toxicol. 128-129: 43–52.
Roland, K., P. Kestemont, R. Loos, S. Tavazzi, B. Paracchini, C. Belpaire, M. Dieu, M. Raes and F. Silvestre. 2014. Looking for protein expression signatures in European eel peripheral blood mononuclear cells after *in vivo* exposure to perfluorooctane sulfonate and a real world field study. Sci. Total Environ. 468-469: 958–967.
Schaerlaekens, D.G., W. Dekker, H. Wickström, F.A.M. Volckaert and G.E. Maes. 2011. Extracting a century of preserved molecular and population demographic data from archived otoliths in the endangered European eel (*Anguilla anguilla* L.). J. Exp. Mar. Biol. Ecol. 398: 56–62.
Tsukamoto, T. and J. Aoyama. 1988. Evolution of freshwater eels of the genus *Anguilla*: a possible scenario. Environ. Biol. Fish. 52: 139–148.
Ulrik, M.G., J.M. Pujolar, A.L. Ferchaud, M.W. Jacobsen, T.D. Als, P.A. Gagnaire, J. Frydenberg, P.K. Bocher, B. Jonsson, L. Bernatchez and M.M. Hansen. 2014. Do North Atlantic eels show parallel patterns of spatially varying selection? BMC Evol. Biol. 14: 138.
van Ginneken, V. and G.E. Maes. 2005. The European eel (*Anguilla anguilla*), its lifecycle, evolution and reproduction: a literature review. Rev. Fish Biol. Fish. 15: 367–398.
Wirth, T. and L. Bernatchez. 2001. Genetic evidence against panmixia in the European eel. Nature 409: 1037–1040.
Wirth, T. and L. Bernatchez. 2003. Decline of North Atlantic eels: a fatal synergy? Proc. R. Soc. Lond. B Biol. Sci. 270: 681–688.

Spawning Ground and Larval Segregation of the Atlantic Eels

Takaomi Arai

Introduction

The spawning areas of the American eel, *Anguilla rostrata* Le Sueur (1817), and the European eel, *A. anguilla* Linnaeus (1758) are known to be neighboring in the Sargasso Sea (Schmidt 1922, 1925) (Fig. 1). American eel leptocephali drift coastward through the Gulf Stream from the Sargasso Sea, whereas European eels are further transported by the North Atlantic Current across the Atlantic Ocean (Schmidt 1922, 1925; Power and McCleave 1983; Boëtius 1985; Kleckner and McCleave 1985) (Fig. 1). The eels grow in the inland freshwaters of eastern North America and Western Europe (Schmidt 1925; Tesch 1977; Kleckner et al. 1983; Kleckner and McCleave 1988) and are distinct, genetically different species (Camparini and Rodinó 1980; Avise et al. 1986), rather than environmentally determined, eco-phenotypic species (Tucker 1959).

McCleave (1993) suggested that the eastward flowing countercurrent from the Sargasso Sea was also involved in the migration of the European eel leptocephali. The geographical segregation of the two species is thought to be due to the difference in the duration of the pelagic migration as a result

Environmental and Life Sciences Programme, Faculty of Science, Universiti Brunei Darussalam, Jalan Tungku Link, Gadong, BE 1410, Brunei Darussalam.
E-mail: takaomi.arai@ubd.edu.bn

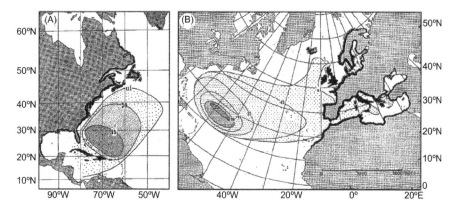

Fig. 1. Larval distributions of the two species of Atlantic eels based on the collections of leptocephali analysed by Johannes Schmidt from 1904 to 1922, shown with shaded ovals for the different sizes of larvae. The black lines on the coastlines show the edges of the species ranges of (A) the American eel, *Anguilla rostrata*, with larval size ranges of 15 mm (darkest shading), 30 mm, and upper limit (ul), and (B) the European eel, *Anguilla anguilla*, with larval size ranges of 10, 15, 25, 45 mm, and upper limit labelled; the small oval with black bars in the 10 mm area shows the estimated location of the spawning area of *A. anguilla*, the location where the newly hatched larvae of 5–7 mm were collected. Taken from Miller et al. (2014), with permission from John Wiley and Sons (License No. 3552730880010).

of differences in the timing of metamorphosis from leptocephalus to glass eel (Schmidt 1922, 1925). Schmidt (1922, 1925) proposed that the American eel larvae grow more rapidly and are ready to metamorphose off the North American coast some 12 months after hatching, while this process takes much longer for the European eel, which cannot metamorphose until they are near Europe, some three years after hatching. On the basis of the data of Schmidt (1925), Harden Jones (1968) and Tesch (1977) speculated that the growth rate of *A. rostrata* is higher than that of *A. anguilla*. However, after re-examining Schmidt's data, Boëtius and Harding (1985) concluded that the growth rate of *A. rostrata* (5.6 mm total length [TL] per month) is not significantly different from that of *A. anguilla* (5.3 mm TL per month). Kleckner and McCleave (1985) and Wippelhauser et al. (1985) estimated a growth rate of 0.24 mm TL per day (7.2 mm TL per month) for *A. rostrata* by regressing total length on the numerical date of collection for a large number of specimens. Thus, the growth rates estimated by different authors were not consistent and the timing of metamorphosis of leptocephalus remained undetermined. It is still impossible to substantiate Schmidt's hypothesis that the growth rate and the timing of metamorphosis are the principal factors in determining the segregation of migrating American and European eel larvae.

Research efforts have accumulated a considerable amount of information about the life history of *Anguilla rostrata* and *A. anguilla*, such as spawning areas, birth dates, larval ages, and growth rates (Schmidt 1922, 1925; Power and McCleave 1983; Boëtius 1985; Kleckner and McCleave 1985; Lecomte-Finiger

1992; Arai et al. 2000a; Wang and Tzeng 1998). However, the mechanisms of segregation of the two species are still speculative, as is the mechanism driving larval migration mechanism. Differing estimates of age at recruitment of *A. rostrata* and *A. anguilla* have also been reported by authors. According to Schmidt (1922), *A. rostrata* larvae remain in the ocean for 10–12 months, whereas Kleckner and McCleave (1985) estimated the time span to be 8–12 months (Fig. 2). Likewise, *A. anguilla* larvae were estimated in one study to spend 2–3 years in the ocean (Schmidt 1922), while Boëtius and Harding (1985) revised the duration upward to 1 year after a re-examination of Schmidt's data (Fig. 2). Shorter durations of recruitment to estuaries were estimated from otolith studies on glass eels (8–9 months, Lecomte-Finiger 1992, 1994; Désaunay et al. 1996a; 7–9 months, Arai et al. 2000a; 13–15 months, Wang and Tzeng 1998) than from historical estimates based on ocean current speed estimates or larval sizes (Fig. 2).

Atlantic eels, especially the European eel, have been studied for well over a century and a large amount of biological information has been amassed. Nevertheless, knowledge concerning many biological processes remains

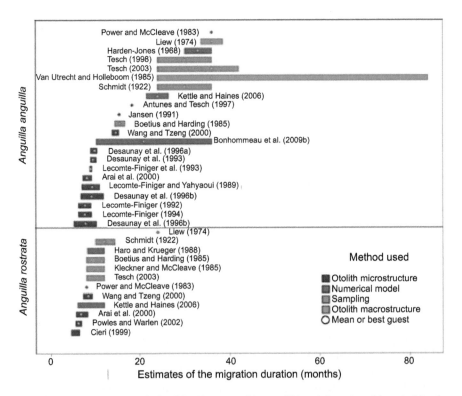

Fig. 2. Oceanic migration periods of the European (*A. anguilla*) and American (*A. rostrata*) eels given in the different studies. Taken from Bonhommeau et al. (2010), with permission from John Wiley and Sons (License No. 3552390937684).

speculative. Several studies have been published about the spawning grounds of Atlantic eels (e.g., Schmidt 1922, 1925; Tesch 1977; McCleave 2003; van Ginnenken and Maes 2005; Miller et al. 2014). Recently, Miller et al. (2014) re-examined previous studies on anguillid leptocephali in the Sargasso Sea and the wider North Atlantic Ocean basin, and presented new analyses using a newly established database containing all available collection of data on the Atlantic eel leptocephali. The authors examined a new, comprehensive database that includes 22,612 *A. anguilla* and 9,634 *A. rostrata* leptocephali, and provides a detailed view of the spatial distributions, temporal distributions and size of the larvae across the Atlantic basin and in the Mediterranean Sea. Their review was made much more comprehensive through the inclusion of data from published studies and unpublished data collected from museums, universities, and individual scientists (Miller et al. 2014). They found that the historical perspective of surveys over the 100-year span along with additional inspection of the now combined dataset provides a better understanding of the early life history of these two species, and can set the stage for more detailed analyses and guide future sampling for leptocephali. In this chapter, I summarize the present status of knowledge regarding the spawning ground and spawning ecology of the Atlantic eels *Anguilla anguilla* and *A. rostrata*. In addition, larval transportation, recruitment and segregation mechanisms are also discussed.

Discovery of the spawning ground

There are two species of freshwater eels in the Atlantic Ocean, the European eel, *Anguilla anguilla*, and the American eel, *A. rostrata*. These species have separate freshwater habitats, along the west coast of Eurasia for the former and along the east coast of the North American continent for the latter (Schmidt 1909; Boëtius 1985). Both *A. anguilla* and *A. rostrata* spawn in an overlapping area of the Sargasso Sea (Schmidt 1922; Schoth and Tesch 1982; McCleave and Kleckner 1987). Schmidt (1922, 1925) collected over 10,000 European eel larvae and over 2,000 American eel larvae during his expedition. The spawning areas of the Atlantic eels have been estimated mainly on the basis of their larval distribution in the Atlantic Ocean. Unlike the Japanese eel, the eggs and spawning adults of the Atlantic eels have not yet been observed in the Sargasso Sea. Schmidt (1923), based on his analysis of the distribution of the smallest larvae,suggested that the European eel spawns only in the Sargasso Sea, located in the southwestern North Atlantic Ocean (Fig. 1). Schmidt (1925) also found that the American eel spawned in an overlapping area to the west, but only had records of 22 larvae that were < 10 mm long.

The peak of European eel spawning is in April, and the spawning area is centered to the northeast of the spawning area of the American eel, which has its spawning peak in February (Schmidt 1925). The times and locations of eel spawning have been supported and refined through analyses of the available historical data by Boëtius and Harding (1985); Kleckner and McCleave (1982, 1985) and McCleave et al. (1987). After Schmidt, intensive surveys were conducted in the 1980s and a number of leptocephali that were < 10 mm TL were collected (Tesch 1982; Schoth and Tesch 1982; Wippelhauser et al. 1985; Castonguay and McCleave 1987; McCleave and Kleckner 1987; Kleckner and McCleave 1988; Tesch and Wegner 1990). During these surveys, more than 700 American eel leptocephali and more than 1600 European eel leptocephali less than 10 mm in TL were collected (McCleave et al. 1987). All specimens of the American eel leptocephali less than 7 mm TL (188 specimens) were collected within a broad ellipse extending eastward from the Bahamas to approximately 58° W longitude. All specimens of the European eel leptocephali less than 7 mm TL (226 specimens) were also collected within a narrow overlapping ellipse. The distribution of the American and the European eel larvae less than 7.5 mm TL was determined to be limited to the north by the boundary between the warm saline surface water of the southern Sargasso Sea and a mixed convergence zone of water to the south. Leptocephali less than 7 mm TL are believed to hatch after three weeks, thus this figure is the estimated length at hatching, based on a growth curve developed from artificial maturation experiments in the laboratory (Yamamoto and Yamauchi 1974; Yamauchi et al. 1976). From these various observations, it has been determined that the European eel spawns primarily from March to June within a narrow geographic ellipse, the long axis of which extends east–west from approximately 48° W to 74° W longitude between 23° N and 30° N latitude and that the American eel spawns primarily from February to April within a broader oval between approximately 52° W and 79° W longitude and 19° N and 29° N latitude (McCleave et al. 1987). Thus, spawning of the European and American eel species are partially sympatric in both space and time (McCleave et al. 1987).

Continental separation of the two species is probably ensured by initial distributional bias from partially allopatric spawning and by differing rates of development (Tesch 1977). Differences between the leptocephali of the two eel species in terms of vertical migration can partially explain how *Anguilla rostrata* leaves the Gulf Stream to recruit to the North American coast, while *Anguilla anguilla* presumably remains in the Gulf stream and is carried on to Europe (Castonguay and McCleave 1987). Social cues unique to each species and the existence of a species-specific pheromone (McCleave 1987) may help prevent interbreeding between the two eel species; observations of spawning behaviour in hormone-treated European eels reinforces the theory that spawning is triggered by pheromones (van Ginnenken and Maes 2005).

Based on the distribution of newly hatched leptocephali, it is believed that adults of both species spawn in, and to the south of, a persistent, meandering, near-surface frontal zone that stretches east–west across the Sargasso Sea (Voorhis and Bruce 1982). This is the so-called subtropical convergence zone (STCZ), a region where the colder waters of the northern Sargasso Sea meet the warmer waters of the southern Sargasso. These fronts act as boundaries for many organisms, and some feature of the frontal zone or the southern waters, such as odour or temperature, may serve as a signal to the migrating eels to cease migration and initiate spawning (Kleckner et al. 1983; McCleave 1987; McCleave et al. 1987). For *Anguilla* larvae, leptocephali are much more abundant on the south face of the front that separates the two general water masses in the STCZ (Kleckner and McCleave 1988; Tesch and Wegner 1990). It is thus hypothesised that differences in species composition are caused by a marked decrease of primary production south of the front (Kleckner et al. 1983; Miller 1995).

There is speculation that the homing mechanism of adult eels may be based on a similar mechanism to that found in the Atlantic salmon, through imprinting on olfactory cues and taste perceptions of the waters of the southern Sargasso Sea. It has been demonstrated that the olfactory senses of sexually immature eels are highly developed and that they are capable of detecting chemical compounds such as b-phenylethanol (Teichmann 1959). In one experiment, the estuarine migration of anosmic and controlled silver-phase American eels was examined during the fall spawning migration. Control eels moved more rapidly, using tidal properties to leave the estuary, while in contrast, anosmic eels took a longer time to leave the estuary and were incapable of using tidal-stream transport for movement out of the estuary (Barbin et al. 1998). From these observations it can be concluded that olfaction plays an important role in the initial migration of adult eels. Another possibility is that a temperature gradient in the surface waters of the frontal zone as high as 2°C per km (Voorhis 1969) could act as a triggering or orientation mechanism. Based on telemetry observations of diurnal migration patterns of migrating silver eels with corresponding temperature fluctuations, however, it seems unlikely that temperature acts as an orientation cue (Tesch 1978, 1989). Finally, some researchers have suggested that the Earth's magnetic field may play a role in eel migration (van Ginneken et al. 2005).

Oceanic migration of Atlantic eels

According to Schmidt (1922), *Anguilla rostrata* and *A. anguilla* larvae spend in the open sea for 10–12 months and 2–3 years, respectively (Fig. 2). Following a re-examination of Schmidt's data, Boëtius and Harding (1985) and Kleckner and McCleave (1985) revised Schmidt's estimation, to 8–12 months for *A. rostrata* and to a little over 1 year for *A. anguilla*. Arai et al. (2000a) suggested that the oceanic migration periods of both species were shorter still, more in the range of 6–8 months for *A. rostrata* and 7–9 months for *A. anguilla*. These estimates overlap with those of Wang and Tzeng (1998), who estimated 7–9 months for *A. rostrata* and Lecomte-Finiger (1992), who estimated 7–9 months for *A. anguilla* based on an analysis of otolith microstructure. The longer estimates of oceanic migration calculated by Boëtius and Harding (1985) and Kleckner and McCleave (1985) might be based on the estimations of somatic growth rate and total length of larvae. *A. rostrata* leptocephali were assumed to be transported clockwise via the Gulf Stream, from the Sargasso Sea to the continental shelves. *A. anguilla* leptocephali appear to be further carried by the North Atlantic Current (also clockwise) (Fig. 1) (Schmidt 1922, 1925; Power and McCleave 1983; Boëtius 1985; Kleckner and McCleave 1985). McCleave (1993) proposed an alternative transportation route for *A. anguilla* leptocephali, one that took them from the Sargasso Sea using eastward countercurrent (an anti-clockwise route). The distance from the spawning areas to the coastal waters are approximately 1,000–5,000 km for *A. rostrata* and 4,000–8,000 km for *A. anguilla*. *Anguilla* leptocephali have been estimated to cover up to 300 km in a week (Wegner 1982), and the total migration period is calculated at a mere 23–117 days for *A. rostrata* and 93–187 days for *A. anguilla*. The situation may not be quite so straight forward; as the leptocephali would have to find their way to a coast through a complicated maze of coastal currents and eddies while making diel vertical migrations (Tesch 1980; Castonguay and McCleave 1987).

The mismatch between the estimated hatching dates of glass eels compared to the spawning times known from the larval sampling surveys, along with the variation among studies, casts clear doubts on the reliability of otolith rings as a source for accurate estimates of the ages of glass eels (McCleave et al. 1998; McCleave 2008). Umezawa and Tsukamoto (1991) found that *Anguilla japonica* glass eels reared in water at a temperature of 12°C deposited narrower daily otolith increments (0.22–0.29 µm) than those reared in water at a temperature of 25°C (1.93–2.02 µm). Water temperatures below 10°C can cause the deposition of otolith material to cease in anguillid glass eels (Fukuda et al. 2009), and there is often an opaque zone in the outer part of the otoliths of the glass eels, in which increments are difficult to see (Antunes and Tesch 1997). Whether rings are formed throughout the later larval stage or during early parts of the glass eel stage is uncertain, especially in the cold waters where the European eel migrates (McCleave 2008; Miller 2009).

In *Anguilla anguilla*, Antunes and Tesch (1997) demonstrated the formation of a diffuse (unclear) incremental area in the otolith during the leptocephalus and metamorphosis stages and were reluctant to use otolith increments for age determination in that species. However, Lecomte-Finiger (1992), Arai et al. (2000a), Wang and Tzeng (1998) and Kuroki et al. (2008) observed clear concentric rings throughout the otolith of *A. anguilla*, including the leptocephalus and metamorphosis stages. The discrepancy may result simply from a technical problem, e.g., over-etching. McCleave et al. (1998) noted that the leptocephalus stage of *A. anguilla* experienced low temperatures and poor nutritive conditions in the open ocean of the North Atlantic and speculated that they could be highly stressed, possibly resulting in otolith increments being deposited at a lower than daily frequency. Although there remains a lack of substantial data on otolith increment deposition in *A. anguilla*, Umezawa and Tsukamoto (1991) observed stable periodicity in increment formation in *A. japonica* even under stressful conditions (near the lower limit of endurable temperatures and prolonged starvation). Under extremely adverse conditions where otolith increments are not deposited daily, normal development and body growth cannot be expected. Obtained through artificial fertilization, *A. japonica* leptocephali of age 30 days did not show normal body growth, having lesser body depth than naturally occurring specimens of the same age. In spite of their abnormal development, the former deposited 25–27 increments, with 3–5 days cessation of increment formation just before death through extreme infirmity (Shinoda et al. 2004). In *A. anguilla* collected in an estuary, otolith structure was similar to those reported in other *Anguilla* spp. (Cheng and Tzeng 1996; Arai et al. 1999, 2001a, 2002), suggesting the same daily periodicity of otolith increment deposition in the estuary specimens. In the field, moreover, abundant food in the form of "marine snow" exists for leptocephali (Otake 1996). However, further validation of otolith daily increments for every developmental stage in the Atlantic eels, is necessary for a more detailed understanding of the early life history of *Anguilla* spp. and to explain the discrepancy between otolith analyses and larval sampling surveys.

General circulation models could provide insights for better understanding of larval migration in the Atlantic Ocean. Kettle and Haines (2006) used Lagrangian simulations over a 4-year period (1993–1996) to estimate the duration and pathways of *A. anguilla* larvae, and found a 2-year duration. However, they used fixed-depth passively drifting particles to analyse migration routes and duration (Kettle and Haines 2006). This assumption remains questionable, as leptocephali show vertical diurnal migrations (Castonguay and McCleave 1987) and might be able to select faster current velocities at different depths.

Bonhommeau et al. (2009a) simulated the drift of particles released in the Sargasso Sea and calculated the minimum time required for eel larvae to achieve their migration. The aim was to evaluate the minimum bound constrained by physical oceanography to passively drift (at fixed depth with no active behaviour) from the Sargasso Sea to the 20° W meridian (Fig. 1). Their results showed that the migration up to the 20° W meridian, i.e., 1000–1500 km from the European shelves, requires at least 10 months and 19 days, and covers in excess of 8000 km. This is the shortest migration duration achieved by the fastest particle among several million released in the Sargasso Sea. It has been suggested that larval duration during oceanic migration shorter than 1 year is highly unlikely for a significant portion of the particles (Bonhommeau et al. 2009a). Further, Bonhommeau et al. (2009b) estimated that the mean duration of migration of European eel leptocephali is 21 months and that less than 0.2% of eel larvae may typically survive the trans-Atlantic migration. Those simulations of the oceanic migration of eel larvae using general ocean circulation models demonstrate the robust potential of these approaches for unraveling migration pathways (Bonhommeau et al. 2010). The approach suggests and appears to validate the estimate of more than 1.5 years for European eel migrations.

In 1991, the spawning area of the Japanese eel was determined to be located in the western North Pacific (Tsukamoto 1992). More recently, the eggs and mature adults of the Japanese eels have been collected (Tsukamoto et al. 2011).

These findings are the first of their kind for any of the 19 freshwater eels. There has been intensive, virtually annual scrutiny of the spawning grounds of the Japanese eel over the past 20 years or so since the discovery of the spawning ground. However, such intensive research has not been conducted for the Atlantic eel species. During the oceanic life stage, it has been suggested that the recruitment variability of the Japanese eel is affected by ocean–atmospheric forcing (Miller et al. 2009). In particular, the latitudinal shifts of spawning locations relative to larval transport by the North Equatorial Current (NEC) are considered to be an important determinant of recruitment success (Kimura et al. 2001). If the eels can travel westward using the NEC and enter the Kuroshio Current, they have a significantly enhanced probability of recruitment success. In contrast, recruitment is reduced if the eels are entrained into the south-flowing Mindanao Current or the mesoscale eddies east of Taiwan (Kim et al. 2007). Similar phenomena were also noted for the two Atlantic species (Knights 2003; Friedland et al. 2007; Bonhommeau et al. 2008a,b). Studies on the distributions of the leptocephali of both species indicate that they spawn south of the isotherm of the Sargasso Sea (Kleckner and McCleave 1988; Tesch and Wegner 1990; McCleave 1993). Frontal jets are also formed in association with these fronts (Eriksen et al. 1991), which appear to transport a variety of leptocephali species eastwards (Miller and McCleave 1994). Therefore, changes in the latitude or intensity of these fronts may affect both the spawning location and the subsequent transport of the leptocephali to continental habitats (Friedland et al. 2007). Seasonal or interannual differences of current speeds or pathways flowing towards North American coasts in both the North Atlantic Current and in currents towards Europe (Krauss 1995), or differences in larval behaviour, may also influence the rapidity of *A. anguilla* migration across the Atlantic Ocean. The disparity in estimated hatching dates and oceanic migration periods between field and modelling surveys and otolith analyses have been suggested as one possible explanation. However, there may not have been mismatches caused by annual environmental fluctuation around the spawning grounds in the Sargasso Sea or in the Atlantic Ocean.

Segregation mechanism

Although Schmidt (1922) reported that there were two Atlantic eel species, *Anguilla anguilla* and *A. rostrata*, Tucker (1959) proposed that there was only one species and that the differences between the two morphologies—in terms of the number of vertebrae in eels and myomeres in leptocephali—could be due to the existence of two different eco-phenotypes of Atlantic eels, as well as due to differences in water temperatures in the spawning areas that are related to the geographic or seasonal differences of where and when the two species spawn (Smith 1989; McCleave 2003). Subsequent molecular genetic studies have clearly shown that the two species are genetically distinct (e.g., Avise et al. 1986; Lehmann et al. 2000; Inoue et al. 2010), although hybrids between the two species have been found in Iceland (Avise 2003; Albert et al. 2006; Pujolar et al. 2014). Recently, similar overlapping spawning areas have been discovered for the Japanese eel, *A. japonica* and the tropical giant mottled eel, *A. marmorata*, in the Pacific Ocean (Tsukamoto et al. 2011), and they are also genetically distinct species. Although it is still unclear what the segregation mechanisms are, studying them is important for understanding the biogeography and evolutionary pathways of eels.

Atlantic eel leptocephali migrate from the southern Sargasso Sea to American continental waters to the west and north, where the migration of the American eel ends. European eel leptocephali, on the other hand, are transported further on to the European continental shelves far to the northeast and east, when the clockwise route is used. Schmidt (1925) suggested that their separate ranges were the consequence of differences in the larval durations and growth rates of the two species. However, Boëtius and Harding (1985) suggested that there was no clear difference in growth rates during the leptocephalus stage of both species after a re-examination of the specimens collected by Schmidt (1925). Castonguay (1987) calculated a faster growth rate in leptocephali based on microstructure increments compared to what had been calculated by previous approaches (Boëtius and Harding 1985; Wippelhauser et al. 1985). Further otolith studies supported Schmidt's hypothesis, by showing that *A. anguilla* began metamorphosing approximately 50–150 days later than *A. rostrata* (Arai et al. 2000a; Wang and Tzeng 1998). Metamorphosis from leptocephalus to glass eel stops the passive drift of eel larvae, presumably established in the phylogeny of these species, and may be involved in their geographical segregation in the Atlantic Ocean. The mean age at recruitment determined for the glass eels in Iceland was similar between the two species (337 ± 42 and 319 ± 36 days for *A. anguilla* and *A. rostrata*, respectively) (Kuroki et al. 2008). In addition, mean age at metamorphosis (278 ± 37 and 254 ± 48 days) and total age (372 ± 51 and 353 ± 43 days) were also similar between the two species. The distance from the Sargasso Sea to Iceland is farther than the distance from the Sargasso Sea to North America (approximately 3,000 km from Newfoundland or Labrador in Canada to Iceland), which would account for the greater ages of *A. rostrata* in Iceland, but the distance to Iceland is not so different from the distance to some parts of Europe. The distance from the Sargasso Sea to Iceland is longer than the distance to some parts of Europe, but perhaps shorter and more direct than the distance to other locations in Europe. Thus, their oceanic migration periods were almost the same when recruited to Iceland. However, the ages of *A. rostrata* in Iceland were more than those in North America, and the *A. anguilla* collected in Iceland were roughly intermediate in age between the rest of Europe and North Africa (Kuroki et al. 2008). These findings support the hypothesis that the timing of metamorphosis is a key factor in determining the place of recruitment of glass eels and maintaining the geographic separation between the two species.

A recent study suggests that different species of anguillid leptocephali may have slightly different growth rates (Arai et al. 2001b; Kuroki et al. 2006), and thus if *A. rostrata* has a faster growth rate as suggested by Arai et al. (2000a) and Wang and Tzeng (2000), their leptocephali would be able to metamorphose into glass eels earlier than those of *A. anguilla*, and would detrain from the Gulf Stream to recruit to North America, while the leptocephali of *A. anguilla* would continue to be transported further by the Gulf Stream and the North Atlantic Drift, eventually reaching Europe or North Africa. *A. rostrata* also appears to have smaller maximum leptocephalus and glass eel sizes than *A. anguilla*, indicating earlier metamorphosis.

Data on the maximum total length of leptocephali of the two Atlantic eel species indicates that there have been no *A. rostrata* leptocephali larger than 70 mm collected (Boëtius and Harding 1985; Kleckner and McCleave 1985), but that *A. anguilla* leptocephali can reach sizes up to 85 mm, with peak abundances at approximately 70–75 mm (Tesch 1980; Boëtius and Harding 1985; Bast and Strehlow 1990). Similarly, the glass eels of the two species have different size ranges: *A. rostrata* glass eels may range from less than 50 mm in the southern part of their range to 65 mm in the northern part (Haro and Krueger 1988; Wang and Tzeng 1998; Sullivan et al. 2006). Glass eels of *A. anguilla* range from approximately 55–80 mm, with some seasonal variation in length (Boëtius and Boëtius 1989; Désaunay and Guérault 1997; Wang and Tzeng 2000). The mean growth rate of each leptocephalus, using otolith age determination, was estimated to be 0.35–0.45 mm/day for *A. rostrata* and 0.21–0.38 mm/day for *A. anguilla* (Arai et al. 2000a; Wang and Tzeng 1998). The slower growth rate during the leptocephalus stage and the longer duration of the stage for *A. anguilla* as compared to *A. rostrata*, seems to contribute to their longer transportation distance and to the drifting further northeast and east for the former. Further information on larval distribution in the Atlantic Ocean, larval age and larval growth is necessary for understanding the mechanisms controlling the larval migration of Atlantic eels.

Conclusion

Research on the Atlantic eels *Anguilla anguilla* and *A. rostrata* extends back over a century, which is the longest period of study for any of the 19 eel species/subspecies. The spawning areas of these Atlantic eels overlap in the Sargasso Sea (Schmidt 1922, 1925). Field observations and leptocephali sampling have revealed that spawning occurs from March to June for the European eel and February to April for the American eel. However, no one has yet discovered their eggs and spawning adults, unlike those of the Japanese eel *A. japonica*, which were recently discovered in the Pacific Ocean (Tsukamoto et al. 2011). As such, different approaches have been applied in order to understand the mechanisms of larval migration, recruitment and segregation, focusing primarily on (1) larval distribution patterns, (2) otolith microstructure and microchemical analyses, and (3) particle drift simulations. However, the results of the different approaches are quite different, especially for the European eel. For example, analysis of the daily otolith increment rate in European eels suggested the migration from the spawning area to their freshwater habitats took less than a year, whereas other studies indicate that it may take as long as 1–1.5 years to complete their oceanic migration. Even in the American eel, otolith daily ages are several months shorter than those of larval distribution patterns and physical models. Regarding otolith daily age analyses, differences also occurred among researchers, although the analyses have been widely applied for early life history analysis in the genus *Anguilla*. In some species of the genus such as *A. japonica* (Tsukamoto 1989), *A. celebesensis* (Arai et al. 2000b) and *A. marmorata* (Sugeha et al. 2001), daily otolith age deposition for their otoliths has been verified. However, factors such as low water temperatures and undernourishment have been known to narrow and even prevent increment deposition (Umezawa and Tsukamoto 1991; Fukuda et al. 2009). No research has yet been carried out to verify otolith daily increment deposition during the leptocephalus stage of the Atlantic eels. Thus, verification of the otolith characteristics through further analysis is required. Nevertheless, differences in the larval migration period and growth rate between these Atlantic eels suggest that larval duration or timing of metamorphosis would determine when they detrain from oceanic currents and recruit to freshwater habitats. The American eel has a faster growth rate than the European eel, and their leptocephali would be able to metamorphose into glass eels earlier than *A. anguilla* leptocephali. Moreover, American eel leptocephali leave the Gulf Stream to recruit to North America, while the leptocephali of *A. anguilla* are transported further by the Gulf Stream and the North Atlantic Drift, eventually reaching coastal Europe or North Africa. *A. rostrata* also appears to have smaller maximum leptocephali and glass eel sizes than does *A. anguilla*, indicating earlier metamorphosis.

Numerous studies have revealed much about the biology and ecology of the Atlantic eels. However, many biological and ecological processes still remain a mystery, lacking both concrete hypotheses and results. Further field studies, in combination with modelling approaches and otolith analyses after validation of daily age deposition for leptocephali and glass eels, are needed in order to understand the migration and segregation mechanisms of the Atlantic eels.

Keywords: *A. anguilla*, *A. rostrata*, Sargasso Sea, Atlantic Ocean, segregation, spawning, oceanic migration period, larval distribution, otolith, circulation model

References

Albert, V., B. Jönsson and L. Bernatchez. 2006. Natural hybrids in Atlantic eels (*Anguilla anguilla*, *A. rostrata*): evidence for successful reproduction and fluctuating abundance in space and time. Mol. Ecol. 15: 1903–1916.
Antunes, C. and F.W. Tesch. 1997. A critical consideration of the metamorphosis zone when identifying daily rings in otoliths of European eel, *Anguilla anguilla* (L.). Ecol. Freshw. Fish 6: 102–107.

Arai, T., T. Otake, D.J. Jellyman and K. Tsukamoto. 1999. Differences in the early life history of Australasian shortfinned eel *Anguilla australis* from Australia and New Zealand, as revealed by otolith microstructure and microchemistry. Mar. Biol. 135: 381–389.

Arai, T., T. Otake and K. Tsukamoto. 2000a. Timing of metamorphosis and larval segregation of the Atlantic eels *Anguilla rostrata* and *A. anguilla*, as revealed by otolith microstructure and microchemistry. Mar. Biol. 137: 39–45.

Arai, T., D. Limbong and K. Tsukamoto. 2000b. Validation of otolith daily increments in the tropical eel *Anguilla celebesensis*. Can. J. Zool. 78: 1078–1084.

Arai, T., D. Limbong, T. Otake and K. Tsukamoto. 2001a. Recruitment mechanisms of tropical eels, *Anguilla* spp., and implications for the evolution of oceanic migration in the genus *Anguilla*. Mar. Ecol. Prog. Ser. 216: 253–264.

Arai, T., J. Aoyama, S. Ishikawa, M.J. Miller, T. Otake, T. Inagaki and K. Tsukamoto. 2001b. Early life history of tropical *Anguilla* leptocephali in the western Pacific Ocean. Mar. Biol. 138: 887–895.

Arai, T., M. Marui, M.J. Miller and K. Tsukamoto. 2002. Growth history and inshore migration of the tropical eel, *Anguilla marmorata* in the Pacific. Mar. Biol. 140: 309–316.

Avise, J.C. 2003. Catadromous eels of the North Atlantic: a review of molecular genetic findings relevant to natural history, population structure, speciation, and phylogeny. pp. 31–48. *In*: K. Aida, K. Tsukamoto and K. Yamauchi (eds.). Eel Biology. Springer, Tokyo, Japan.

Avise, J.C., G.S. Helfman, C. Saunders and L.S. Hales. 1986. Mitochondrial DNA differentiation in North Atlantic eels: population genetic consequences of an unusual life history pattern. *In*: Proceedings of the National Academy of Science, USA 83: 4350–4354.

Barbin, G.P., S.J. Parker and J.D. McCleave. 1998. Olfactory clues play a critical role in the estuarine migration of silver-phase American eels. Environ. Biol. Fish. 53: 283–291.

Bast, H.D. and B. Strehlow. 1990. Length composition and abundance of eel larvae, *Anguilla anguilla* (Anguilliformes: Anguillidae), in the Iberian Basin (northeastern Atlantic) during July–September 1984. Helgol. Wiss. Meeresunters 44: 353–361.

Boëtius, J. 1985. Greenland eels, *Anguilla rostrata* Lesueur. Dana 4: 41–48.

Boëtius, J. and E.F. Harding. 1985. A re-examination of Johannes Schmidt's Atlantic eel investigations. Dana 4: 129–162.

Boëtius, I. and J. Boëtius. 1989. Ascending elvers, *Anguilla anguilla*, from five European localities. Analyses of pigmentation stages, condition, chemical composition and energy reserves. Dana 7: 1–12.

Bonhommeau, S., E. Chassot and E. Rivot. 2008a. Fluctuations in European eel (*Anguilla anguilla*) recruitment resulting from environmental changes in the Sargasso Sea. Fish. Oceanogr. 17: 32–44.

Bonhommeau, S., E. Chassot, B. Planque, E. Rivot, A.H. Knap and O. Le Pape. 2008b. Impact of climate on eel populations of the Northern Hemisphere. Mar. Ecol. Prog. Ser. 373: 71–80.

Bonhommeau, S., B. Blanke, A.M. Tréguier, N. Grima, E. Rivot, Y. Vermard, E. Greiner and O. Le Pape. 2009a. How fast can the European eel larvae cross the Atlantic Ocean? Fish. Oceanogr. 18: 371–385.

Bonhommeau, S., O. Le Pape, D. Gascuel, B. Blanke, A.M. Tréguier, N. Grima, Y. Vermard, M. Castonguay and E. Rivot. 2009b. Estimates of the mortality and the duration of the trans-Atlantic migration of European eel leptocephali using a particle tracking model. J. Fish Biol. 74: 1891–1914.

Bonhommeau, S., M. Castonguay, E. Rivot, R. Sabatié and O. Le Pape. 2010. The duration of migration of Atlantic *Anguilla* larvae. Fish Fisheries 11: 289–306.

Camparini, A. and E. Rodinó. 1980. Electrophoretic evidence for two species of *Anguilla* leptocephali in the Sargasso Sea. Nature (London) 287: 435–437.

Castonguay, M. and J.D. McCleave. 1987. Vertical distributions, diel and onto genetic vertical migrations, and net avoidance of leptocephali of *Anguilla* spp. and other common species in the Sargasso Sea. J. Plankt. Res. 9: 195–214.

Cheng, P.W. and W.N. Tzeng. 1996. Timing of metamorphosis and estuarine arrival across the dispersal range of the Japanese eel *Anguilla japonica*. Mar. Ecol. Prog. Ser. 131: 87–96.

Cieri, M.D. 1999. Migrations, growth and early life history of the American eel (*Anguilla rostrata*). Ph.D. thesis, University of Maine, Orono, ME, USA.

Désaunay, Y. and D. Guérault. 1997. Seasonal and long-term changes in biometrics of eel larvae: a possible relationship between recruitment variation and North Atlantic ecosystem productivity. J. Fish Biol. 51: 317–339.

Désaunay, Y., D. Guérault and R. Lecomte-Finiger. 1996a. Variation of the oceanic larval migration of *Anguilla anguilla* (L.) glass eels from a two year study in the Vilaine Estuary (France). Arch. Polish Fish. 4: 195–210.

Désaunay, Y., R. Lecomte-Finiger and D. Guérault. 1996b. Mean age and migration patterns of *Anguilla anguilla* (L.) glass eels from three French estuaries (Somme, Vilaine and Adour rivers). Arch. Polish Fish. 4: 187–194.

Eriksen, C.C., R.A. Weller, D.L. Rudnick, R.T. Pollard and L.A. Regier. 1991. Ocean frontal variability in the frontal air–sea interaction experiment. J. Geophys. Res. 96: 8569–8591.

Friedland, K.D., M.J. Miller and B. Knights. 2007. Oceanic changes in the Sargasso Sea and declines in recruitment of the European eel. ICES J. Mar. Sci. 64: 519–530.

Fukuda, N., M. Kuroki, A. Shinoda, Y. Yamada, A. Okamura, J. Aoyama and K. Tsukamoto. 2009. Influence of water temperature and feeding regime on otolith growth in *Anguilla japonica* glass eels and elvers: does otolith growth cease at low temperatures? J. Fish Biol. 74: 1915–1933.

Harden Jones, F.R. 1968. Fish Migration. Arnold, London.

Haro, A.J. and W.H. Krueger. 1988. Pigmentation, size, and migration of elvers (*Anguilla rostrata* (Lesueur)) in a coastal Rhode Island stream. Can. J. Zool. 66: 2528–2533.

Inoue, J.G., M. Miya, M.J. Miller, T. Sado, R. Hanel, K. Hatooka, J. Aoyama, Y. Minegishi, M. Nishida and K. Tsukamoto. 2010. Deep-ocean origin of the freshwater eels. Biol. Let. 6: 363–366.

Jansen, S. 1991. Mikrostructuruntersuchungen an Otolithendes Europaischen Aales *Anguilla anguilla* (Linnaeus, 1758). Diplomarbeit, Fachbereich Biologie, Universitat Hamburg. 65pp.

Jespersen, P. 1942. Indo-Pacific leptocephali of the Genus *Anguilla*. Dana Rep. 22: 1–128.

Kettle, A.J. and K. Haines. 2006. How does the European eel (*Anguilla anguilla*) retain its population structure during its larval migration across the North Atlantic Ocean? Can. J. Fish. Aquat. Sci. 63: 90–106.

Krauss, W. 1995. Currents and mixing in the Irminger Sea and in the Iceland Basin. J. Geophys. Res. 100: 10,851–10,871.

Kim, H., S. Kimura, A. Shinoda, T. Kitagawa, Y. Sasai and H. Sasaki. 2007. Effect of El Nino on migration and larval transport of the Japanese eel (*Anguilla japonica*). ICES J. Mar. Sci. 64: 1387–1395.

Kimura, S., T. Inoue and T. Sugimoto. 2001. Fluctuation in the distribution of low salinity water in the North Equatorial Current and its effect on the larval transport of the Japanese eel. Fish. Oceanogr. 10: 51–60.

Kleckner, R.C. and J.D. McCleave. 1982. Entry of migrating American eel leptocephali into the Gulf Stream system. Helgol. Meeresunter. 35: 329–339.

Kleckner, R.C. and J.D. McCleave. 1985. Spatial and temporal distribution of American eel larvae in relation to North Atlantic Ocean current systems. Dana 4: 67–92.

Kleckner, R.C. and J.D. McCleave. 1988. The northern limit of spawning by Atlantic eels (*Anguilla* spp.) in the Sargasso Sea in relation to thermal fronts and surface water masses. J. Mar. Res. 46: 647–667.

Kleckner, R.C., J.D. McCleave and G.S. Wippelhauser. 1983. Spawning of American eel, *Anguilla rostrata*, relation to thermal fronts in the Sargasso Sea. Environ. Biol. Fish. 9: 289–293.

Knights, B. 2003. A review of the possible impacts of long-term oceanic and climate changes and fishing mortality on recruitment of anguillid eels of the Northern Hemisphere. Sci. Total Environ. 310: 237–244.

Kuroki, M., J. Aoyama, M.J. Miller, S. Wouthuyzen, T. Arai and K. Tsukamoto. 2006. Contrasting patterns of growth and migration of tropical anguillid leptocephali in the western Pacific and Indonesian Seas. Mar. Ecol. Prog. Ser. 309: 233–246.

Kuroki, M., M. Kawai, B. Jónsson, J. Aoyama, M.J. Miller, D.L.G. Noakes and K. Tsukamoto. 2008. Inshore migration and otolith microstructure/microchemistry of anguillid glass eels recruited to Iceland. Environ. Biol. Fish. 83: 309–325.

Lecomte-Finiger, R. 1992. Growth history and age at recruitment of European glass eels (*Anguilla anguilla*) as revealed by otolith microstructure. Mar. Biol. 114: 205–210.

Lecomte-Finiger, R. 1994. The early life of the European eel. Nature 370: 424.

Lecomte-Finiger, R. and A. Yahyaoui. 1989. Otolith microstructure analysis in the study of the early life history of the European eel *Anguilla anguilla*. Proceedings of the EIFAC, Working Party on Eels, Porto (Portugal).

Lecomte-Finiger, R., Y. Désaunay, D.D. Guérault and P. Grellier. 1993. The immigration of *Anguilla Anguilla* (L.) glass eels in coastal waters: question about the determinism of the otolith structure. Report of the eight session of the Working Party on Eel, Olsztyn, Poland.

Lehmann, D., H. Hettwer and H. Taraschewski. 2000. RAPD-PCR investigations of systematic relationships among four species of eels (Teleostei: Anguillidae), particularly *Anguilla anguilla* and *A. rostrata*. Mar. Biol. 137: 195–204.

Liew, P.K.L. 1974. Age determination of American eels based on the structure of their otoliths. pp. 124–136. *In*: T.B. Bagenal (ed.). Ageing of Fish. Unwin Brothers Limited, Surrey.

McCleave, J.D. 1987. Migration of *Anguilla* in the ocean: signpost for adults! signposts for leptocephali? pp. 102–117. *In*: W.F. Herrnkind and A.B. Thistle (eds.). Signpost in the Sea: Proceedings of a Multidisciplinary Workshop on Marine Animal Orientation and Migration. Florida State University, Tallahassee FL.

McCleave, J.D. 1993. Physical and behavioral controls on the oceanic distribution and migration of leptocephali. J. Fish Biol. 43 (suppl. A): 243–273.

McCleave, J.D. 2003. Spawning areas of the Atlantic eels. pp. 141–155. *In*: K. Aida, K. Tsukamoto and K. Yamauchi (eds.). Eel Biology. Springer, Tokyo, Japan.

McCleave, J.D. 2008. Contrasts between spawning times of *Anguilla* species estimated from larval sampling at sea and from otolith analysis of recruiting glass eels. Mar. Biol. 155: 249–262.

McCleave, J.D. and R.C. Kleckner. 1987. Distribution of leptocephali of the catadromous *Anguilla* species in the western Sargasso Sea in relation to water circulation and migration. Bull. Mar. Sci. 41: 789–806.

McCleave, J.D., R.C. Kleckner and M. Castonguay. 1987. Reproductive sympatry of American and European eels and implications for migration and taxonomy. Am. Fish. Soc. Symp. 1: 286–297.

McCleave, J.D., P.J. Brickley, K.J. O'Brien, D.A. Kistner, M.W. Wong, M. Gallagher and S.M. Watson. 1998. Do leptocephali of the European eel swim to reach continental caters? Status of the question. J. Mar. Biol. Assoc. UK 78: 285–306.

Miller, M.J. 1995. Species assemblages of leptocephali in the Sargasso Sea and Florida Current. Mar. Ecol. Prog. Ser. 121: 11–26.

Miller, M.J. 2009. Ecology of anguilliform leptocephali: remarkable transparent fish larvae of the ocean surface layer. Aqua-BioSci. Monogr. 2: 1–94.

Miller, M.J. and J.D. McCleave. 1994. Species assemblages of leptocephali in the subtropical convergence zone of the Sargasso Sea. J. Mar. Res. 52: 743–772.

Miller, M.J., S. Kimura, K.D. Friedland, B. Knights, H. Kim, D.J. Jellyman and K. Tsukamoto. 2009. Review of ocean–atmospheric factors in the Atlantic and Pacific oceans influencing spawning and recruitment of anguillid eels. *In*: A. Haro, T. Avery, K. Beal, J. Cooper, R. Cunjak, M. Dadswell, R. Klauda, C. Moffitt, R. Rulifson and K. Smith (eds.). Challenges for Diadromous Fishes in a Dynamic Global Environment. American Fisheries Society Symposium Publication, Bethesda, MD.

Miller, M.J., S. Bonhommeau, P. Munk, M. Castonguay, R. Hanel and J.D. McCleave. 2014. A century of research on the larval distributions of the Atlantic eels: a re-examination of the data. Biol. Rev. DOI: 10.1111/brv.12144.

Otake, T. 1996. Possible food sources for Anguilliformes leptocephali. pp. 33–43. *In*: O. Tabeta (ed.). Early Life-history and Prospects of Seed Production of the Japanese Eel *Anguilla japonica*. Koseisha-Koseikaku, Tokyo.

Power, J.H. and J.D. McCleave. 1983. Simulation of the North Atlantic Ocean drift of *Anguilla* leptocephali. Fish. Bull. 81: 483–500.

Powles, P.M. and S.M. Warlen. 2002. Recruitment season, size and age of young American eels (*Anguilla rostrata*) entering an estuary near Beaufort, North Carolina. Fish. Bull. 100: 299–306.

Pujolar, J.M., M.W. Jacobsen, T.D. Als, J. Frydenberg, E. Magnussen, B. Jönsson, X. Jiang, L. Cheng, D. Bekkevold, G.E. Maes, L. Bernatchez and M.M. Hansen. 2014. Assessing patterns of hybridization between North Atlantic eels using diagnostic single-nucleotide polymorphisms. Heredity 112: 1–11.

Schmidt, J. 1909. On the distribution of the fresh-water eels (*Anguilla*) throughout the world. Atlantic Ocean and adjacent regions. I. Meddel fra Kommissionen for Havundersøgelser Serie III: 1–45.

Schmidt, J. 1922. The breeding places of the eel. Phil. Trans. R. Soc. (Ser. B) 211: 178–208.

Schmidt, J. 1923. Breeding places and migration of the eel. Nature 111: 51–54.

Schmidt, J. 1925. The breeding places of the eel. Annu. Rep. Smithson. Inst. 1924: 279–316.

Schoth, M. and F.W. Tesch. 1982. Spatial distribution of 0-group eel larvae (*Anguilla* sp.) in the Sargasso Sea. Helgol. Meeresunter. 35: 309–320.

Shinoda, A., H. Tanaka, H. Kagawa, H. Ohta and K. Tsukamoto. 2004. Otolith microstructural analysis of reared larvae of the Japanese eel *Anguilla japonica*. Fish. Sci. 70: 340–342.

Smith, D.G. 1989. Family Anguillidae: Leptocephali. pp. 898–899. *In*: E.B. Böhlke (ed.). Fishes of the Western North Atlantic. Sears Foundation for Marine Research, New Haven.

Sugeha, H.Y., A. Shinoda, M. Marui, T. Arai and K. Tsukamoto. 2001. Validation of otolith daily increments in the tropical eel *Anguilla marmorata*. Mar. Ecol. Prog. Ser. 220: 291–294.

Sullivan, M.C., K.W. Able, J.A. Hare and H.J. Walsh. 2006. *Anguilla rostrata* glass eel ingress into two U.S. east coast estuaries: patterns, processes and implications for adult abundance. J. Fish Biol. 69: 1081–1101.

Teichmann, H. 1959. Uber die leistung des Gurichsinnes beim Aal (*Anguilla anguilla* L.) Z. Vergl. Physiol. 42: 206–254.

Tesch, F.W. 1977. The Eel. Biology and management of anguillid eels. Chapman and Hall, London.

Tesch, F.W. 1978. Telemetric observations on the spawning migration of the eel (*Anguilla anguilla*) west of the European continental shelf. Environ. Biol. Fish. 3: 203–209.

Tesch, F.W. 1980. Occurrence of eel *Anguilla anguilla* larvae west of the European continental shelf 1971–1977. Environ. Biol. Fish. 5: 53–55.

Tesch, F.W. 1982. The Sargasso Sea eel expedition 1979. Helgoländer Meeresunters 35: 263–277.

Tesch, F.W. 1989. Changes in swimming depth and direction of silver eels (*Anguilla anguilla* L.) from the continental shelf to the deep sea. Aquat. Living Resour. 2: 9–20.

Tesch, F.W. 1998. Age and growth rates of North Atlantic eel larvae (*Anguilla* spp.), based on published length data. Helgol. Meeresunter. 52: 75–83.

Tesch, F.W. 2003. The Eel, 3rd edn. Blackwell Publishing, Oxford, U.K.

Tesch, F.W. and G. Wegner. 1990. The distribution of small larvae of *Anguilla* sp. related to hydrographic conditions 1981 between Bermuda and Puerto Rico. International Revue der Gesamten Hydobiologie 75: 845–858.

Tsukamoto, K. 1989. Otolith daily growth increments in the Japanese eel. Bull. Jpn. Soc. Sci. Fish. 55: 1017–1021.

Tsukamoto, K. 1992. Discovery of the spawning area for the Japanese eel. Nature 356: 789–791.

Tsukamoto, K., S. Chow, T. Otake, H. Kurogi, N. Mochioka, M.J. Miller, J. Aoyama, S. Kimura, S. Watanabe, T. Yoshinaga, A. Shinoda, M. Kuroki, M. Oya, T. Watanabe, K. Hata, S. Ijiri, Y. Kazeto, K. Nomura and H. Tanaka. 2011. Oceanic spawning ecology of freshwater eels in the western North Pacific. Nature Commun. 2: 179.

Tucker, D.W. 1959. A new solution to the Atlantic eel problem. Nature 183: 495–501.

van Ginnenken, V. and G.E. Maes. 2005. The European eel (*Anguilla anguilla*, Linnaeus), its lifecycle, evolution and reproduction: a literature review. Rev. Fish. Biol. Fisheries 15: 367–398.

van Ginneken, V., B. Muusze, J. Klein Breteler, D. Jansma and G. van den Thillart. 2005. Microelectronic detection of activity level and magnetic orientation of yellow European eel (*Anguilla anguilla* L.) on a pond. Environ. Biol. Fish. 72: 313–320.

van Utrecht, W.L. and M.A. Holleboom. 1985. Notes on eel larvae (*Anguilla anguilla* Linnaeus, 1758) from the central and eastern North Atlantic and on glass eels from European continental shelf. Bijdragen tot de Dierkunde 55: 249–262.

Umezawa, A. and K. Tsukamoto. 1991. Factors influencing otolith increment formation in Japanese eel, *Anguilla japonica* T. and S., elvers. J. Fish Biol. 39: 211–223.

Voorhis, A.D. 1969. The horizontal extent and persistence of thermal fronts in the Sargasso Sea. Deep-Sea Res. 16(Supplement): 331–337.

Voorhis, A.D. and J.G. Bruce. 1982. Small-scale surface stirring and frontogenesis in the subtropical convergence of the western North Atlantic. J. Mar. Res. 40 (Supplement): 801–821.

Wang, C.H. and W.N. Tzeng. 1998. Interpretation of geographic variation in size of American eel *Anguilla rostrata* elvers on the Atlantic coast of North America using their life history and otolith ageing. Mar. Ecol. Prog. Ser. 168: 35–43.

Wang, C.H. and W.N. Tzeng. 2000. The timing of metamorphosis and growth rates of American and European eel leptocephali: a mechanism of larval segregative migration. Fish. Res. 46: 191–205.

Wegener, G. 1982. Main hydrographic features of the Sargasso Sea in Spring 1979. Helgöl. Meeresunter. 35: 385–400.

Wippelhauser, G.S., J.D. McCleave and R.C. Kleckner. 1985. *Anguilla rostrata* leptocephali in the Sargasso Sea during February and March 1981. Dana 4: 93–98.

Yamamoto, K. and K. Yamauchi. 1974. Sexual maturation of Japanese eel and production of eel larvae in the aquarium. Nature 251: 220–222.

Yamauchi, K., M. Nakamura, H. Takahashi and K. Takano. 1976. Cultivation of larvae of Japanese eel. Nature 263: 412.

Spawning Ground of the Japanese Eel *Anguilla japonica*

Takaomi Arai

Introduction

At the beginning of the 20th century, Schmidt (1922) conducted numerous expeditions and discovered that the spawning areas for both the European eel (*Anguilla anguilla*, Anguillidae) and the American eel (*A. rostrata*, Anguillidae) were located far offshore in the Sargasso Sea of the Atlantic Ocean. Approximately 70 years after Schmidt's (1922) discovery, the spawning area of the Japanese eel (*A. japonica*, Anguillidae) was found in the North Equatorial Current to the west of the Mariana Islands, approximately 3000 km from their growth habitats in east Asia in 1991 (Tsukamoto 1992). Since this discovery, intensive investigations have been conducted almost every year for the last 20 years. The preleptocephali have been collected just after hatching (Tsukamoto 2006), and the eggs and maturing eels have been collected in the spawning area (Tsukamoto et al. 2011). These findings have not been reported for 18 of the 19 anguillid eel species and such findings and studies would further contribute towards understanding the life history, migration route and behaviour, and ecological involvement with other species for the other 18 eel species.

Environmental and Life Sciences Programme, Faculty of Science, Universiti Brunei Darussalam, Jalan Tungku Link, Gadong, BE 1410, Brunei Darussalam.
E-mail: takaomi.arai@ubd.edu.bn

The population size of wild glass eels has linearly decreased from over 200 tonnes in the early 1960s to 20 tonnes at present. A shortage of Japanese eel fry has become a very serious problem for fish cultures in recent years (Arai 2014a). Eel stocks all over Europe are also declining (Dekker 2003a), and eel fishery yields have decreased in most European countries. The population of the European eel is considered to be outside of the safe biological limits, and current fisheries are not sustainable (Dekker 2003b). The European eel was even recently categorised as critically endangered by the European Union (EU) and the United Nations (UN) (CITES 2007). Since the early 1980s, glass eel recruitment has decreased and dropped to 1% of the levels encountered in the 1970s. The causes of decline in stock and recruitment are not well understood. Overfishing, habitat loss, migration barriers, increased natural predation, parasitism, ocean climate variation, and pollution might be responsible (Knights 2003; Marcogliese and Casselman 2009; Bonhommeau et al. 2008; Friedland et al. 2007). In 2014, the Japanese and the American eels were both added to the IUCN's list with an endangered classification and critically endangered classification, respectively (Jacoby and Gollock 2014), suggesting that these eels also have a high risk of extinction.

To address the rapid decline in eel resources, artificially induced breeding techniques for the eel have been intensively studied in Japan. A challenging research project for the artificial production of glass eels commenced in the late 1960s. Since then, through a continuous process of trial and error, the production of second-generation larvae was finally achieved in 2010 (Ijiri et al. 2011). However, the techniques for producing glass eels in captivity have not yet been firmly established. The quality of eggs obtained through controlled maturation is still highly variable, and the survival rates of the larvae are usually extremely low. In addition, the growth of the larvae is slower in captivity than in the wild (Tanaka et al. 2003). Therefore, further studies will be required to improve the culture conditions and environment for enhancing the eel stock in the future.

In this chapter, the present status of the knowledge regarding the spawning ground and spawning ecology of the Japanese eel *Anguilla japonica* is discussed. This paper also discusses and evaluates the evidence from intensive investigations of the spawning ground of the Japanese eel, which has accrued for the past 20 years since its discovery in 1991. How these efforts have contributed to eel stock enhancement rather than to biological interests is also examined. Most of the content in this chapter is cited and referenced from the latest studies by Arai (2014a,b).

History of the discovery of the spawning ground

The search for the spawning area of the Japanese eel began in the 1930s, approximately 10 years after Schmidt (1922, 1925) discovered the spawning area of Atlantic eels (Tsukamoto et al. 2003). At the beginning of the search, information regarding the life history and ecology of the Japanese eel was scant. Matsui (1957) hypothesised that the Japanese eel spawned in an area east of Taiwan and south of Okinawa because the adult Japanese eel reached the southern most limit of its continental distribution near Taiwan (Fig. 1). The first collected Japanese eel leptocephalus was fully grown, at a total length (TL) of 57 mm, which was just before metamorphosis at the Bashi Strait just off the southern tip of Taiwan in 1967 (Matsui et al. 1968) (Fig. 1). Kuo (1971) suggested that the Japanese eel of Taiwan would spawn locally in the waters off southwestern Taiwan, because the elver catches were higher on the west coast of Taiwan than on the east coast. Therefore, the spawning ground proposed by Matsui (1957) was plausible only if the east coast of Taiwan was supplied with eel larvae that were transported by the Kuroshio Current from south to north. In the 1970s, a total of 55 leptocephali (approximately 50-50 mm TL) were collected predominantly in the waters east of Taiwan (Tanaka 1975; Takai and Tabeta 1976). However, in the 1980s, smaller leptocephali (approximately 40–50 mm TL) were collected in more southern waters east of Luzon Island, off of the Philippines (Kajihara 1988; Tsukamoto et al. 1989). Subsequent research efforts collected a total of 28 smaller Japanese eel larvae (approximately 20–31 mm TL) in a more eastern area (Ozawa et al. 1989, 1991). Based on the collected data and the current patterns in this area, the spawning ground of the Japanese

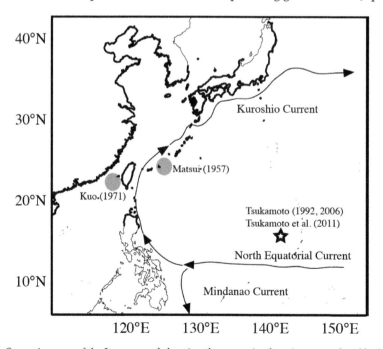

Fig. 1. Spawning area of the Japanese eel showing the spawning locations speculated by Matsui (1957) and Kuo (1971) (grey circle) and estimated by Tsukamoto (1992, 2006) and Tsukamoto et al. (2011) (star) along with the major current systems in the western North Pacific. Areas covered by thick lines indicate the geographic range of the Japanese eel.

eel was estimated to occur somewhere around the Mariana Islands (Fig. 1). In 1991, a large collection of 958 leptocephali (approximately 10–20 mm TL), including the smallest leptocephalus (7.7 mm TL) ever collected, revealed that the spawning ground of the Japanese eel was located at approximately 15° N, 140° E in the North Equatorial Current to the west of the Mariana Islands (Tsukamoto 1992) (Fig. 1). These research efforts suggested that the estimated spawning ground of the Japanese eel has historically moved from north (off Taiwan) to south (off the Philippines), and then from west to east, west of the Mariana Islands in the western North Pacific (Fig. 1).

Since the discovery of the spawning ground of the Japanese eel, further intensive investigations have been conducted almost every year for more than 20 years in order to determine the precise location where the actual spawning occurs. In 2005, a total of 130 preleptocephali (4.2–6.5 mm TL just after hatching) and 60 leptocephali (11.7–18.4 mm TL) were collected at 14° N, 142° E in the region of the North Equatorial Current (Tsukamoto 2006). Back-calculated hatching dates using the daily incremental growth rate of the otolith, which was observed for these preleptocephali and other leptocephali specimens, demonstrated that the Japanese eel had hatched at approximately the time of the new moon during those years (Tsukamoto 2006). In 2009, thirty-one Japanese eel eggs and three female eels with functional ovaries were collected in the spawning ground (Tsukamoto et al. 2011).

The spawning area was determined to be just south of a weak salinity front that was typically present in the region as a result of tropical rainfall (Kimura et al. 1994, 2001; Kimura and Tsukamoto 2006). The salinity front was hypothesised to act as a cue for the eels to help them find the spawning area. Although Tsukamoto et al. (2003) and Tsukamoto (2006) proposed that the Japanese eel might spawn near seamounts as a landmark for forming spawning aggregations, spawning adult males and a large number of the Japanese eel preleptocephali were collected further off from the seamounts (Chow et al. 2009). If the Japanese eel spawns around seamounts, then it should be possible to constantly collect eggs and small larvae without any annual variation. However, only 31 eggs were collected during 20 years of intensive surveys, despite the fact that the fecundity (number of eggs) of anguillid eels, including the Japanese eel, was reported to be higher than one million per eel (MacNamara and McCarthy 2012). Therefore, these findings suggested that Japanese eel spawning can occur in the open water of the spawning ground with no relation to seamounts.

Spawning ecology

Although the investigation of the spawning ground of the Japanese eel has a long history, little information is available on the spawning ecology of the matured (silver) eel. Thus, it is unknown whether the silver eels migrate in schools or as single individuals on the way to the spawning ground after maturation in their growth habitats. In addition, it is also unknown whether spawning occurs in large or small aggregations in the spawning area. If they do not migrate in schools, it has been hypothesised that they might use pheromones to locate each other within the spawning area (Sorensen and Winn 1984; McCleave and Kleckner 1985; Huertas et al. 2008). *A. anguilla* was found to have high olfactory sensitivity to substances released by conspecifics (Huertas et al. 2008). Both the bile fluid and skin mucus were implicated as the routes of release of these odorants but other routes, such as urine, could not be excluded (Huertas et al. 2008). Furthermore, the nature of these odorants depended on both the sex and the reproductive status of the donor; exposure to water conditioned by mature con-specifics caused a stimulation of sexual maturation in immature eels (Huertas et al. 2008). These results would be consistent with a role of chemical communication in eel reproduction.

There were no observations of the spawning behaviour of the other anguillid species in their spawning areas. Information regarding some aspects of the spawning behaviour of anguillid eels was obtained from laboratory observations of artificially matured silver eels. In 1980, a male *A. anguilla* injected with pituitary extract swam close to an artificially matured female and the release of sperm was first observed in 1980 (Boëtius and Boëtius 1980). Similar behaviour of artificially matured males swimming next to the females and nudging them under the head as gametes were released was also observed with *A. rostrata* (Sorensen and Winn 1984). The spawning behaviour of artificially matured *A. anguilla* was also observed in a group of two females and three males (van Ginneken et al. 2005). During a 283 minute (min) observation of the two females, van Ginneken et al. (2005) found female-female interaction: 'lethargic behaviour' (33.6%) vs. 'cruising together' (66.4%). In the period when males and females were together (188 min), 'approaching the head region of the female' (57.7%), 'touching the operculum' (39.4%), or 'approaching the urogenital area' (2.9%) by the males (total 725 seconds (s)) were observed. Sperm release in the presence of a female took 115 s of the total approaching time of 725 s (15.9%), while in the case of male-male interaction this was only 15 s of the total period of 116 s (12.9%). The induced spawning behaviour of eels was collective and simultaneous, corresponding to spawning in a group (van Ginneken et al. 2005). The study was the first to observe and record group spawning behaviour in freshwater eels. These results suggested that anguillid eels would form mixed-sex spawning aggregations, rather than showing paired spawning in an ambient environment.

Migration route of adult eels to the spawning ground

The migration route of juvenile Japanese eels is well understood. After hatching, the larvae (leptocephali) drift from their spawning grounds with the North Equatorial Current (NEC) and then the Kuroshio Current (KC) for 4–6 months in order to reach the coasts of East Asia. They then metamorphose into glass eels and become pigmented elvers when they enter river/estuarine habitats in the Philippines, Taiwan, China, Korea, and Japan (Cheng and Tzeng 1996; Arai et al. 1997; Tesch 2003). The larval migration of Japanese eels is thought to be controlled by environmental factors in the region of the NEC, such as the salinity front at the bifurcation of the NEC, allowing the Ekman transport of leptocephali operating in an Ekman layer approximately 70 m deep (Kimura et al. 2001). The salinity front is thought to be critical to the spawning migration of Japanese eels (Kimura and Tsukamoto 2006), and its interannual variability associated with El Niño/Southern Oscillation (ENSO) events probably leads to a reduced larval transport into the KC, causing poor recruitment in eastern Asia (Kimura et al. 2001).

However, because of the long distance migration route(s), no study has found the spawning migration route of eels from their growth habitats to spawning areas. American eels migrate from a wide range of latitudes; however McCleave and Kleckner (1985) examined the hypothesis that one general compass orientation direction could allow anguillid eels from different latitudes to reach the spawning area in the Sargasso Sea. Two migration routes to the Sargasso Sea were proposed for the European eel: *A. anguilla*: a northern migration route via the central Atlantic Ocean (Tesch 1986) and a southern route via the Azores Current (Fricke and Kaese 1995). The former was suggested by the direct compass course to the south-west from Europe to the Sargasso Sea and the latter was the result of a numerical simulation. The duration of the migration estimated by the simulation was 4–6 months and was congruent with the actual departure times of *A. anguilla* silver eels in Europe (September to November). Knights (2003) suggested that the deep boundary currents of the North Atlantic subtropical gyre could be followed by silver eels migrating on each side of the Atlantic basin using periodic deep diving, which might help them in locating the spawning area. In case of *A. japonica*, a few speculative ideas on the migration route of the silver eels have been proposed (Fig. 3). One idea is the direct compass course, in which fish from different areas of their species range migrate directly towards the spawning area, traversing beneath the strong Kuroshio Current. This possibility was proposed by Matsui (1972), but at the time the spawning area was hypothesised to be closer to the continental growth habitats, closer to the Okinawa Islands. Matsui (1972) also speculated that silver eels of *A. japonica* might use deep currents, such as the southward flowing deep boundary current that is often present under the Kuroshio in the East China Sea or further to the north-east (Nakamura et al. 2008), to avoid swimming against the Kuroshio. The others are routes along the surface currents: an along-Kuroshio route, which could include a southward migration along the Ogasawara Islands (Tsukamoto 2009), and an anti-Kuroshio route (Yokose 2008). The anti-Kuroshio route hypothesis proposed that *A. japonica* migrate upstream in the Kuroshio and then further upstream in the NEC, while being led by olfactory attractants specific to the spawning area (Yokose 2008). In anguillid eels, olfaction has been shown to play a critical role in orientation during their spawning migration (Sébert et al. 2008). Westin (1990) reported that silver eels, originating from glass eels imported from France for stocking in Sweden, showed a lower swimming speed as compared to indigenous silver eels. The imported eels also failed to find their way out of the Baltic Sea. Swimming against the KC and NEC would be highly energy consuming, but if the fish swam within the peripheral areas at the edge of the current, it might be possible to migrate along an anti-Kuroshio route (Yokose 2008), which would essentially retrace their larval migration route using their olfaction (Fig. 3).

How have recent spawning ground investigations of the Japanese eel contributed to the stock enhancement?

Although the intensive spawning ground investigations of the Japanese eel over the last 20 years provided updated biological information for the eel, the eel stock is decreasing linearly with the increase of the aquaculture demand for human consumption (Fig. 2). The reasons for investigating the eel's spawning ground other than the biological interests are to facilitate efforts to prevent further declines in its population (Tsukamoto et al. 2011) and to provide some clues to help advance the establishment of commercial glass eel production, i.e., to understand the environmental conditions of the spawning ground for implementation in commercial production (Ijiri et al. 2011).

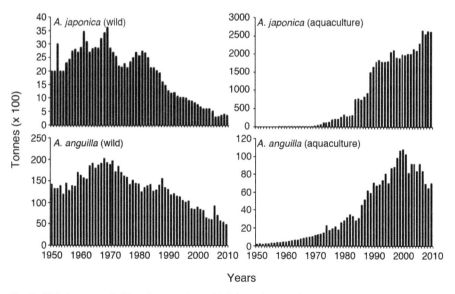

Fig. 2. Global capture (left) and aquaculture (right) production from 1950–2010 for the Japanese eel (top) and the European eel (bottom). Sources: FAO. ©2010–2012. FAO FishFinder—Web Site. FAO FishFinder Contacts. In: FAO Fisheries and Aquaculture Department (online) http://www. fao.org/fishery/fishfinder/contacts/en.

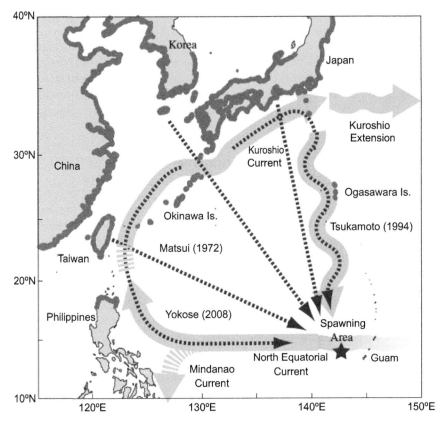

Fig. 3. Hypothesised migration routes by silver eels from their growth habitats to the spawning area of *Anguilla japonica* in the western North Pacific, as proposed by Matsui (1972), Yokose (2008) and Tsukamoto (2009). Taken from Tsukamoto (2009), with permission from John Wiley and Sons (License No. 3523550529064).

The fluctuation of the Japanese eel population has garnered particular attention (Miller et al. 2009) because of its high economic value (FAO 2010), complex life history (Arai and Chino 2012), and its declining recruitment since the 1970s (Tatsukawa 2003). A similar declining trend has as also been reported for the European and American eels (Dekker 2003c). The reasons for the recruitment declines of these temperate anguillid eels are not clear but may have been caused by overfishing, habitat degradation, pollution, parasites, virus, and global climate change (Knights 2003; Marcogliese and Casselman 2009; Bonhommeau et al. 2008; Friedland et al. 2007). In addition, their effects would be different depending on the eel life histories.

During the oceanic life stage, it has been suggested that the recruitment variability of the Japanese eel is affected by ocean–atmospheric forcing (Miller et al. 2009). In particular, the latitudinal shifts of spawning locations in relation to larval transport by the North Equatorial Current (NEC) are considered to be an important determinant of recruitment success (Kimura et al. 2001). If the eels can travel westward using the NEC and enter the Kuroshio Current, they have a significantly enhanced probability of recruitment success. In contrast, if the eels are entrained into the south-flowing Mindanao Current or mesoscale eddies east of Taiwan, recruitment is reduced (Kim et al. 2007). Specifically, when precipitation is low during some ENSO years, the salinity front (and therefore the spawning location) may move considerably southward, thereby increasing the possibility that the eel larvae will enter the Mindanao Current (Kimura et al. 2001; Kimura and Tsukamoto 2006). In addition, the bifurcation latitude of the NEC varies both seasonally and interannually (Qiu and Lukas 1996), which potentially also affects the recruitment variability of the Japanese eel (Zenimoto et al. 2009). In particular, ENSO events shift the bifurcation latitude of the NEC northward, which results in more NEC water flowing into the Mindanao Current and hampers eel recruitment (Zenimoto et al. 2009). Another possible climatic effect is the change in ocean productivity, which may be critical for feeding success and for larval survival during their migration route (Miller et al. 2009; Bonhommeau et al. 2008).

Spawning ground investigations can provide exact locations and this information may be useful for estimations of the possible transportation route of the leptocephali. This information may provide further data to predict the possible recruitment mechanism for glass eels. However, the spawning ground was reported to shift in association with ocean climate change (Kimura and Tsukamoto 2006). Although the NEC region is the only spawning area of the Japanese eel, the 2002 expedition indicated that the larval distribution of the species was likely related to a salinity front generated by two distinct waster masses in the NEC. This investigation was conducted during an El Niño event, and the salinity front had moved. Smaller larvae (less than 10 mm TL) were collected just south of the salinity fronts, where these larvae were never collected during typical years (Kimura and Tsukamoto 2006). Because the latitudinal location of the Japanese eel spawning events was found to shift by months or years (Tsukamoto 2009), the determination of the exact spawning ground might be difficult. Further, the examined spawning locations were not so different from the first spawning ground discovery in 1991 (Tsukamoto 1992), even though these investigations were conducted almost annually during this period over the past 20 years (Tsukamoto et al. 2011) through snapshot findings (Tsukamoto 2006). Such ocean climate effects leading to fluctuations of the Japanese eel population during the oceanic life stage could be estimated using ocean climate data, satellite tracked buoy data and a coupled ocean-atmosphere circulation model using the data from the first discovery.

Information regarding the environmental conditions in the spawning ground may help advance the establishment of artificial breeding techniques for the production of commercial glass eels in captivity. Water temperature, spawning depth and diet during hatching and the early developmental stages of the larvae in the spawning ground are accessible and constitute fundamental information for improving and developing eel culture techniques.

Knowing the optimal temperature range will lead to the efficient production of healthy Japanese eel larvae and healthy larvae are essential for producing glass eels. Tanaka (1996) reported a higher hatching rate (60–70%) at a relatively high temperature (25–28°C); however the survival rate for pre-feeding larvae was low (0%, 8 days after hatching). Chang et al. (2004) indicated that the Japanese eel eggs and yolk sac (prefeeding) larvae could adapt to a wide range of temperatures (3–32°C) and suggested that the optimal temperature range might be 24–26°C for incubating eggs and 26–28°C for prefeeding larvae. Recently, Okamura et al. (2007) provided an answer for producing healthy larvae. Their results demonstrated that many larvae become deformed at 19–22°C, whereas much fewer become deformed at 25–28°C. Assuming that the occurrence rate of deformed fish is a more effective index for determining the optimum temperature for eel larvae, the optimal temperature range is approximately 25–28°C (Okamura et al. 2007). This result is in good agreement with the environmental conditions likely experienced by wild larvae, as a temperature range of 25–28°C corresponds to depths of 150–200 m in the spawning ground (Tsukamoto et al. 2011). At these depths in this area, the specific gravity of the sea water ranges from 1.022–1.023, which is almost the same for artificially fertilised eggs (Okamura et al. 2007). These results suggested that the optimal water temperatures for the Japanese eel eggs and pre-feeding larvae in the wild and in captivity are approximately 25–28°C. As a result, the experimental studies in captivity revealed the optimal water temperature for incubating eggs and rearing larvae of the Japanese eel. This information may be more useful for predicting where to find wild eggs and larvae, which would better help to determine the spawning layer in the spawning ground rather than extracting information from the spawning ground investigation field data for the eel.

Despite intensive research on wild and captive eels, no resource has provided access to all of the life cycle stages of the Japanese eel since the production of preleptocephali 25 years ago (Tanaka et al. 2001). Therefore, the transition from the preleptocephalus (newly hatched larva) to the leptocephalus stage (typical leaf-like eel larva) has remained the missing link in the eel life cycle. Yamamoto and Yamauchi (1974) first succeeded in obtaining fertilised eggs and larvae of the Japanese eel, and preleptocephalus larvae were reared for 2 weeks, reaching 7 mm TL (Yamauchi et al. 1976). Thereafter, eel larvae could be successfully obtained; however suitable larval feeds were not identified. As a result, the preleptocephalus larvae could not survive beyond the depletion of their yolk and oil droplet stores. In other eel species, for example, the European eel (Prokhorchik 1986) and New Zealand eels *A. dieffenbachii* and *A. australis* (Lokman and Young 2000), experimentally produced larvae survived only for a few days and like the Japanese eel did not develop into leptocephali. Tanaka et al. (2001) finally found that a slurry-type diet made from shark egg powder was a suitable feed for captive-bred eel larvae. The larvae were successfully reared with this diet in aquaria for 100 days and raised to 22.8 mm TL. The age, TL and body proportions of the reared specimens overlapped with those of wild leptocephali (Tanaka et al. 2001). Although preleptocephalus larvae can be reared using this diet, it remains inadequate because the leptocephali reared in this manner cannot be raised to the subsequent stage. Soon after the findings on this diet by Tanaka et al. (2001), the diet was improved by supplementation with krill hydrolysate, soybean peptide, vitamins and minerals. Leptocephali that fed on this new diet grew to 50 to 60 mm in TL and began to metamorphose into glass eels (Tanaka et al. 2003). The artificially produced glass eels could then be grown and artificially matured (Ijiri et al. 2011). Thereafter, a second generation of larvae was produced in 2010 (Ijiri et al. 2011).

While there have been a few studies, the feeding ecology of leptocephali, including the anguillid eel is still not well understood. From the analyses of the gut contents, gut pigment content and nitrogen stable isotopes, the most likely food source of congrid leptocephali (*Conger myriaster* and *C. japonicus*, Congridae) has as been inferred to be particulate organic matter (POM) (Otake et al. 1993). Mochioka and Iwamizu (1996) examined the gut contents from five families of eels (Congridae, Muraenidae, Muraenesocidae, Nettastomatidae and Ophichthidae) and reported that the larvacean houses, and their faecal pellets were commonly found in their guts. These studies indicate that the trophic level of leptocephali should be low considering their gut contents, e.g., larvacean houses and POM. Kimura and Tsukamoto (2006) reported that the carbon isotope ratios of the Japanese eel leptocephali corresponded to the carbon stable isotope ratios of POM, which confirmed that POM was the most likely food source of leptocephali. A likely reflection of this complex composition of POM was observed in a recent study using DNA barcoding for the qualitative analysis of the diet of small European eel leptocephali. This result suggested that gelatinous zooplankton such as Hydrozoa, Thaliacea and Ctenophoramay somehow contribute to the diet of the early larvae of that species (Riemann et al. 2010). However, the comparisons of POM stable isotope ratios and those of the Japanese eel leptocephali indicated that leptocephali appeared to feed more on POM rather than directly feeding on zooplankton or secondary consumers (Miyazaki et al. 2011) even if their DNA sequences were detected in the gut contents of the European eel leptocephali by Riemann et al. (2010). More recently, the nitrogen isotope enrichment values of reared larvae have suggested that the primary food source of wild larvae was also consistent only with POM, such as marine snow and discarded appendicularian houses (Miller 2012). These results all led to the conclusion that leptocephali appeared to feed on POM and did not directly feed on zooplankton or secondary consumers.

Although the findings using shark egg powder, krill hydrolysate, soybean peptide, vitamins and minerals as optimal diets for larvae in captivity (Tanaka et al. 2001, 2003) have provided a significant breakthrough for cultivating eel larvae, these diets could not be found in the NEC in the spawning ground and were highly different from those of wild larvae (POM; Kimura and Tsukamoto 2006; Miyazaki et al. 2011; Miller 2012). Judging from these results and the present state of eel farming for the Japanese eel, the eels can be reared throughout their lives from eggs to adults in captivity.

Other environmental conditions such as salinity, light intensity (wavelength), water circulation and water quality, may also be important factors affecting the survival rate in captivity. These factors may directly determine the location of the spawning ground. In addition to water temperature and diet, other factors at optimal conditions must be considered while determining rearing conditions (Tanaka et al. 2001, 2003; Tanaka 2003; Kagawa et al. 2005; Okamura et al. 2009a,b). Therefore, in order to determine the optimal conditions for the various ambient factors involved in artificial eel culturing, field investigations of the spawning ground of the Japanese eel might be less relevant and not as important a link to eel propagation in captivity.

Future perspectives for research: Factors improved by improving artificial breeding techniques and the conservation of wild stocks for their enhancement

The intensive spawning ground investigations of the Japanese eel after the first discovery of its spawning ground in 1991 have provided valuable biological information, such as information on wild egg collection and the wild spawning conditions of the adult specimens collected in the spawning ground. These findings were the first for a member of the 19 species of the freshwater eel. However, this exertion and long-term (20 years) expedition might be less influential for improving and developing artificial breeding and culturing techniques along with stock enhancement for the Japanese eel than previously thought.

Although the glass eel has been successfully produced artificially, and has been grown and matured in captivity, the quality of eggs obtained through this type of controlled maturation is still highly variable, and the survival rates of the larvae are usually extremely low for the Japanese eel (Tanaka et al. 2003; Ijiri et al. 2011). Therefore, further studies should focus on developing better maturation induction procedures and rearing regimes for the larvae to establish techniques for the consistent mass production of glass eels. Unlike anadromous salmon, the spawning grounds of all catadromous eels are located in the open ocean, generally more than thousands of kilometers away from coastal and inland farming facilities. It is almost impossible to establish eel propagation facilities and eel farming industries in the open ocean, but even consistent mass collections of wild eggs will succeed in the future. However, the spawning ground may be shifted by oceanic climate conditions on a monthly and yearly basis, which might prove to be a challenge for the collection of wild eggs for propagation. Furthermore, the wild eggs and spawning condition adult eels in the open ocean might be considered to be common property. Therefore, international fishery management for the usage of precious resources must be considered. The Japanese eel is distributed widely in the East Asian countries of Japan, China, Korea and Taiwan as well as in the Philippines and Southeast Asian countries as a panmictic population (Sang et al. 1994). These facts suggest that further studies will be specifically required to improve and develop techniques for the consistent mass production of glass eels in captivity to improve Japanese eel propagation and stock enhancement, which is in contrast to further intensive spawning ground investigations.

The world-wide decline of the anguillid eels is one of the major concerns for fishery management. Similar to the American eel and the European eel, the Japanese eel has experienced a sharp decline across its range over the last 30–40 years (ICES 2006; Aprahamian et al. 2007; Castonguay et al. 2007; Dekker et al. 2007; Tatsukawa 2003; Arai 2014a,b) (Fig. 2). There were dramatic declines in the populations of these eels, which have plummeted to approximately 1%–10% of their levels from the 1970–1980s (Dekker 2007; Casselman 2003; Tatsukawa 2003), and the effective population size of the Japanese eel has also decreased (Tzeng et al. 2004). Current populations are far below the biologically safe limit, and extinctions in the near future are a real possibility. The reasons for the population decline are still uncertain, but they may be related to changes in oceanic currents and global climate, degradation of habitat, pollution, exotic parasite infection, and over-fishing. Population levels of the American, European and Japanese eels are on the verge of collapse, and the need for management and conservation is urgent because eel fishing is performed at almost every life stage, from glass eel to silver eel.

As a consequence, in case of the European eel, the European Commission has agreed to an Eel Recovery Plan (ERP), the aim of which is to return the European eel stock to sustainable levels of adult abundance and glass eel recruitment (Svedäng and Gipperth 2012). In October 2003, the European Commission presented the Council and the European Parliament with a recovery plan for the European eel. The European Council then asked the Commission to come forward with proposals for the long-term management of the eel. The Parliament adopted a solution in November 2005 calling on the Commission to propose legislation regarding eel recovery measures (Svedäng and Gipperth 2012). In June 2007, the European eel was listed in Appendix II of the CITES (the Convention on International Trade in Endangered Species of Wild Fauna and Flora). Appendix II "includes species not necessarily threatened with extinction, but in which trade must be controlled to avoid utilisation incompatible with their survival" (CITES 2007; WWF 2007). The EU passed Council Regulation (EC) number 1100/2007 for the purpose of establishing measures for the recovery of the stock of the European eel in 2007 (Svedäng and Gipperth 2012). According to this EU regulation, national recovery plans should be developed to meet the stipulated objectives. Regulation EC 1100/2007 recognises several possible measures to reduce eel mortality: reducing/closing fisheries (both commercial and recreational); facilitating migration and improving river habitats; and transporting silver eels from inland waters to waters from which they can escape freely to the Sargasso Sea (Svedäng and Gipperth 2012). A critical feature of these plans is to permit the escapement to sea of at least 40% of the virgin silver eel biomass, which would have occurred prior to anthropogenic influences. The principle behind this conservation measure is that escape is one of the key stages of the European eel's catadromous life cycle that can be controlled, and it is hypothesised that positive alterations to this will, in turn, have a proportionally positive effect on recruitment (the number of juveniles migrating to the freshwater habitat for maturation) (Bilotta et al. 2011). Although stock assessment and management of the European eel has received increasing attention from both the scientific community and the fisheries agencies in recent years (ICES 2006), such assessment and management of the Japanese eel has not been well studied with a concrete policy and destination for stock enhancement. However, in 2014 the Japanese and the American eels were added to the IUCN's list with an endangered classification and critically endangered classification, respectively (Jacoby and Gollock 2014), suggesting that these eels also have a high risk of extinction. Please note that despite the high demand for the product, the peak capture of Japanese eels is less than the lowest captures of European eels (Halpin 2007). This fact indicates the relatively low virgin biomass of Japanese eels. Landings of Japanese eels have declined since the 1970s, with a brief increase in the 1980s from an increased effort in South Korea. Although landings have decreased, the consumption of eels has increased (Fig. 2), most markedly in Japan, where more than 50% of the world production of eels is consumed (Ringuet et al. 2002). More than 90% of the world production of eels is cultured in East Asian countries, in particular in Japan, Taiwan and China (Ringuet et al. 2002). In November 2007, Taiwan banned the exports of glass eels. However further strict regulations for the Japanese eel trade and fishery, especially in the glass and silver eel stages and for the conservation of wild eel stocks, which have all been discussed at length by EU countries, are indispensable for eel stock enhancement in Asian countries. There is a need to define a joint assessment of wild eel stocks. This step includes the quantitative parameters of the world populations (distribution, structure and abundance) for each biological stage and the qualitative parameters (e.g., the fecundity of spawners) to define appropriate stock enhancement targets. It is also necessary to improve the monitoring of eel stocks to better appreciate the efficiency of stock enhancement programmes.

Conclusion

The intensive spawning ground investigation of the Japanese eel has played an important role in expanding our basic knowledge of freshwater eel biology. During the past 20 years, a significant amount of human power, research funding, cruise costs and plenty of time have been expended almost every year on research cruises that spanned thousands of kilometers. After the first discovery of the spawning ground in 1991, important data were collected only for wild eggs and mature adults in the spawning ground. The outcomes of these expeditions do not necessarily contribute towards the development and improvement of artificial breeding techniques for the complete propagation and stock enhancement of the Japanese eel. In addition to the Japanese eel, other anguillid eels, such as the American and European eels, are also threatened. Recruitment remains in decline and stock recovery will be a long-term process for biological reasons; therefore, studies aimed at eel stock management in coastal and inland waters should be the current focus. Furthermore, the present eel aquaculture completely (100%) depends on wild glass eels or elvers as fry. More than 90% of the world production of eels is cultured in East Asia, in particular in Japan, Taiwan and China (Ringuet et al. 2002). Rearing eel larvae to the glass eel stage has not been successfully completed commercially. Therefore, in order to enhance eel stocks and their commercial use for human consumption, studies related to the establishment of commercial glass eel production are urgently required and should be more focused on this goal. Because eels are not protected under local or international law, the Japanese eel and European eels are currently seriously threatened with extinction.

Keywords: *A. japonica*, Mariana Islands, North Equatorial Current, Kuroshio Current, egg, new moon, salinity front, spawning ground investigation, stock enhancement

References

Aprahamian, M.W., A.M. Walker, B. Williams, A. Bark and B. Knights. 2007. On the application of models of European eel (*Anguilla anguilla*) production and escapement to the development of Eel Management Plans: the River Severn. ICES J. Mar. Sci. 64: 1472–1482.

Arai, T. 2014a. How have spawning ground investigations of the Japanese eel *Anguilla japonica* contributed to the stock enhancement? Rev. Fish Biol. Fisheries 24: 75–88.

Arai, T. 2014b. Do we protect freshwater eels or do we drive them to extinction? Springer Plus 3: 534.

Arai, T. and N. Chino. 2012. Diverse migration strategy between freshwater and seawater habitats in the freshwater eels genus *Anguilla*. J. Fish Biol. 81: 442–455.

Arai, T., T. Otake and K. Tsukamoto. 1997. Drastic changes in otolith microstructure and microchemistry accompanying the onset of metamorphosis in the Japanese eel *Anguilla japonica*. Mar. Ecol. Prog. Ser. 161: 17–22.

Arai, T., A. Kotake, M. Ohji, N. Miyazaki and K. Tsukamoto. 2003a. Migratory history and habitat use of Japanese eel *Anguilla japonica* in the Sanriku Coast of Japan. Fish. Sci. 69: 813–818.

Arai, T., A. Kotake, M. Ohji, M.J. Miller, K. Tsukamoto and N. Miyazaki. 2003b. Occurrence of sea eels of *Anguilla japonica* along the Sanriku Coast of Japan. Ichthyol. Res. 50: 78–81.

Arai, T., A. Kotake, P.M. Lokman, M.J. Miller and K. Tsukamoto. 2004. Evidence of different habitat use by New Zealand freshwater eels, *Anguilla australis* and *A. dieffenbachii*, as revealed by otolith microchemistry. Mar. Ecol. Prog. Ser. 266: 213–225.

Arai, T., A. Kotake and T.K. McCarthy. 2006. Habitat use by the European eel *Anguilla anguilla* in Irish waters. Estuar. Coast. Shelf Sci. 67: 569–578.

Arai, T., A. Kotake and M. Ohji. 2008. Variation in migratory history of Japanese eels, *Anguilla japonica*, collected in the northern most part of its distribution. J. Mar. Biol. Assoc. UK 88: 1075–1080.

Arai, T., N. Chino and A. Kotake. 2009. Occurrence of estuarine and sea eels *Anguilla japonica* and a migrating silver eel *Anguilla anguilla* in Tokyo Bay area, Japan. Fish. Sci. 75: 1197–1203.

Bilotta, G.S., P. Sibley, J. Hateley and A. Don. 2011. The decline of the European eel *Anguilla anguilla*: quantifying and managing escapement to support conservation. J. Fish Biol. 78: 23–28.

Boëtius, I. and J. Boëtius. 1980. Experimental maturation of female silver eels, *Anguilla anguilla*. Estimates of fecundity and energy reserves for migration and spawning. Dana 1: 1–28.

Bonhommeau, S., E. Chassot and E. Rivot. 2008. Fluctuations in European eel (*Anguilla anguilla*) recruitment resulting from environmental changes in the Sargasso Sea. Fish. Oceanogr. 17: 32–44.

Brannon, E.L., D.F. Amend, M.A. Cronin, J.E. Lannan, S. LaPatra, W.J. McNeil, R.E. Noble, C.E. Smith, A.J. Talbot, G.A. Wedemeyer and H. Westers. 2004. The controversy about salmon hatcheries. Fisheries 29: 12–31.

Casselman, J.M. 2003. Dynamics of resources of the American eel, *Anguilla rostrata*: declining abundance in the 1990s. pp. 255–274. *In*: K. Aida, K. Tsukamoto and K. Yamauchi (eds.). Eel Biology. Springer-Verlag, Tokyo.

Castonguay, M., P.V. Hodson, C. Moriarty, K.F. Drinkwater and B.M. Jessop. 2007. Is there a role of ocean environment in American and European eel decline. Fish. Oceanogr. 3: 197–203.

Chang, S.L., G.H. Kou and I.C. Liao. 2004. Temperature adaptation of the Japanese eel (*Anguilla japonica*) in its early stages. Zool. Stud. 43: 571–579.

Cheng, P.W. and W.N. Tzeng. 1996. Timing of metamorphosis and estuarine arrival across the dispersal range of the Japanese eel *Anguilla japonica*. Mar. Ecol. Prog. Ser. 131: 87–96.

Chilcote, M.W. 2003. Relationship between natural productivity and the frequency of wild fish in mixed spawning populations of wild and hatchery steelhead (*Oncorhynchus mykiss*). Can. J. Fish. Aquat. Sci. 60: 1057–1067.

Chino, N. and T. Arai. 2009. Relative contribution of migratory type on the reproduction of migrating silver eels, *Anguilla japonica*, collected off Shikoku Island, Japan. Mar. Biol. 156: 661–668.

Chino, N. and T. Arai. 2010a. Migratory history of the giant mottled eel (*Anguilla marmorata*) in the Bonin Islands of Japan. Ecol. Freshw. Fish 19: 19–25.

Chino, N. and T. Arai. 2010b. Occurrence of marine resident tropical eel *Anguilla bicolor bicolor* in Indonesia. Mar. Biol. 157: 1075–1081.

Chino, N. and T. Arai. 2010c. Habitat use and habitat transitions in the tropical eel, *Anguilla bicolor bicolor*. Environ. Biol. Fish. 89: 571–578.

Chow, S., H. Kurogi, N. Mochioka, S. Kaji, M. Okazaki and K. Tsukamoto. 2009. Discovery of mature freshwater eels in the open ocean. Fish. Sci. 75: 257–259.

Ciccotti, E. and G. Fontennelle. 2000. Aquaculture of European eel (*Anguilla anguilla*) in Europe: a review. pp. 9–11. *In*: Abstracts of the 3rd East Asian Symposium on Eel Research—Sustainability of Resources and Aquaculture of Eels. Keelung: Taiwan Fisheries Research Institute.

Daverat, F., K.E. Limberg, I. Thibault, J.C. Shiao, J.J. Dodson, F. Caron, W.N. Tzeng, Y. Iizuka and H. Wickström. 2006. Phenotypic plasticity of habitat use by three temperate eel species, *Anguilla anguilla*, *A. japonica* and *A. rostrata*. Mar. Ecol. Prog. Ser. 308: 231–241.

Dekker, W. 2003a. On the distribution of the European eel (*Anguilla anguilla*) and its fisheries. Can. J. Fish. Aquat. Sci. 60: 787–799.

Dekker, W. 2003b. Did lack of spawners cause the collapse of the European eel, *Anguilla anguilla*? Fish. Manage. Ecol. 10: 365–376.

Dekker, W. 2003c. Worldwide decline of eel resources necessitates immediate action. Fisheries 28: 28.

Dekker, W., M. Pawson and H. Wickström. 2007. Is there more to eels than slime? An introduction to the papers presented at the ICES Theme Session in September 2006. ICES J. Mar. Sci. 64: 1366–1367.

FAO (Food and Agriculture Organization of the United Nations). 2010. The state of world fisheries and aquaculture. Food and Agriculture Organization of the United Nations, Rome.

Fricke, H. and R. Kaese. 1995. Tracking of artificially matured eels (*Anguilla anguilla*) in the Sargasso Sea and the problem of the eel's spawning site. Naturwissenschaften 82: 32–36.

Friedland, K.D., M.J. Miller and B. Knight. 2007. Oceanic changes in the Sargasso Sea and declines in recruitment of the European eel. ICES J. Mar. Sci. 64: 519–530.

Gousset, B. 1990. European eel (*Anguilla anguilla* L.) farming technologies in Europe and in Japan: application of a comparative analysis. Aquaculture 87: 209–235.

Gousset, B. 1992. Eel Culture in Japan. Bulletin No. 10. L' Institut Oceanographique. Monaco.

Halpin, P. 2007. *Unagi: Freshwater "Eel". Anguilla japonica, A. anguilla, A. rostrata*. Seafood Watch, Seafood Report. Monterey Aquarium, Monterey.

Heinsbroek, L.T.N. 1991. A review of eel culture in Japan and Europe. Aquacult. Fish. Manage. 22: 57–72.

Hilborn, R. 1992. Hatcheries and the future of salmon in the Northwest. Fisheries 17: 5–8.

Hiroi, O. 1998. Historical trends of salmon fisheries and stock conditions in Japan. North Pacific Anadromous Fish Commission Bulletin 1: 23–27.

Huertas, M., A.V.M. Canário and P.C. Hubbard. 2008. Chemical communication in the genus *Anguilla*: a minireview. Behaviour 145: 1389–1407.

ICES. 2006. Report of the Joint EIFAC/ICES Working Group on Eels. ICES Document CM 2006/ACFM: 16: 367pp.

Ijiri, S., K. Tsukamoto, S. Chow, H. Kurogi, S. Adachi and H. Tanaka. 2011. Controlled reproduction in the Japanese eel (*Anguilla japonica*), past and present. Aquacult. Europe 36: 13–17.

Jacoby, D. and M. Gollock. 2014. *Anguilla anguilla*. The IUCN Red List of Threatened Species. Version 2014.3. <www.iucnredlist.org>. Downloaded on 07 December 2014.

Kaeriyama, M. 1989. Aspects of salmon ranching in Japan. Physiol. Ecol. Japan Spec. 1: 625–638.

Kaeriyama, M. 1998. Dynamics of chum salmon, *Oncorhynchus keta*, populations released from Hokkaido, Japan. North Pacific Anadromous Fish Comm. Bull. 1: 90–102.

Kaeriyama, M. 1999. Hatchery programmes and stock management of salmonid populations in Japan. pp. 103–131. *In*: B.R. Howell, E. Moksness and T. Svåsand (eds.). Stock Enhancement and Sea Ranching. Blackwell, Oxford.

Kagawa, H., H. Tanaka, H. Ohta, T. Unuma and K. Nomura. 2005. The first success of glass eel production in the world: Basic biology on fish reproduction advances new applied technology in aquaculture. Fish Physiol. Biochem. 31: 193–199.

Kajihara, T. 1988. Distribution of *Anguilla japonica* leptocephali in western Pacific during September 1986. Bull. Jpn. Soc. Sci. Fish. 54: 929–933.

Kim, H., S. Kimura, A. Shinoda, T. Kitagawa, Y. Sasai and H. Sasaki. 2007. Effect of El Niño on migration and larval transport of the Japanese eel (*Anguilla japonica*). ICES J. Mar. Sci. 64: 1387–1395.

Kimura, S. and K. Tsukamoto. 2006. The salinity front in the North Equatorial Current: a landmark for the spawning migration of the Japanese eel (*Anguilla japonica*) related to the stock recruitment. Deep-Sea Res. II 53: 315–325.

Kimura, S., K. Tsukamoto and T. Sugimoto. 1994. A model for the larval migration of the Japanese eel: roles of the trade winds and salinity front. Mar. Biol. 119: 185–190.

Kimura, S., T. Inoue and T. Sugimoto. 2001. Fluctuation in the distribution of low salinity water in the North Equatorial Current and its effect on the larval transport of the Japanese eel. Fish. Oceanogr. 10: 51–60.

Knights, B. 2003. A review of the possible impacts of long-term oceanic and climate changes and fishing mortality on recruitment of anguillid eels of the Northern Hemisphere. Sci. Total Environ. 310: 237–244.

Kobayashi, T. 1980. Salmon propagation in Japan. pp. 91–107. *In*: J.E. Thorpe (ed.). Salmon Ranching. Academic Press, London.

Kotake, A., T. Arai, T. Ozawa, S. Nojima, M.J. Miller and K. Tsukamoto. 2003. Variation in migratory history of Japanese eels, *Anguilla japonica*, collected in coastal waters of the Amakusa Islands, Japan, inferred from otolith Sr/Ca ratios. Mar. Biol. 142: 849–854.

Kotake, A., A. Okamura, Y. Yamada, T. Utoh, T. Arai, M.J. Miller, H.P. Oka and K. Tsukamoto. 2005. Seasonal variation in migratory history of the Japanese eel, *Anguilla japonica*, in Mikawa Bay, Japan. Mar. Ecol. Prog. Ser. 293: 213–221.

Kuo, H. 1971. Onshore migration of the elvers of Japanese eel in Taiwan. Aquaculture 8: 52–56.

Lamson, H.M., J.C. Shiao, Y. Iizuka, W.N. Tzeng and D.K. Cairns. 2006. Movement patterns of American eels (*Anguilla rostrata*) between salt- and freshwater in a coastal watershed, based on otolith microchemistry. Mar. Biol. 149: 1567–1576.

Levin, P.S., R.W. Zabel and J.G. Williams. 2001. The road to extinction is paved with good intentions: negative association of fish hatcheries with threatened salmon. Proc. Royal Soc. London Ser. B. 268: 1153–1158.

Liao, I.C., Y.K. Hsu and W.C. Lee. 2002. Technical innovations in eel culture systems. Rev. Fish. Sci. 10: 433–450.

Lokman, P.M. and G. Young. 2000. Induced spawning and early ontogeny of New Zealand freshwater eels (*Anguilla dieffenbachii*) and (*A. australis*). N.Z. J. Mar. Freshwat. Res. 34: 135–145.

MacNamara, R. and T.K. McCarthy. 2012. Size-related variation in fecundity of European eel (*Anguilla anguilla*). ICES J. Mar. Sci. 69: 1333–1337.

Marcogliese, L.A. and J.M. Casselman. 2009. Long-term trends in size and abundance of juvenile American eel (*Anguilla rostrata*) in the St. Lawrence River. Am. Fish. Soc. Symp. 58: 197–206.

Matsui, I. 1952. Studies on the morphology, ecology and pond-culture of the Japanese eel (*Anguilla japonica* Temminck and Schlegel). J. Shimonoseki Coll. Fish. 2: 1–245.

Matsui, I. 1957. On the records of a leptocephalus and catadromous eels of *Anguilla japonica* in the waters around Japan with a presumption of their spawning places. J. Shimonoseki Univ. Fish. 7: 151–167.

Matsui, I. 1972. Eel Biology: Biological Study. Kouseisha-Kouseikaku, Tokyo.

Matsui, I., T. Takai and A. Kataoka. 1968. Anguillid leptocephalus found in the Japan Current and its adjacent waters. J. Shimonoseki Univ. Fish. 17: 17–23.

Mayama, H. 1985. Technical innovations in chum salmon enhancement with special reference to fry condition and timing of release. National Oceanographic and Atmospheric Administration (NOAA), National Marine Fisheries Service Technical Report 27: 83–86.

Mayama, H. and Y. Ishida. 2003. Japanese studies on the early ocean life of juvenile salmon. North Pacific Anadromous Fish. Comm. Bull. 3: 41–67.

McCleave, J.D. and R.C. Kleckner. 1985. Oceanic migrations of Atlantic eels (*Anguilla* spp.): adults and their offspring. Contribut. Mar. Sci. 27: 316–337.

McDowall, R.M. 1988. Diadromy in Fishes. Croom Helm, London.

Miller, M.J., S. Kimura, K.D. Friedland, B. Knights, H. Kim, D.J. Jellyman and K. Tsukamoto. 2009. Review of ocean-atmospheric factors in the Atlantic and Pacific oceans influencing spawning and recruitment of anguillid eels. *In*: A. Haro, T. Avery, K. Beal, J. Cooper, R. Cunjak, M. Dadswell, R. Klauda, C. Moffitt, R. Rulifson and K. Smith (eds.). Challenges for Diadromous Fishes in a Dynamic Global Environment. American Fisheries Society Symposium Publication. Bethesda, MD.

Miller, M.J., Y. Chikaraishi, N.O. Ogawa, Y. Yamada, K. Tsukamoto and N. Ohkouchi. 2012. A low trophic position of Japanese eel larvae indicates feeding on marine snow. Biol. Lett., http://dx.doi.org/10.1098/rsbl.2012.0826.

Miyazaki, S., H.Y. Kim, K. Zenimoto, T. Kitagawa, M.J. Miller and S. Kimura. 2011. Stable isotope analysis of two species of anguilliform leptocephali (*Anguilla japonica* and *Ariosoma major*) relative to their feeding depth in the North Equatorial Current region. Mar. Biol. 158: 2555–2564.

Mochioka, N. and M. Iwamizu. 1996. Diet of anguilloid larvae: leptocephali feed selectively on larvacean houses and fecal pellets. Mar. Biol. 125: 447–452.

Morita, K., S.H. Morita and M. Fukuwaka. 2006. Population dynamics of Japanese pink salmon (*Oncorhynchus gorbuscha*): are recent increases explained by hatchery programmes or climatic variations? Can. J. Fish. Aquat. Sci. 63: 55–62.

Nagata, M. and M. Kaeriyama. 2003. Salmonid status and conservation in Japan. pp. 89–98. *In*: P. Gallaugher and L. Wood (eds.). Proceedings of the World Summit on Salmon. Simon Fraser University, Vancouver, Canada.

Nakamura, H., A. Nishina, H. Ichikawa, M. Nonaka and H. Sasaki. 2008. Deep countercurrent beneath the Kuroshio in the Okinawa Trough. J. Geophys. Res. 113: C06030.

Nickelson, T. 2003. The influence of hatchery coho salmon (*Oncorhynchus kisutch*) on the productivity of wild coho salmon populations in Oregon coastal basins. Can. J. Fish. Aquat. Sci. 60: 1050–1056.

Okamura, A., Y. Yamada, N. Horie, T. Utoh, N. Mikawa, S. Tanaka and K. Tsukamoto. 2007. Effects of water temperature on early development of Japanese eel *Anguilla japonica*. Fish. Sci. 73: 1241–1248.

Okamura, A., Y. Yamada, N. Mikawa, N. Horie, T. Utoh, T. Kaneko, S. Tanaka and K. Tsukamoto. 2009a. Growth and survival of eel leptocephali (*Anguilla japonica*) in low-salinity water. Aquaculture 296: 367–372.

Okamura, A., Y. Yamada, T. Horita, N. Horie, N. Mikawa, T. Utoh, S. Tanaka and K. Tsukamoto. 2009b. Rearing eel leptocephali (*Anguilla japonica* Temminck and Schlegel) in a planktonkreisel. Aquacult. Res. 40: 509–512.

Otake, T., K. Nogami and K. Maruyama. 1993. Dissolved and particulate organic matter as possible food sources for eel leptocephali. Mar. Ecol. Prog. Ser. 92: 27–34.

Ozawa, T., O. Tabeta and N. Mochioka. 1989. Anguillid leptocephali from the western North Pacific east of Luzon, in 1988. Bull. Jpn. Soc. Sci. Fish. 55: 627–632.

Ozawa, T., F. Kakizoe, O. Tabeta, T. Maeda and Y. Yuwaki. 1991. Japanese eel leptocephali from three cruises in the western North Pacific. Bull. Jpn. Soc. Sci. Fish. 57: 1877–1881.

Prokhorchik, G.A. 1986. Postembryonic development of European eel, *Anguilla anguilla*, under experimental conditions. J. Ichthyol. 26: 121–127.

Qiu, B. and R. Lukas. 1996. Seasonal and interannual variability of the North Equatorial Current, the Mindanao Current, and the Kuroshio along the Pacific western boundary. J. Geophys. Res. 101: 12315–12330.

Riemann, L., H. Alfredsson, M.M. Hansen, T.D. Als, T.G. Nielsen, P. Munk, K. Aarestrup, G.E. Maes, H. Sparholt, M.I. Petersen, M. Bachler and M. Castonguay. 2010. Qualitative assessment of the diet of European eel larvae in the Sargasso Sea resolved by DNA barcoding. Biol. Lett. 6: 819–822.

Ringuet, S., F. Muto and C. Raymakers. 2002. Eels, their harvest in Europe and Asia. Traffic Bull. 19: 2–27.

Salo, E.O. 1991. Life history of chum salmon (*Oncorhynchus keta*). pp. 231–309. *In*: C. Groot and L. Margolis (eds.). Pacific Salmon Life Histories. UBC Press, Vancouver, BC.

Sang, T.K., H.Y. Chang, T.C. Che and C.F. Hui. 1994. Population structure of the Japanese eel, *Anguilla japonica*. Mol. Biol. Evol. 11: 250–260.

Satoh, H. 1979. Try for perfect culture of the Japanese eel. Iden. 33: 23–30.

Schmidt, J. 1922. The breeding places of the eel. Phil. Trans. Roy. Soc. London 211: 179–208.

Schmidt, J. 1925. The breeding places of the eel. Smithsonian Inst. Ann. Rep. 1924: 279–316.

S´ebert, M.E., F.A. Weltzien, C. Moisan, C. Pasqualini and S. Dufour. 2008. Dopaminergic systems in the European eel: characterization, brain distribution and potential role in migration and reproduction. Hydrobiologia 602: 27–46.

Sorensen, P.W. and H.E. Winn. 1984. The induction of maturation and ovulation in American eels and the relevance of chemical and visual cues to their spawning behavior. J. Fish Biol. 25: 261–268.

Svedäng, H. and L. Gipperth. 2012. Will regionalization improve fisheries management in the EU? An analysis of the Swedish eel management plan reflects difficulties. Mar. Pol. 36: 801–808.

Takai, T. and O. Tabeta. 1976. Anguillid leptocephali. pp. 12–13. *In*: T. Ishii (ed.). Preliminary Report of the Hakuho Maru Cruise KH-75-1. Ocean Research Institute, University of Tokyo, Tokyo.

Tanaka, S. 1975. Collection of leptocephali of the Japanese eel in waters south of the Okinawa Islands. Bull. Jpn. Soc. Sci. Fish. 41: 129–136.

Tanaka, H. 1996. Rearing of eel larvae. pp. 119–127. *In*: O. Tabeta (ed.). Early Life History and Prospects of Seed Production of the Japanese Eel *Anguilla Japonica*. Kouseisya-Kouseikaku Press, Tokyo.

Tanaka, H. 2003. Techniques for larval rearing. pp. 427–434. *In*: K. Aida, K. Tsukamoto and K. Yamauchi (eds.). Eel Biology. Springer-Verlag, Tokyo.

Tanaka, H., H. Kagawa and H. Ohta. 2001. Production of leptocephali of Japanese eel (*Anguilla japonica*) in captivity. Aquacult. 201: 51–60.

Tanaka, H., H. Kagawa, H. Ohta, T. Unuma and K. Nomura. 2003. The first production of glass eel in captivity: fish reproductive physiology facilitates great progress in aquaculture. Fish Physiol. Biochem. 28: 493–497.

Tatsukawa, K. 2003. Eel resources in East Asia. pp. 293–298. *In*: K. Aida, K. Tsukamoto and K. Yamauchi (eds.). Eel Biology. Springer, Tokyo.

Tesch, F.W. 1986. Der Aal als Konkurrent von anderen Fischarten und von Krebsen. Öʻsterreichs Fischerei 39: 5–20.

Tesch, F.W. 2003. The Eel Biology and Management of Anguillid Eels. Chapman and Hall, London.

Tsukamoto, K. 1992. Discovery of the spawning area for the Japanese eel. Nature 356: 789–791.

Tsukamoto, K. 2006. Spawning of eels near a seamount. Nature 439: 929.

Tsukamoto, K. 2009. Oceanic migration and spawning of anguillid eels. J. Fish Biol. 74: 1833–1852.

Tsukamoto, K. and T. Arai. 2001. Facultative catadromy of the eel, *Anguilla japonica*, between freshwater and seawater habitats. Mar. Ecol. Prog. Ser. 220: 365–376.

Tsukamoto, K., A. Umezawa, O. Tabeta, N. Mochioka and T. Kajihara. 1989. Age and birth date of *Anguilla japonica* leptocephali collected in western North Pacific in September 1986. Bull. Jpn. Soc. Sci. Fish. 55: 1023–1028.

Tsukamoto, K., S. Chow, T. Otake, H. Kurogi, N. Mochioka, M.J. Miller, J. Aoyama, S. Kimura, S. Watanabe, T. Yoshinaga, A. Shinoda, M. Kuroki, M. Oya, T. Watanabe and K. Hata et al. 2011. Oceanic spawning ecology of freshwater eels in the western North Pacific. Nature Commun. 2: 179.

Tsukamoto, K., T. Otake, N. Mochioka, T.W. Lee, H. Fricke, T. Inagaki, J. Aoyama, S. Ishikawa, S. Kimura, M.J. Miller, H. Hasumoto, M. Oya and Y. Suzuki. 2003. Seamounts, new moon and eel spawning: the search for the spawning site of the Japanese eel. Environ. Biol. Fish. 66: 221–229.

Tzeng, W.N. 2004. Modern research on the natural life history of the Japanese eel *Anguilla japonica*. J. Fish. Soc. Taiwan 31: 73–84.

van Ginneken, V., G. Vianen, B. Muusze, A. Palstra, L. Verschoor, O. Lugten, M. Onderwater, S. van Schie, P. Niemantsverdriet, R. van Heeswijk, E. Eding and G. van den Thillart. 2005. Gonad development and spawning behaviour of artificially-matured European eel (*Anguilla anguilla* L.). Anim. Biol. 55: 203–218.

Wang, Y., C. Zhao, Z. Shi, Y. Tan, K. Zhang and T. Li. 1980. Studies on the artificial inducement of reproduction in common eel. J. Fish. China 4: 147–156.

Waples, R.S. 1999. Dispelling some myths about hatcheries. Fisheries 24: 12–21.

Westin, L. 1990. Orientation mechanisms in migrating European silver eel (*Anguilla anguilla*): temperature and olfaction. Mar. Biol. 106: 175–179.

Yamamoto, K. and K. Yamauchi. 1974. Sexual maturation of Japanese eel and production of eel larvae in the aquarium. Nature 251: 220–222.

Yamauchi, K., M. Nakamura, H. Takahashi and K. Takano. 1976. Cultivation of larvae of Japanese eel. Nature 263: 412.

Yokose, H. 2008. Geological approach to the spawning sites of the Japanese eel. Kaiyo Monthly Special Issue 48: 45–58.

Zenimoto, K., T. Kitagawa, S. Miyazaki, Y. Sasai, H. Sasaki and S. Kimura. 2009. The effects of seasonal and interannual variability of oceanic structure in the western Pacific North Equatorial Current on larval transport of the Japanese eel *Anguilla japonica*. J. Fish Biol. 74: 1878–1890.

Spawning Ground of Tropical Eels

Takaomi Arai

Introduction

Schmidt (1922) discovered that the spawning area for both the European eel *Anguilla anguilla* and the American eel *Anguilla rostrata* is located far offshore in the Sargasso Sea of the Atlantic Ocean thousands of kilometers from their growth habitats in Europe and North America, indicating that these two species of Atlantic freshwater eels make remarkably long spawning migrations. The European eel is widely distributed in North Africa and Europe. It has a longer larval period as compared to other anguillid eels (Arai et al. 2001a). More recently, the spawning area of the Japanese eel, *Anguilla japonica*, was discovered far offshore in the Philippine Sea of the western North Pacific (Tsukamoto 1992) where all of the oceanic stages of this species were first collected, including spawning adults, eggs, and recently hatched larvae (Tsukamoto et al. 2011). The Japanese eel spawns in summer, with juvenile recruits being transported back to the coasts of northeastern Asian countries in winter. Eel species in both the Atlantic and Pacific spawn in similar westward-flowing currents at the southern edges of the subtropical gyres in both oceans; therefore, their larvae (leptocephali) can be passively transported to coastal areas. The long migrations to these spawning areas have fascinated scientists because each eel must migrate thousands of kilometers back to the same area to spawn. The discoveries of the spawning areas of temperate eels

Environmental and Life Sciences Programme, Faculty of Science, Universiti Brunei Darussalam, Jalan Tungku Link, Gadong, BE 1410, Brunei Darussalam.
E-mail: takaomi.arai@ubd.edu.bn

have stimulated numerous studies regarding the life history and freshwater ecology of these eels and indicate that temperate eels have well-defined spawning seasons.

Nineteen species/subspecies of freshwater eels have been reported worldwide, 13 of which inhabit tropical regions (Arai 2016) that are globally distributed in temperate, tropical, and subtropical areas. In tropical areas, seven species occur in the western Pacific around Indonesia (Ege 1939; Castle and Williamson 1974; Arai et al. 1999a). In general, freshwater eels are divided into temperate and tropical eels, based on their major distributions (Arai 2016) and ecological properties. Recent molecular phylogenetic researches on freshwater eels have revealed that tropical eels are the most basal species originating in the Indonesian region and that freshwater eels radiated out from the tropics to colonise the temperate regions (Minegishi et al. 2005), suggesting that tropical freshwater eels must be more closely related to the ancestral form than their temperate counterparts. Thus, studying the biological aspects of tropical eels provides clues for understanding the nature of primitive forms of catadromous migration in freshwater eels and how the large-scale migration of temperate eel species became established. The drastic decline of glass eel recruitment in Europe and East Asia in recent times has caused serious problems for maintaining sustainable levels of adult abundance (Arai 2014a,b). Tropical eels are becoming the major target species in order to satisfy the high demand for eel products. The tropical eel *Anguilla marmorata* is very common in the Indo-Pacific Ocean and has a more-extensive distribution than temperate freshwater eels (Ege 1939; Jespersen 1942; Arai 2016). However, remarkably little is known about the spawning area, spawning ecology, and life histories of the many tropical eel species that are found in the Indo-Pacific region.

Early in the last century, the Danish Foundation's Oceanographical Expedition Round the World conducted from 1928 to 1930, collected many leptocephali off of the west coast of Sumatra, an island that is part of Indonesia, indicating that *Anguilla bicolor bicolor* spawned there near the Mentawai Trench (Jespersen 1942). More recently, it was thought that *Anguilla celebesensis* spawned in Tomini Bay of northeastern Sulawesi Island, also part of Indonesia (Aoyama et al. 2003). These studies suggest that, in contrast to the long migrations made by temperate eels, tropical eels make much shorter migrations in order to spawn. However, the natural reproductive ecology and spawning patterns of tropical and temperate eels have remained a mystery, and it has thus remained difficult to determine the nature of the migrations of freshwater eels.

In this chapter, I review the current state of knowledge about the spawning grounds of tropical eels. The unique spawning strategy suggests the nature of primitive forms of spawning ecology and migration mechanisms in freshwater eels.

Spawning grounds off the coast of Sumatra Island of Indonesia

Investigations of the spawning grounds of the Atlantic and Japanese eels have a long history and have been intensively conducted around the respective areas. However, research on tropical eels is largely limited to waters around Indonesia. According to Jespersen (1942), the spawning grounds of the tropical eels *A. bicolor bicolor* and *A. marmorata*, which inhabit Java and the North Sulawesi islands, respectively, are potentially located off of the southwestern coast of Sumatra (for *A. bicolor bicolor*) and the Celebes, Sulu and Molucca seas (for *A. marmorata*), that is, close to their native areas (Fig. 1). The occurrence of leptocephali of various sizes, including preleptocephalus to metamorphosing

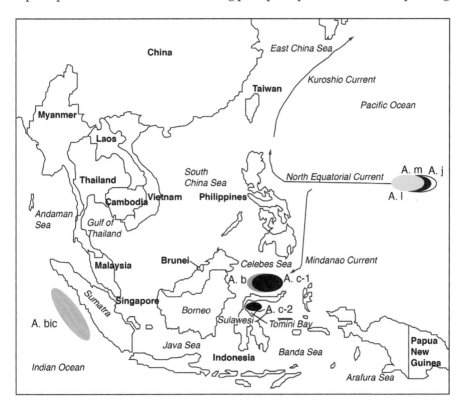

Fig. 1. Map of the locations where the tropical eel leptocephali of *Anguilla celebesensis*, *A. borneensis*, *A. marmorata*, *A. luzonensis* and *A. bicolor bicolor*, were collected and a possible spawning area of *A. bicolor bicolor* off the coast of Sumatra Island (Jespersen 1942), *A. celebesensis* in Celebes Sea and Tomini Bay (Jespersen 1942; Aoyama et al. 2003), *A. borneensis* in the Celebes Sea (Jespersen 1942; Aoyama et al. 2003), *A. marmorata* in the western North Pacific (Arai et al. 2002; Tsukamoto et al. 2011) and *A. luzonensis* in the western North Pacific (Arai et al. 2003; Kuroki et al. 2012). A spawning area of a temperate species of *A. japonica* in the western North Pacific is also illustrated. A. c-1: spawning area of *A. celebesensis* in Celebes Sea, A. c-2: spawning area of *A. celebesensis* in Tomini Bay, A. b: spawning area of *A. borneensis*, A. m: spawning area of *A. marmorata*, A. l: spawning area of *A. luzonensis*, A. bic: spawning area of *A. bicolor bicolor*, A. j: spawning area of *A. japonica*.

stages, in waters off the coast of Sumatra (Jespersen 1942), provide evidence of the small-scale nature of their migration from the spawning area to growth habitats. This situation is quite different from that of temperate eels, which must travel thousands of kilometers between their spawning grounds and growth habitats (Arai 2014c). However, unlike in case of the Japanese eels, eggs, recently hatched leptocephali and adult eels have not yet been discovered. Therefore, the exact location of the spawning grounds of tropical eels is yet to be determined.

The reports by Dr. Johannes Schmidt made in the early 1920s describe the initial discovery of the spawning area of Atlantic eels in the Sargasso Sea of the Atlantic Ocean (Schmidt 1922, 1925). Inspired by Schmidt's studies, further attempts were made to discover the spawning grounds of anguillid eels in the Pacific and Indian oceans. From 1928 to 1930, the Carlsberg Foundation's Oceanographical Expedition Round the World collected a total of 1,225 anguillid eels ranging from fairly small leptocephali to metamorphosing larvae and glass eels were collected in the eastern Indian Ocean (Jespersen 1942) (Fig. 1). Most of the leptocephali and at least 950 leptocephali collected by Jespersen (1942) were identified as the shortfin type of anguillid species. As *Anguilla bicolor bicolor* is the species known to inhabit the eastern Indian Ocean, it was therefore believed that this species spawns near the Mentawai Trench (Ege 1939). However, almost all of the small leptocephali (10–20 mm) could not be categorized as being shortfin or longfin types, and at least eight large-sized specimens were tentatively identified as *Anguilla celebesensis*, which is found in the western Pacific region (Ege 1939). These problems were due to the undeveloped morphological characters of the small leptocephali and the critical overlap of the taxonomic characteristics among tropical anguillid species (Castle 1963; Aoyama et al. 1999). It is difficult to definitively identify most of anguillid leptocephali using morphological features due to overlapping ranges of total myomeres and other traits. Modern molecular genetic techniques, however, can be used to distinguish the different species of anguillid eels at the leptocephalus stage. As a result, the locations of the spawning areas of anguillid eels off the coast of Sumatra are still poorly understood.

To solve the problems of identifying anguillid leptocephali, a genetic approach has been adopted (Aoyama et al. 1999, 2003). A research cruise conducted in the eastern Indian Ocean (Aoyama et al. 2007) sampled the same area approximately 60 years after the cruise by Jespersen (1942). The BJ-03-2 research cruise of the R/V Baruna Jaya VII, sponsored by the Research Center for Oceanography of the Indonesian Institute of Sciences, was carried out from 5 to 20 June, 2003. A total of 43 mostly large (45.0–55.5 mm TL) anguillid leptocephali were collected. Forty-one of the specimens were identified as *A. bicolor bicolor*, one specimen was identified as *A. marmorata* (46.8 mm TL), and one specimen was identified as *A. interioris* (the smallest specimen, at 12.4 mm TL) (Aoyama et al. 2007). The total number of specimens collected was much lower than the number collected by Jespersen (1942), and fewer small-sized specimens were collected. The differences in collection and specimen sizes can be attributed to the different times of year when the samples were taken, as well as cruise efforts.

The leptocephali samples collected by Jespersen (1942) and Aoyama et al. (2007) consisted primarily of *A. bicolor bicolor*. Jespersen (1942) reported a number of leptocephali in the waters off west Sumatra, and 85% of the individuals ranging from 20 to 56 mm could be distinguished as the shortfin type, which were probably *A. bicolor bicolor* based on the distribution range of adult shortfinned eel. In addition to a wide range of sizes of leptocephali collected by Jespersen (1942) during September–November 1929, some specimens were undergoing metamorphosis, and several glass eels were collected as well. Both the smallest leptocephali (10–20 mm, N = 72; only collected October 6 to November 4), which could not be determined as being either shortfinned or longfinned, and the metamorphosing leptocephali (45–55 mm, N = 54) were collected in a narrow band of water offshore of the Mentawai Islands. Based on the abundance of leptocephali and the presence of some relatively small leptocephali, Jespersen (1942) concluded that *A. bicolor bicolor* spawned in waters over the Mentawai Trench.

Spawning of tropical eels has been suggested to occur over several months (Arai et al. 2001b), but relatively few small leptocephali were collected in the region off the western coast of Sumatra by Aoyama et al. (2007), although their sample size was much lower than that of Jespersen (1942). Typically, when sampling for anguillid leptocephali with plankton nets near their spawning areas during the spawning season, many small-sized leptocephali—less than 10–20 mm TL—have been collected for both temperate (McCleave 2003; Tsukamoto et al. 2003; Tsukamoto 2006) and tropical eel species (Miller et al. 2002; Aoyama et al. 2003; Miller 2003). However, relatively few small leptocephali identified as *A. bicolor bicolor* were collected by Aoyama et al. (2007) leading those authors to propose that the spawning ground of *A. bicolor bicolor* may not be limited to the Mentawai Trench. Aoyama et al. (2007) suggested that the leptocephali of *A. bicolor bicolor* found in waters around Sumatra may have been transported from farther offshore by the South Equatorial Countercurrent (SECC) or by the Southwest Monsoon Current to the north. The primary spawning area of *A. bicolor bicolor* might therefore be farther offshore than previously thought by Jespersen (1942) (Aoyama et al. 2007). However, the research period of the BJ-03-2 cruise was only two weeks, considerably less time than that of the Carlsberg cruise. Consequentially, speculation about the location of the spawning grounds of *A. bicolor bicolor* continues to this day.

The spawning ground of the Japanese eel was determined to be just south of a weak salinity front that is typically present in the North Equatorial Current region as a result of tropical rainfall (Kimura et al. 1994, 2001; Kimura and Tsukamoto 2006). The salinity front was hypothesised to act as a cue for the eels to help them find the spawning area. However, the latitudinal location (north-south) of the salinity front is highly dependent on the appearance of an El Niño (Kimura and Tsukamoto 2006). During an El Niño, the area of low-salinity water in the western equatorial Pacific shrinks because of low precipitation, causing the salinity front to shift considerably farther southward. Thus, the location of the spawning area may change with the appearance of an El Niño. Moreover, the spawning area of *A. bicolor bicolor* may be subjected to oceanographic variations that change its location both spatially and temporally. The Indian Ocean is strongly affected by monsoons, with the magnitude of monsoons varying from year to year. Furthermore, an El Niño may affect monsoon intensity, frequently and consequentially influence precipitation rates and water temperature around the spawning ground. Additional intensive research cruises are needed to determine the exact spawning area of *A. bicolor bicolor* and other tropical eels. Finally, a single *A. interioris* specimen of 12.4 mm TL was collected, which suggests that the species may also spawn around Mentawai Trench.

Spawning grounds around Sulawesi Island of Indonesia

Jespersen (1942) also found small anguillid leptocephali ranging in size from 19 to 33 mm TL in the Celebes and Sulu seas during the Carlsberg Foundation's Oceanographic Expedition indicating that the spawning grounds of *A. celebesensis* may be located in the Celebes Sea off of north Sulawesi Island.

   Another research cruise to collect anguillid leptocephali was undertaken from 14 January to 10 March 2000 in the Celebes Sea and Sulu seas by the R/V Hakuho Maru, formerly belonging to the Ocean Research Institute of the University of Tokyo (Limbong et al. 2000). A total of 12 *Anguilla* leptocephali, including six specimens smaller than 10 mm TL (Limbong et al. 2000) were collected from the Celebes Sea (Fig. 1). In addition, four small recently hatched leptocephali (6.3–7.5 mm TL) and one large leptocephalus (47.4 mm TL) were collected off the northeastern tip of Sulawesi Island, at the eastern edge of the Celebes Sea (Limbong et al. 2000). A subsequent cruise of the R/V Baruna Jaya VII (Research Center for Oceanography, Indonesian Institute of Sciences), made in the waters around Sulawesi from 8 to 30 May 2001, partially overlapped the sampling area of the Hakuho Maru cruise (Aoyama et al. 2003). Between the two research cruises, 67 leptocephali and one glass eel were collected. Genetic analysis of the leptocephali determined that there were 12 *A. borneensis* (34.0–50.7 mm TL and one glass eel of 47.8 mm), 41 *A. celebesensis* (13.0–47.8 mm), three *A. borneensis* (8.5, 13.0, 35.4 mm), four *A. bicolor pacifica* (42.6–49.2 mm), and one *A. interioris* (48.9 mm), all taken from the waters around Sulawesi Island (Aoyama et al. 2003). Jespersen (1942) also collected a number of leptocephali in the same area, but could not make precise species identifications. The small leptocephali of *A. borneensis* that were 8.5 mm and 13.0 mm, 16 and 26 days after hatching, respectively (Aoyama et al. 2003), were collected in the Celebes Sea to the east of Borneo, whereas one specimen of the same species (35.4 mm) was collected to the south in Makassar Strait, where water from the Celebes Sea is transported (Aoyama et al. 2003).

The freshwater growth habitat of *A. borneensis* is limited to the east-central part of Borneo (Ege 1939; Tesch 2003), which suggests that this species spawns in the Celebes Sea and then migrates back to its growth habitat adjacent to the spawning area (Fig. 2). Another tropical anguillid species, *Anguilla celebesensis*,

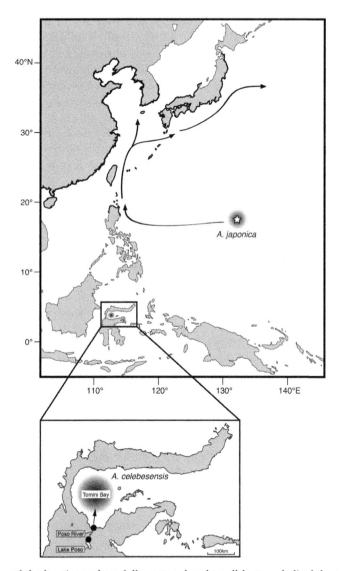

Fig. 2. Map of the locations where fully matured and small leptocephali of the tropical eel *Anguilla celebesensis* were collected, a possible spawning area of *Anguilla celebesensis* in Tomini Bay (Aoyama et al. 2003; Arai 2014c), central Sulawesi of Indonesia and the short-distance migratory route of *A. celebesensis* from its growth habitat. The general location of the offshore spawning areas with the migratory route and the distribution range (black lines on coastlines) of *Anguilla japonica* is also illustrated, so as to provide a comparison between the spawning migration of *A. celebesensis* and those of a temperate eel species.

has a wider distribution that extends from Luzon in the Philippines to across Sulawesi (Ege 1939; Tesch 2003). Interestingly, the small leptocephali of this species collected approximately 25 days after hatching (Aoyama et al. 2003), were found in two different seasons and in two different areas separated by the northern Sulawesi Island; a 12.3 mm specimen was found in the Celebes Sea in February 2000, while a 13 mm specimen was found in Tomini Bay in May 2001 (Fig. 1). The nine *A. celebesensis* leptocephali collected in Tomini Bay ranged in total length from 13 to 48.9 mm (Aoyama et al. 2003). These findings indicate that individuals of *A. celebesensis* inhabiting the watershed of Tomini Bay spawn over a relatively long period and that their leptocephali are retained in Tomini Bay because it is semi-enclosed and its waters apparently do not mix very well with those of other areas (Fig. 2) (Aoyama et al. 2003). Aoyama et al. (2003) suggested that these eels are probably geographically isolated from those in the Celebes Sea. The results also indicate that *A. celebesensis* in Sulawesi might have at least two spawning sites, despite it being such a geographically restricted area (Fig. 2).

Distribution of tropical eel leptocephali in the western Indian Ocean

The research cruises undertaken to investigate the distribution of tropical leptocephali by Jespersen covered not only areas around Sumatra in the eastern Indian Ocean and Sulawesi in the Celebes Sea but also areas around Madagascar off the southeast coast of Africa in the western Indian Ocean between 1929 December and January 1930 (Fig. 1). In contrast to the research cruises made in Indonesia, however, only 17 specimens of *A. bicolor bicolor*, *A. marmorata* and *A. mossambica* were collected, ranging in size from 45 to 54 mm TL (Jespersen 1942). No further study has been conducted in this area, as there was in Indonesia. The results suggest that the spawning grounds of the three species, along with that of *A. bengalensis labiata*, which is found in the western Indian Ocean, have not yet been discovered, and the lack of data discourages even speculative theories. Réveillac et al. (2009), on the basis of an analysis of otolith microstructure, suggested that the spawning grounds of *A. mossambica* might be located west of the Mascarene Ridge in the Indian Ocean.

Inference of the spawning ground of *Anguilla marmorata* through otolith analyses

As only a handful of research cruises have been dispatched to investigate the locations of the spawning grounds of tropical eels, only a small amount of information is currently available regarding their spawning grounds. However, recent studies on otolith microstructure and the microchemistry of leptocephali and glass eels of tropical anguillids have provided valuable new information about possible locations of their spawning grounds. It has been shown that otoliths in *Anguilla celebesensis* (Arai et al. 2000) and *A. marmorata* are incrementally deposited on a daily basis (Sugeha et al. 2001). Thus, otolith information in combination with oceanographic features, such as water current and numerical simulations, may aid in surmising their spawning grounds.

A total of 28 leptocephali (20 specimens of *Anguilla bicolor pacifica* and eight specimens of *A. marmorata*) were collected in the western Pacific Ocean (Fig. 1) (Arai et al. 2001). The age of *Anguilla bicolor pacifica* leptocephali ranged widely, from 40 to 128 days (Arai et al. 2001), while the ages of *A. marmorata* leptocephali were separated into two different latitudinal ranges in the Northern and Southern hemispheres (Fig. 1) (Arai et al. 2001). The former ranged from 38 to 58 days, and the latter ranged from 40 to 69 days, plus one specimen collected in the southern most region that was 99 days old (Arai et al. 2001). The estimated hatch dates, calculated from their sampling date and ages, were between April and July 1995 for *Anguilla bicolor pacifica*, and between May and July 1995 for *A. marmorata* (May to June 1995 in the Northern hemisphere specimens and May to July 1995 in the Southern hemisphere specimens) (Arai et al. 2001).

The geographic distribution of *Anguilla bicolor pacifica* and *A. marmorata* leptocephali indicated that these two species are widely dispersed throughout the western Pacific Ocean and that there are at least two separate spawning areas for each species (Arai et al. 2001). Various sizes (27.6–54.1 mm TL) of *A. bicolor pacifica* leptocephali were found in regions of the North Equatorial Current (westward flow, approximately 8–20° N) the North Equatorial Countercurrent (eastward flow, approximately 0–8° N) and the South Equatorial Current (westward flow, approximately 15° S-4° N) (Fig. 1) (Arai et al. 2001). Three specimens were collected in the North Equatorial Current region, with the remainder collected between 8° N and 5° S in the region of the North Equatorial Countercurrent and the South Equatorial Current. *A. marmorata* leptocephali clearly showed a patchy distribution in both the Northern (15–16° N, 22.0–39.6 mm TL) and Southern hemispheres (3–15° S, 25.2–47.3 mm TL) (Arai et al. 2001). All Northern hemisphere specimens occurred in the region of the North Equatorial Current, whereas the Southern hemisphere specimens occurred in the region of the South Equatorial Current or the South Equatorial Countercurrent, which is often present from approximately 5–10° S, near to where several specimens were collected (Arai et al. 2001).

Leptocephali of both species were present in the far western region of the North Equatorial Current (Fig. 1) (Arai et al. 2001). The westward flow of the North Equatorial Current has been consistently observed at these latitudes during the late summer and early fall (Reverdin et al. 1994; Kaneko et al. 1998; Kawabe and Taira 1998; Wijffels et al. 1998) suggesting that the leptocephali in this region were being transported westward into the northward flowing Kuroshio Current or into the southward flowing Mindanao Current (Lukas et al. 1991) and that they therefore might have been distributed to various areas such as the southern islands of Japan, Taiwan or the Philippines, in a manner similar to that of the leptocephali of the Japanese eel (Tsukamoto 1992). The specimens of *Anguilla bicolor pacifica* sampled in this area were large and probably at least three months old, making it impossible to determine where they were spawned, although this species has been reported from the Mariana Islands in the east and is present in the Philippines, Indonesia and Papua New Guinea (Ege 1939). In contrast, the much smaller *A. marmorata* leptocephali ranging in age from 35 to 58 days, were also found in this region, which suggests that this species may spawn within the North Equatorial Current west of the Mariana Islands, as does *A. japonica* (Tsukamoto 1992). Similarly, the ages of *A. japonica* leptocephali previously collected in the same area of the North Equatorial Current ranged from 27 to 42 days (Tsukamoto 1992). The presence of a similar size range of *A. marmorata* leptocephali—as small as 9–10 mm TL in the western NEC area to the west of the Mariana Islands, based on samples taken over three different years (1991, 1994, 1995)—indicates that this area is a spawning area for this species (Miller et al. 2002). A recent molecular study on the population structure of *A. marmorata* demonstrated that this species was divided into distinct North Pacific and South Pacific populations (Minegishi et al. 2008); thus this area is a possible spawning site for the North Pacific population. The distribution of *Anguilla marmorata* leptocephali collected in five different years in the NEC indicates that a spawning area of this species is also located to the west of the Mariana Islands, in an area that overlaps with the spawning grounds of the Japanese eel (Kuroki et al. 2006). Recently, one spawning-condition female *Anguilla marmorata* was collected from a location in the western North Pacific (Tsukamoto et al. 2011, Fig. 1) that corresponded to the spawning area theorized by Arai et al. (2002) and Kuroki et al. (2006), based on otolith analyses.

Estimation of spawning ground of *Anguilla celebesensis* and *A. luzonensis* by otolith analyses

An interesting finding was that the age at metamorphosis and the age at recruitment of *Anguilla celebesensis* that recruit to the Philippines and to northern Indonesia appear to be different (Arai et al. 2003). *A. celebesensis* (but now thought to be *A. luzonensis* based on molecular analysis; see below) collected from the Cagayan River in the Philippines on 24 September 1994, which metamorphosed and recruited on average (mean ± SD) at 124 ± 12.0 days and 157 ± 13.7 days, respectively, had early life-history parameters 26 to 45 days longer than those of *A. celebesensis* collected from the Poigar River on 7 July 1997 (90 ± 13.6 and 112 ± 14.2 days, respectively) and the Poso River on 15 July 1999 (98 ± 7.2 and 122 ± 7.2 days, respectively) (Arai et al. 2003). These findings are similar to those of a previous study (Marui et al. 2001) of *A. celebesensis* glass eels sampled from the Poigar River on 13 October 1996 (metamorphosis: 87 ± 15.7 days; recruitment: 112 ± 15.0 days; N = 14) and of three glass eels collected at the southern tip of Mindanao Island in the Philippines on 30 January 1998 (98, 95, 120 days and 127, 138, 151 days). In fact, the average age at metamorphosis of glass eels that recruited to the Poigar River in 1997 ranged from 84 to 95 days throughout the year (Arai et al. 2001a), suggesting that *A. celebesensis* glass eels recruiting to the Poigar River may consistently have a shorter larval duration than those that recruit to the Philippines. The *A. celebesensis* glass eels from both the Poso and Poigar rivers in Indonesia have early life-history parameters that are shorter than those of any other anguillid glass eels studied (see Arai et al. 2001b; Marui et al. 2001), which suggests that the distance between their spawning areas and the Poso and Poigar rivers in Indonesia may be much shorter than the distance between the spawning area and the recruitment area of *A. celebesensis* in the Philippines. The patterns of surface-water circulation around the Philippines and northern Indonesia make it difficult to develop a hypothesis as to how the three samples of glass eels could have originated from the same spawning population (Fig. 2).

The surface-water circulation on the eastern side of the Philippines and the Celebes Sea is primarily influenced by the westward-flowing North Equatorial Current (NEC) that predominates between approximately 5° N and 17° N in the western North Pacific (Reverdin et al. 1994; Kaneko et al. 1998; Kawabe and Taira 1998) because this water flows both north and south (Toole et al. 1990) when it reaches the east coast of the Philippines (Fig. 1). The northern part of the NEC flows northward past the top of the Philippines and enters the northward flow of the Kuroshio Current northeast of Taiwan, while the southern part of the NEC turns south to create the southward flow of the Mindanao Current (Wijffels et al. 1995; Qu et al. 1998). The Mindanao Current flows south along the eastern side of the southern Philippines and reaches the Mindanao Eddy region at the mouth of the Celebes Sea, with some of it entering the Celebes Sea (Lukas et al. 1991; Miyama et al. 1995), while another part enters the North Equatorial Countercurrent, which flows to the east. Therefore, the only place where *Anguilla celebesensis* could spawn and recruit to all three areas sampled during the present study would be the western NEC region, which has recently been reported to be a spawning area for *A. marmorata* (Miller et al. 2002).

However, if all *Anguilla celebesensis* were spawned in this region, based on the distances from the NEC to the sampling locations in both areas, the glass eels recruiting to northern Indonesia should be older than those in the northern Philippines, but this was not the case. The age data from the glass eels sampled in northern Indonesia and the Philippines in this and previous studies, along with the ocean current patterns of the region suggest that these glass eels came from at least two, or more likely three, spawning areas. Although it is theoretically possible that the *A. celebesensis* glass eels from the Poigar and Poso rivers on opposite sides of Sulawesi Island could have come from a spawning area near the mouth of the Celebes Sea, recent data on the distribution of leptocephali of this species indicate that spawning occurs in both the Celebes

Sea and in Tomini Bay, where the Poso River reaches the sea (Aoyama et al. 2003). That the life-history parameters of the *Anguilla celebesensis* glass eels collected from the Cagayan River in the northern Philippines (which metamorphosed and recruited on average at 124 ± 12.0 days and 157 ± 13.7 days, respectively) do not differ significantly from the corresponding parameters (120 ± 13.0 days and 154 ± 13.5 days) for *A. marmorata* collected in the Cagayan River (Arai et al. 1999b) suggests that the *A. celebesensis* recruited to the northern Philippines might have been transported from somewhere in the same region of the NEC as the spawning area of *A. japonica* (Tsukamoto 1992) and that the northern population of *A. marmorata* (Minegishi et al. 2008), which has been identified on the basis of the consistent presence of leptocephali there (Miller et al. 2002), some of which have been genetically identified (Aoyama et al. 1999). Both species have been collected in relatively large numbers (800 to 1000 elvers) in the Cagayan River estuary during certain years (Tabeta et al. 1976). Therefore, although *A. celebesensis* has not been genetically identified from the NEC in the summer or early fall, it is possible that it was overlooked among the *A. marmorata* leptocephali because of the similar morphology of these two species. It is also possible that *A. celebesensis* has a different spawning season in the NEC than does *A. marmorata*, which also appears to have a very different pattern of recruitment that includes areas from southern Japan to northern Indonesia (Yamamoto et al. 2001; Miller et al. 2002; Minegishi et al. 2008). Another alternative is that this apparent northern population of *A. celebesensis* in the Philippines (and possibly Taiwan) may spawn over deep waters in the northern South China Sea; but, it is difficult to evaluate this hypothesis at present. However, if *Anguilla celebesensis* does spawn in the NEC along with the other two species, this would be an extremely interesting discovery because it would indicate that three species with apparently different recruitment patterns all utilize the same westward flowing current and then differentially recruit to very specific areas.

Regarding *A. celebesensis* from the Cagayan River estuary, the specimens are thought to be misidentified and are likely *A. luzonensis* (Kuroki et al. 2012). Arai et al. (2003) studied the otolith microstructure of *A. celebesensis* glass eels from the Cagayan River, using morphological characteristics such as total number of vertebrae, position of the dorsal fin (ano-dorsal vertebrae), and pigmentation patterns as a basis for identification (Tabeta et al. 1976). However, the northern Philippines is farther north than the known distribution range of *A. celebesensis* (Ege 1939), so it now appears fairly certain that the glass eels originally identified as *A. celebesensis* from this area were in fact *A. luzonensis*, as no *A. celebesensis* from the northern Philippines have been identified using DNA analysis whereas *A. luzonensis* glass eels and yellow eels are known to be present there (Teng et al. 2009; Watanabe et al. 2009).

The genetically identified *A. luzonensis* leptocephali, which ranged in size from 29.2 to 51.2 mm TL, were collected in the western North Pacific in 2002 (n = 2), 2005 (n = 1), 2006 (n = 1), and 2009 (n = 1) (Kuroki et al. 2012). All of these specimens had the typical morphology of longfin leptocephali and were originally thought to be *A. marmorata* or *A. celebesensis* based on their myomere counts and their distribution. The leptocephali were collected in two general areas between 13° N and 17.5° N (Fig. 1), with four being collected offshore (13–17.5° N, 137–141° E) in the region west of the West Mariana Ridge to the west or northwest of Guam (Kuroki et al. 2012). One specimen was collected not very far offshore to the east of Luzon, the Philippines (15° N, 125° E) at the western edge of the NEC region in June 2006 (Kuroki et al. 2012). The smallest leptocephalus (29.2 mm) was collected at 13° N in April 2009, which was two degrees further south from where the others were captured (Kuroki et al. 2012). The otolith microstructure of three of the leptocephali was examined, and the estimated ages of these *A. luzonensis* leptocephali were 103, 115, and 138 days old (Kuroki et al. 2012).

The leptocephali of *A. luzonensis* were collected offshore or at the edge of the subtropical gyre of the western North Pacific (Kuroki et al. 2012). The 29.2 mm *A. luzonensis* leptocephalus collected at 13° N, 140° E in 2009 suggests that the spawning area of this species is relatively close to where it was collected (Kuroki et al. 2012). The offshore presence of these leptocephali supports the hypothesis that this species would have a spawning area similar to the three other anguillid eels in the region. An offshore spawning area for *A. luzonensis* is also supported by the early-life history data obtained from the misidentified *A. luzonensis* glass eels that were collected from the Cagayan River (Arai et al. 2003).

Reproductive ecology suggests spawning grounds

The natural reproductive ecology and spawning patterns of both the tropical and temperate eel species remain a mystery, and it is thus extremely difficult to determine the nature of the migrations of freshwater eels. Arai (2014c) found that tropical freshwater eels in Lake Poso, located in central Sulawesi, Indonesia (Fig. 1), had higher gonadosomatic index values than the temperate eels that were collected in coastal waters, preparing for spawning migration and showed histologically fully developed gonads. The results suggested that, in contrast to the long-distance migrations made by the Atlantic and Japanese eels, freshwater eels originally migrated only short distances, perhaps less than 100 km to local spawning areas adjacent to their freshwater growth habitats (Arai 2014c).

A total of 41 *Anguilla celebesensis* and 64 *A. marmorata* were collected at the inlet and outlet of Lake Poso, and 13 *A. marmorata* were collected by eel traps from the mouth of the Poso River between February 2008 and July 2009 (Fig. 1) (Arai 2014c). The lake is connected to the sea (Tomini Bay) by a single river (the Poso), which drains the water from an elevation of 512 m past a waterfall and along a 40-km stretch down to the sea. The GSI of *A. celebesensis* and *A. marmorata* ranged from 4.6 to 11.2 and from 0.0 to 6.4, respectively (Arai 2014c).

The findings indicate that freshwater eels living in tropical regions must have life-history characteristics that differ markedly from those of their temperate relatives, which have a single spawning site for each species, long spawning migrations in both the North Atlantic Ocean (Schmidt 1922) and the North Pacific Ocean (Tsukamoto 1992), and a distinct spawning season. Temperate anguillids have clear seasonal patterns of downstream migration, spawning in the open ocean, and recruit of glass eels (Tesch 2003). However, Arai (2014c) clearly demonstrated that the spawning areas of *Anguilla celebesensis* are a relatively short distance from their freshwater habitats. Migrating silver eels of a few species have been caught incidentally along continental margins of both the Atlantic and Pacific oceans (Wenner 1973; Ernst 1977; Bast and Klinkhardt 1988; Sasai et al. 2001; Kotake et al. 2005; Chino and Arai 2009) but, until recently, never close to their spawning areas. Female European eels migrating downstream typically have GSI values > 1.2 (Vøllestad and Jonsson 1986; Durif et al. 2005) but less than 3.0 (Svedang and Wickström 1997; Durif et al. 2005). The levels of female maturation (GSI) of Japanese eels as they began their spawning migration in coastal areas ranged from 1.0 to 4.0, similar to that of eels that were collected in the East China Sea and other coastal areas of Japan (Sasai et al. 2001; Kotake et al. 2005). Arai (2014c) found that the GSI value exceeded 4.0 for all of the *Anguilla celebesensis* specimens. The GSI values of migrating *A. celebesensis* collected from weirs near the outlet of Lake Poso ranged from 3.3 to 11.4 (mean ± SD: 6.9 ± 1.8) (Hagihara et al. 2012). Gonadal histological analysis also indicated that the eels were physiologically in final preparation for spawning.

Before beginning their oceanic migrations, freshwater eels metamorphose from yellow (immature) to silver in freshwater and estuarine habitats. This process begins weeks or months prior to migration (Fontaine et al. 1995). In temperate eels, the process of maturation is not completed during the early stages of the migration out of freshwater, estuarine and coastal habitats because their GSI values are still quite low, and the gonads are also not yet fully developed, indicating that the process of gonadal maturation mostly occurs during spawning migrations and when the eels reach the offshore spawning area. However, the spawning area of *Anguilla celebesensis* is thought to be located in Tomini Bay, which is near to their growth habitats (Fig. 2, Aoyama et al. 2003). The Poso River flows into the southern part of the bay and is the largest drainage and potential source of silver eels close to the bay (Fig. 2). Such a considerably short spawning migration of less than 100 km may induce the final stage of maturation, whereas eels are still in terrestrial water for a short period of time before reaching the spawning area. The GSI values of *Anguilla marmorata* collected from both Lake Poso and River Poso were greater than those of temperate silver eels, but less than those of *Anguilla celebesensis* (Arai 2014c). The GSI values of migrating *A. marmorata* ranged from 1.8 to 5.7 (mean ± SD: 3.1 ± 0.8) around Lake Poso and less than those of *A. celebesensis* (Hagihara et al. 2012). The spawning area of the Poso population of *Anguilla marmorata* may be located in the western North Pacific (Arai et al. 2002; Kuroki et al. 2006; Tsukamoto et al. 2011).

Spawning grounds of tropical anguillid eels

The spawning area of the European eel *Anguilla anguilla* was found to be located surprisingly far offshore in the Sargasso Sea of the Atlantic Ocean, and their reproduction took place 5000 to 8000 km away from the European coast (Schmidt 1922). The spawning grounds of the American eel were also located in the same sea, but the distance between the spawning grounds and the growth habitats was slightly less, ranging from 1000 km to 5000 km (Schmidt 1922). The spawning area of the Japanese eel, located in the Pacific Ocean to the west of the Mariana Islands, is approximately 2000 to 3000 km from their growth habitats in East Asia. The spawning grounds of these temperate species are located thousands of kilometers away from their growth habitats as they are long-distance spawners. The long distance migrations of these temperate eels consist of two different strategies; the larval migration towards their freshwater growth habitats and the migration of adults back to the spawning area. These strategies have been poorly understood due to the critical lack of information about the location of the spawning areas and the migration routes for both tropical eels and other temperate Austaralasian eels, despite tropical eels comprising two-thirds of the genus *Anguilla*. According to the distribution of the tropical anguillid leptocephali in the Indo-Pacific region, otolith analyses and gonadal histology, the spawning ecology of tropical eels is quite different from that of the temperate eels.

The Mentawai Trench, situated to the west of Indonesia's Sumatra Island, was initially believed to be the spawning area for *A. bicolor bicolor* in the eastern Indian Ocean based on the results of an analysis of approximately 1300 leptocephali, including small specimens less than 20 mm TL, performed by (Jespersen 1942). However, further analysis of the age and growth patterns of the specimens collected during a second expedition (Aoyama et al. 2007) revealed slow growth and long larval duration of *A. bicolor bicolor* leptocephali (Kuroki et al. 2007), suggesting that the spawning area is more likely to be located somewhat farther offshore than proposed by Jespersen (1942). Nevertheless, if the spawning ground of *A. bicolor bicolor* is located near the Mentawai Trench, the spawning grounds for this short-scale migrator would be located approximately 200 to 1000 km from their growth habitats. The spawning grounds of *A. celebesensis* and *A. margumora* are located near their growth habitats and range from 100 km to 600 km, similar to the Celebes Sea and/or in Tomini Bay in Indonesian waters, and their spawning migration routes are thus relatively short distances. Other tropical species, such as *A. marmorata* and *A. luzonensis* in the North Pacific, spawn in an area that is relatively close to the spawning area of the Japanese eel (Figs. 1, 2). Distances between their spawning grounds and growth habitats range from

2000 km to 3000 km for *A. marmorata* and 2000 km for *A. luzonensis*, longer than those of other tropical species like *A. bicolor bicolor*, *A. celebesensis* and *A. margumora* but shorter than those of the temperate species *A. anguilla*, *A. rostrata* and *A. japonica* as being long-scale spawners (Table 1).

Molecular phylogenetic research on freshwater eels has recently revealed that tropical eels are the most basal species originating in the Indonesian region and that freshwater eels radiated out from the tropics to colonise the temperate regions (Minegishi et al. 2005). Tropical freshwater eels must therefore be more closely related to the ancestral form than their temperate counterparts. Ancestral eels most likely undertook a catadromous migration from a local, short-distance movement in tropical coastal waters using simple migratory mechanisms to subsequently developing the long-distance migration characteristics of present-day temperate eels, with their spawning grounds remaining in tropical areas (low latitude) and occurring in a subtropical gyre in both hemispheres (Fig. 3).

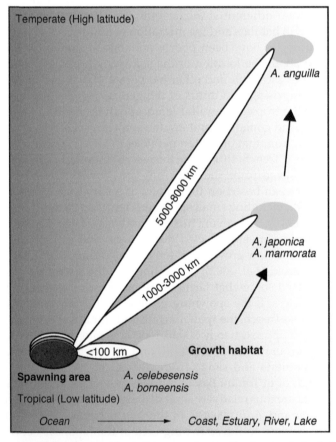

Fig. 3. Scenario of the expansion of catadromous eel migration in ancestral eels from a local short-distance movement in tropical regions to a long-distance migration in temperate regions.

Keywords: tropical eels, short-scale migration, Indonesia, oceanic migration, evolution, GSI, year-round, ancestral, El Niño

References

Aoyama, J., N. Mochioka, T. Otake, S. Ishikawa, Y. Kawakami, P.H.J. Castle, M. Nishida and K. Tsukamoto. 1999. Distribution and dispersal of anguillid leptocephali in the western Pacific Ocean revealed by molecular analysis. Mar. Ecol. Prog. Ser. 188: 193–200.

Aoyama, J., S. Wouthuyzen, M.J. Miller, T. Inagaki and K. Tsukamoto. 2003. Short-distance spawning migration of tropical freshwater eels. Biol. Bull. 204: 104–108.

Aoyama, J., S. Wouthuyzen, M.J. Miller, Y. Minegishi, M. Kuroki, S.R. Suharti, T. Kawakami, K.O. Sumardiharga and K. Tsukamoto. 2007. Distribution of leptocephali of the freshwater eels, genus *Anguilla*, in the waters off west Sumatra in the Indian Ocean. Environ. Biol. Fish. 80: 445–452.

Arai, T. 2014a. How have spawning ground investigations of the Japanese eel *Anguilla japonica* contributed to the stock enhancement? Rev. Fish Biol. Fisheries 24: 75–88.

Arai, T. 2014b. Do we protect freshwater eels or do we drive them to extinction? Springer Plus 3: 534.

Arai, T. 2014c. Evidence of local short-distance spawning migration of tropical freshwater eels, and implications for the evolution of freshwater eel migration. Ecol. Evol. 4: 3812–3819.

Arai, T. 2016. Taxonomy and distribution. pp. 1–20. *In*: T. Arai (ed.). Biology and Ecology of Anguillid Eels. CRC Press, Boca Raton, FL, USA (this book).

Arai, T., J. Aoyama, D. Limbong and K. Tsukamoto. 1999a. Species composition and inshore migration of the tropical eels *Anguilla* spp. recruiting to the estuary of the Poigar River, Sulawesi Island. Mar. Ecol. Prog. Ser. 188: 299–303.

Arai, T., L. Daniel, T. Otake and K. Tsukamoto. 1999b. Metamorphosis and inshore migration of tropical eels, *Anguilla* spp., in the Indo-Pacific. Mar. Ecol. Prog. Ser. 182: 283–293.

Arai, T., L. Daniel and K. Tsukamoto. 2000. Validation of otolith daily increments in the tropical eel, *Anguilla celebesensis*. Can. J. Zool. 78: 1078–1084.

Arai, T., J. Aoyama, S. Ishikawa, T. Otake, M.J. Miller, T. Inagaki and K. Tsukamoto. 2001a. Early life history of tropical *Anguilla* leptocephali in the western Pacific Ocean. Mar. Biol. 138: 887–895.

Arai, T., D. Limbong, T. Otake and K. Tsukamoto. 2001b. Recruitment mechanisms of tropical eels, *Anguilla* spp. and implications for the evolution of oceanic migration in the genus *Anguilla*. Mar. Ecol. Prog. Ser. 216: 253–264.

Arai, T., M. Marui, M.J. Miller and K. Tsukamoto. 2002. Growth history and inshore migration of the tropical eel, *Anguilla marmorata* in the Pacific. Mar. Biol. 140: 309–316.

Arai, T., M.J. Miller and K. Tsukamoto. 2003. Larval duration of the tropical eel *Anguilla celebesensis* from Indonesian and Philippine coasts. Mar. Ecol. Prog. Ser. 251: 255–261.

Bast, H.D. and M.B. Klinkhardt. 1988. Catch of a silver eel (*Anguilla anguilla* (L., 1758)) in the Iberian Basin (Northeast Atlantic) (Teleostei: Anguillidae). Zool. Anz. 221: 386–398.

Castle, P.H.J. 1963. Anguillid leptocephali in the southwest Pacific. Zool. Publ. Victoria Univ. Wellington 33: 1–14.

Castle, P.H.J. and G.R. Williamson. 1974. On the validity of the freshwater eel species *Anguilla ancestralis* Ege from Celebes. Copeia 2: 569–570.

Chino, N. and T. Arai. 2009. Relative contribution of migratory type on the reproduction of migrating silver eels *Anguilla japonica*, collected off Shikoku Island, Japan. Mar. Biol. 156: 661–668.

Durif, C., S. Dufour and P. Elie. 2005. The silvering process of *Anguilla anguilla*: a new classification from the yellow resident to the silver migrating stage. J. Fish Biol. 66: 1025–1043.

Ege, V. 1939. A revision of the genus *Anguilla* Shaw. Dana Rep. 16: 1–256.

Ernst, P. 1977. Catch of an eel (*Anguilla anguilla*) northeast of the Faroe Islands. Ann. Biol. 32: 175.

Fontaine, Y.A., M. Pisam, C. Le Moal and A. Rambourg. 1995. Silvering and gill "mitochondria-rich" cells in the eel, *Anguilla anguilla*. Cell Tiss. Res. 281: 465–471.

Hagihara, S., J. Aoyama, D. Limbong and K. Tsukamoto. 2012. Morphological and physiological changes of female tropical eels, *Anguilla celebesensis* and *Anguilla marmorata*, in relation to downstream migration. J. Fish Biol. 81: 408–426.

Kaneko, I., Y. Takatsuki, H. Kamiya and S. Kawae. 1998. Water property and current distributions along the WHP-P9 section (137–142° E) in the western North Pacific. J. Geophys. Res. 103: 12959–12984.

Kawabe, M. and K. Taira. 1998. Water masses and properties at 165° E in the western Pacific. J. Geophys. Res. 103: 12941–12958.

Kimura, S. and K. Tsukamoto. 2006. The salinity front in the North Equatorial Current a landmark for the spawning migration of the Japanese eel (*Anguilla japonica*) related to the stock recruitment. Deep-Sea Res. II 53: 315–325.

Kimura, S., K. Tsukamoto and T. Sugimoto. 1994. A model for the larval migration of the Japanese eel: roles of the trade winds and salinity front. Mar. Biol. 119: 185–190.

Kimura, S., T. Inoue and T. Sugimoto. 2001. Fluctuation in the distribution of low salinity water in the North Equatorial Current and its effect on the larval transport of the Japanese eel. Fish. Oceanogr. 10: 51–60.

Kotake, A., A. Okamura, Y. Yamada, T. Utoh, T. Arai, M.J. Miller, H.P. Oka and K. Tsukamoto. 2005. Seasonal variation in migratory history of the Japanese eel, *Anguilla japonica*, in Mikawa Bay, Japan. Mar. Ecol. Prog. Ser. 293: 213–221.

Kuroki, M., J. Aoyama, M.J. Miller, S. Wouthuyzen, T. Arai and K. Tsukamoto. 2006. Contrasting patterns of growth and migration of tropical anguillid leptocephali in the western Pacific and Indonesian Seas. Mar. Ecol. Prog. Ser. 309: 233–246.

Kuroki, M., J. Aoyama, S. Wouthuyzen, K. Sumardiharga, M.J. Miller and K. Tsukamoto. 2007. Age and growth of *Anguilla bicolor bicolor* leptocephali in the eastern Indian Ocean. J. Fish Biol. 70: 538–550.

Kuroki, M., M.J. Miller, J. Aoyama, S. Watanabe, T. Yoshinaga and K. Tsukamoto. 2012. Evidence of offshore spawning for the newly discovered anguillid species *Anguilla luzonensis* (Teleostei: Anguillidae) in the western North Pacific. Pac. Sci. 66: 497–507.

Limbong, D., S. Wouthuyzen, J. Aoyama, Y. Kawakami, T. Inagaki, M. Oya, A. Shinoda, J. Inoue, M. Kawai, H. Matsubara, S.Y. Mujiono, M.J. Miller, S. Berhimpon, K. Sumadhiharga and K. Tsukamoto. 2000. Distribution of anguillid leptocephali in the South Pacific, Celebes Sean and Sulu Sea. Proceeding of the 11th JSPS Joint seminar on marine science.

Lukas, R., E.R. Firing, P. Hacker, P.L. Richardson, C.A. Collins, R. Fine and R. Gammon. 1991. Observations of the Mindanao Current during the western Equatorial Pacific Ocean Circulation Study. J. Geophys. Res. 96: 7089–7104.

Marui, M., T. Arai, M.J. Miller, D.J. Jellyman and K. Tsukamoto. 2001. Comparison of the early life history between New Zealand temperate eels and Pacific tropical eels revealed by otolith microstructure and microchemistry. Mar. Ecol. Prog. Ser. 213: 273–284.

McCleave, J.D. 2003. Spawning areas of the Atlantic eels. pp. 141–155. *In*: K. Aida, K. Tsukamoto and K. Yamauchi (eds.). Eel Biology. Springer-Verlag, Tokyo.

Miller, M.J. 2003. The worldwide distribution of anguillid leptocephali. pp. 157–168. *In*: K. Aida, K. Tsukamoto and K. Yamauchi (eds.). Eel Biology. Springer-Verlag, Tokyo.

Miller, M.J., N. Mochioka, T. Otake and K. Tsukamoto. 2002. Evidence of a spawning area of *Anguilla marmorata* in the western North Pacific. Mar. Biol. 140: 809–814.

Minegishi, Y., J. Aoyama, J.G. Inoue, M. Miya, M. Nishida and K. Tsukamoto. 2005. Molecular phylogeny and evolution of the freshwater eels genus *Anguilla* based on the whole mitochondrial genome sequences. Mol. Phyl. Evol. 34: 134–146.

Minegishi, Y., J. Aoyama and K. Tsukamoto. 2008. Multiple population structure of the giant mottled eel *Anguilla marmorata*. Mol. Ecol. 17: 3109–3122.

Miyama, T., T. Awaji, K. Akitomo and N. Imasato. 1995. Study of seasonal transport variations in the Indonesian seas. J. Geophys. Res. 100: 20517–20541.

Qu, T., H. Mitsudera and T. Yamagata. 1998. On the western boundary currents in the Philippine Sea. J. Geophys. Res. 103: 7537–7548.

Réveillac, É., T. Robinet, M.W. Rabenevanana, P. Valade and É. Feunteun. 2009. Clues to the location of the spawning area and larval migration characteristics of *Anguilla mossambica* as inferred from otolith microstructural analyses. J. Fish Biol. 74: 1866–1877.

Reverdin, G., C. Frankignoul, E. Kestenare and M.J. McPhaden. 1994. Seasonal variability in the surface currents of the equatorial Pacific. J. Geophys. Res. 99: 20323–20344.

Sasai, S., J. Aoyama, S. Watanabe, T. Kaneko, M.J. Miller and K. Tsukamoto. 2001. Occurrence of migrating silver eels, *Anguilla japonica*, in the East China Sea. Mar. Ecol. Prog. Ser. 212: 305–310.

Schmidt, J. 1922. The breeding places of the eel. Philos. Trans. R. Soc. Lond. B. Biol. Sci. 211: 178–208.

Schmidt, J. 1925. The breeding places of the eel. Annu. Rep. Smithson. Inst. 1924: 279–316.

Sugeha, H.Y., S. Shinoda, M. Marui, T. Arai and K. Tsukamoto. 2001a. Validation of otolith daily increments in the tropical eels *Anguilla marmorata*. Mar. Ecol. Prog. Ser. 220: 291–294.

Svedäng, H. and H. Wickström. 1997. Low fat contents in females silver eels: indications of insufficient energetic stores for migration and gonadal development. J. Fish Biol. 50: 475–486.

Tabeta, O., T. Tanimoto, T. Takai, I. Matsui and T. Imamura. 1976. Seasonal occurrence of anguillid elvers in Cagayan River, Luzon Island, the Philippines. Bull. Jpn. Soc. Sci. Fish. 42: 421–426.

Teng, H.Y., Y.S. Lin and C.S. Tzeng. 2009. A new *Anguilla* species and a reanalysis of the phylogeny of freshwater eel. Zool. Stud. 48: 808–822.

Tesch, F.W. 2003. The Eel. Biology and management of anguillid eels. Chapman and Hall, London.

Toole, J.M., R.C. Millard, Z. Wang and S. Pu. 1990. Observations of the Pacific North Equatorial Current bifurcation at the Philippine coast. J. Phys. Oceanogr. 20: 307–318.

Tsukamoto, K. 1992. Discovery of the spawning area for the Japanese eel. Nature 356: 789–791.

Tsukamoto, K. 2006. Spawning of eels near a seamount. Nature 439: 929.

Tsukamoto, K., T.W. Lee and H. Fricke. 2003. Spawning area of the Japanese eel. pp. 121–140. *In*: K. Aida, K. Tsukamoto and K. Yamauchi (eds.). Eel Biology. Springer-Verlag, Tokyo.

Tsukamoto, K., S. Chow, T. Otake, H. Kurogi, N. Mochioka, M.J. Miller, J. Aoyama, S. Kimura, S. Watanabe, T. Yoshinaga, A. Shinoda, M. Kuroki, M. Oya, T. Watanabe, K. Hata, S. Ijiri, Y. Kazeto, K. Nomura and H. Tanaka. 2011. Oceanic spawning ecology of freshwater eels in the western North Pacific. Nat. Commun. 2: 179. doi:10.1038/ncomms1174.

Vøllestad, L.A. and B. Jonsson. 1986. Life-history characteristics of the European eel *Anguilla anguilla* in the Imsa River, Norway. Trans. Am. Fish. Soc. 115: 864–871.

Watanabe, S., J. Aoyama and K. Tsukamoto. 2009. A new species of freshwater eel *Anguilla luzonensis* (Teleostei: Anguillidae) from Luzon Island of the Philippines. Fish. Sci. 75: 387–392.

Wenner, C.A. 1973. Occurrence of American eels, *Anguilla rostrata*, in waters overlying the eastern North American continental shelf. J. Fish. Res. Board. Can. 30: 1752–1755.

Wijffels, S., E. Firing and J. Toole. 1995. The mean current structure and variability of the Mindanao Current at 8° N. J. Geophys. Res. 100: 18421–18435.

Wijffels, S.E., M.M. Hall, T. Joyce, D.J. Torres, P. Hacker and E. Firing. 1998. Multiple deep gyres of the western North Pacific: A WOCE section along 149° E. J. Geophys. Res. 103: 12985–13009.

Yamamoto, T., N. Mochioka and A. Nakazono. 2001. Seasonal occurrence of anguillid glass eels at Yakushima Island, Japan. Fish. Sci. 67: 530–532.

7

Early Life History and Recruitment

Donald J. Jellyman[1],* and *Cédric Briand*[2]

Introduction

The glass eel is the stage of arrival of juvenile eels from the sea (Fig. 1). The name refers to the transparency of the small eels which are virtually colourless upon arrival in freshwater apart from some pigmentation around the top of the cranium and the tip of the tail. With residence in freshwater, glass eels become increasingly pigmented and undergo marked changes in their behaviour. Historically the term "elver" has been used to describe this stage of freshwater recruitment (e.g., Schmidt 1925; Ege 1939; Deelder 1958), but the term "glass eel" is now almost universally used to avoid confusion with small eels that have been resident in freshwater for varying periods. A recent review (Fukuda et al. 2013), and also the ICES standard terminology (ICES 2014) concluded that the term 'glass eel' should be used to describe juvenile eels until the dorsolateral pigmentation across the top of the body is complete; this stage also corresponds to the commencement of feeding and growth.

Historically, glass eels were an important traditional food source for European countries (Fig. 2). While salmon and lampreys were the desirable species for the landowners along the Severn River, England, glass eels ("elvers") were a traditional food for the working classes (Hunt 2007). Glass eels were

[1] National Institute of Water and Atmosphere, PO Box 8602, Christchurch 8053, New Zealand.
[2] Institution d'Aménagement de la Vilaine; Patrick Lambert, Irstea, UR EABX, Aquatic Ecosystems and Global Change Research Unit, 50 avenue de Verdun, 33612 Cestas Cedex, France.
E-mail: Cedric.Briand@eptb-vilaine.fr
* Corresponding author: don.jellyman@niwa.co.nz

Fig. 1. Glass eels. Part of the daily catch at the Arzal trapping ladder (Brittany, France), 21 March 2014 (Photo Cédric Briand).

Fig. 2. Packing of glass eels for use as food, Aginaga on the banks of the Oria river, Basque district, Spain. Photo source Estibaliz Diaz (AZTI).

cooked and pressed into "elver cakes" in Ireland (McCarthy 2014), although commercial fishing for glass eels has been illegal since 1842. In Spain, glass eels were traditionally served as ingredients in an expensive soup (Tesch 2003).

Glass eels have been commercially harvested in Europe since the end of the 19th century (Briand et al. 2008). In coastal France, glass eels were once common at every fishmongers stall, and were eaten boiled with salads or fried with garlic and parsley (Feunteun and Robinet 2014). The glass eel fishery developed in conjunction with the extension of rail transport (Briand et al. 2008). Prior to demand from Japan that commenced in 1969, the main market for human consumption was in Spain, where the largest trade was 521 t in 1925 (Briand et al. 2008). The total European eel catch peaked around 1976 when 2700 t were caught (Briand et al. 2008). Average weights of glass eel vary both spatially and temporally, but using a mean weight of 0.4 g, 2700 t then would equate to approximately 6.7 billion glass eels. Largest catches of *A. anguilla* glass eels are made in France, where a variety of fishing techniques are used, including trawling in estuaries (Fig. 3).

Today, the primary value of glass eels is as seed stock for culture (Dekker and Beaulaton in press). Of the approximately 250,000 t of eels produced annually worldwide, 95% are raised in captivity (Ringuet et al. 2002) using a variety of techniques ranging from low density and extensive pond systems,

Fig. 3a. Glass eel fishing boat, Operating with large "pibalour" nets, in the Gironde estuary (France) (Photo Laurent Beaulaton).

Fig. 3b. Glass eel fishing boats, Vilaine estuary, France.

to high density, heated and recirculating systems. All eel culture is dependent upon the supply of wild glass eel as seed stock. With the decrease in *A. japonica* abundance, *A. anguilla* glass eels, which have the largest landings in the world, have provided the largest source of seed stock for Asian aquaculture (Ringuet et al. 2002). However, *A. anguilla* has been listed on the second annex of the CITES, and its export has been banned from 2010. For this reason, there has been increased interest in obtaining glass eels of other temperate and tropical species. At present, at least five temperate species (*A. anguilla, A. japonica, A. rostrata, A. australis, A. dieffenbachii*) and three tropical species (*A. reinhardtii, A. marmorata, A. mossambica*) have been trialled for culture.

In the past decade, significant advances have been made in the laboratory production of glass eels (e.g., Tanaka et al. 2003), and the F2 generation of *A. japonica* has been successfully bred in captivity and is ready to produce the F3 generation (Kuroki et al. 2014b). Current challenges are to improve the quality of eggs, and obtain alternatives for the shark egg yolk extract that forms the basis of the larval food. Despite such advances, a technique needs to be developed for the mass production of glass eels for the aquaculture industry (Kuroki et al. 2014b), while for the foreseeable future, eel culture will continue to be dependent upon the supply of wild glass eels.

Physical appearance

The marine stage of the eel, the leptocephalus, is a distinctive larva with a highly laterally compressed and almost transparent body (Miller 2009). This larval form is well adapted to the pelagic life in the ocean (Hulet 1978; Hulet and Robins 1989; Lecomte-Finiger et al. 2004; Miller et al. 2015). Leptocephali have a low incidence of predation at sea (Miller et al. 2015), possibly because of their transparency and mimicry of gelatinous zooplankton. Metamorphosis to the glass eel stage involves development of most organs (intestine, gills and kidneys), and the development of muscles and bones (Otake 2003).

At the stage of initial arrival in freshwater from the sea, glass-eels are still almost completely transparent with no external pigmentation. A small rosette of pigment is present around the top of the cranium, the so-called tache cérébrale (Strubberg 1913). With residence in fresh water, external pigmentation commences at the tail tip and advances forward to cover the dorsal midline and flanks-sequential changes in pigmentation were first classified into 14 different stages and substages (Strubberg 1913) although more abbreviated classifications exist (e.g., Boetius 1976; Elie et al. 1982). Small interspecific differences in the spread of pigmentation do occur (e.g., Fukuda et al. 2013), and can sometimes be used to differentiate between coexisting species (Robinet et al. 2003), but Strubberg's classification has been widely adopted in order to classify glass eel stages (e.g., Jellyman 1977; Haro and Krueger 1988). Glass eels arriving in freshwater are progressively more pigmented during the arrival season (Jellyman 1977; Haro and Krueger 1988; Tesch 2003; Iglesias et al. 2010). Active migrants ascending trapping ladders are also more pigmented than their estuarine counterparts (Pease et al. 2003; Briand 2009).

The average size and condition of temperate species decline during the recruitment season (e.g., Chisnall et al. 2002; Désaunay and Guérault 1997; Jessop 1998; Sullivan et al. 2009), possibly because early season arrivals have spent more time migrating during the warmer spring and summer periods (Désaunay and Guérault 1997), and also because larger larvae might swim faster and arrive earlier (Wilson et al. 2007). Being the largest and best-conditioned recruits, these early season glass eels are in demand for eel culture as they grow faster and have a higher survival than later arrivals (Tabeta and Mochioka 2003), but also because earlier arrival results in a longer growing season. Glass eels that become sedentary and remain in estuarine areas tend to be smaller than those that actively migrate upstream (Edeline et al. 2006; Bureau du Colombier et al. 2007).

Prior to the development of microchemistry techniques, the identification of eel species was done by meristics. Ege (1939) carried out a comprehensive taxonomic review of the genus *Anguilla* that used features like colouration, the relative length of the dorsal fin and vertebral counts, to distinguish between the species. Although there was some brief controversy that vertebral number might be influenced by water temperature during larval development and hatching (Tucker 1959; see Sinha and Jones 1975 for a review), vertebral counts (and myomere counts of larvae) became the primary tool for species identification. The advent of molecular systematics (allozyme markers, mtDNA and microsatellites) has led to a more robust understanding of speciation in eels, including ancestral radiation from the western Pacific (Aoyama and Tsukamoto 1997), confirmation of multiple spawning areas for *A. marmorata* (Ishikawa et al. 2004), and panmixia (e.g., Wirth and Bernatchez 2003; Als et al. 2011). Many such studies use glass eels, as they are small and readily collected over broad spatial and temporal scales.

Another powerful diagnostic tool for unravelling the life histories of eels has been otolith microchemistry, where the proportional measurements of elements like strontium and calcium can provide information of prior residence in marine/estuarine/freshwater habitats. Otoliths of glass eels have been widely studied in order to provide proportional times spent in such habitats (e.g., Jessop et al. 2002; Tzeng et al. 2003; Cairns et al. 2004; Kotake et al. 2004; Daverat and Tomas 2006; Thibault et al. 2007; Arai et al. 2009). The technology has also been applied to glass eels of both temperate and tropical species (e.g., Shiao et al. 2001, 2003), to determine the time of invasion of fresh water, to examine the core of glass eel otoliths for clues about the spatial homogeneity of spawning grounds (Martin et al. 2010), and also to differentiate between stocked and wild eels in lakes (Shiao et al. 2006).

Likewise otolith microstructure has been a fertile area of research, providing estimates of the duration of larval life, growth rates, and dates of spawning. Underpinning such studies is the pioneering research by Pannella (1971) and Campana and Neilson (1985) that showed the presence of daily growth rings in the otoliths of juvenile fish. McCleave (2008) lists studies on the daily age at recruitment for 12 species or subspecies of *Anguilla*, and results for a selection of temperate and tropical species are given in Table 1. From these data it is apparent that the temperate species have a significantly longer larval life than the tropical species, indicating the associated greater distance between freshwater habitats and spawning areas (Tsukamoto et al. 2002; Aoyama et al. 2003).

Such ageing studies rely on the regular deposition of daily rings, and some validation studies using dye immersion techniques on glass eels have been carried out (e.g., Umezawa and Tsukamoto 1991; Cieri and McCleave 2001) in both the laboratory and the field. However, there are still unresolved issues about the interpretation of daily growth rings, especially in that part of the otolith termed the "transition zone" that is formed during metamorphosis (e.g., Antunes and Tesch 1997). Reviews of the spawning times of temperate and tropical glass eels derived from daily growth rings in otoliths, and comparison with the times estimated from the size of larvae caught at sea, have highlighted mismatches between these data (McCleave 2008; Miller et al. 2014). In the case of *A. anguilla*, otolith analysis indicates year-round spawning which conflicts with the collection of newly hatched larvae in late winter and spring. Bonhommeau et al. (2010) also highlighted such differences for *A. anguilla*, where otolith studies indicate larval durations of 7–9 months, whereas field

Table 1. Ages of glass eels at recruitment (from selected studies).

Species	Location	N	Age at recruitment (days)		Reference
			Range	Mean (SD)	
Temperate species					
A. anguilla	Portugal	15	220–281	249 (22.6)	Arai et al. 2000
	Portugal to Sweden	56	420–468	447 ± 41.67	Wang and Tzeng 2000
	France	64		272.1–317.6	Désaunay et al. 1996b
	France 1990	66		239 ± 31	Désaunay et al. 1996a
	France 1991/92	46		290 ± 33	
A. rostrata	Maine	10	171–252	206 ± 22.3	Arai et al. 2000
	Haiti - Canada	125	220–284	255.3 ± 30.2	Wang and Tzeng 2000
	North Carolina, New Jersey	167		167 ± 16.9 to 209.3 ± 18.1	Powles and Warlen 2002
A. japonica	Taiwan	55		154.7 ± 10.7 to 164.3 ± 14.2	Cheng and Tzeng 1996
	China	66		162.9 ± 8.6 to 178.3 ± 9.9	Cheng and Tzeng 1996
	Japan	10		182.1 ± 12.4	Cheng and Tzeng 1996
A. australis	Australia	15	186–239	208 ± 17.4	Arai et al. 1999a
	Australia (various sites)	150	169–317	229.2 ± 29.4	Shiao et al. 2001, 2002
	New Zealand (W. Coast)	15	214–263	232 ± 19.8	Arai et al. 1999b
	New Zealand (W. Coast)	22		258 ± 19.7	Shiao et al. 2001
	New Zealand (W. Coast)	10	216–326	268 ± 31.3	Marui et al. 2001
A. dieffenbachii	New Zealand (W. Coast)	10	240–332	302 ± 28.2	Marui et al. 2001
	New Zealand (E. Coast)	10	264–330	295 ± 24.5	Marui et al. 2001

Tropical species

A. celebesensis	Sulawesi	189	104–118	109 ± 10.9	Arai et al. 2001, 2003
	Philippines	13	130–177	157 ± 13.7	Arai et al. 1999a
A. marmorata	Sulawesi	68	144–182	155 ± 14.8	Arai et al. 2001, 2003
	Philippines	16	147–219	174 ±19.2	Marui et al. 2001
	Indonesia	18	129–177	152 ± 15.2	Arai et al. 1999a
	Taiwan	15	116–166	144 ± 15.7	Arai et al. 2002
	Japan	15	133–189	154 ± 17.3	Arai et al. 2002
	Reunion, Mauritius, Mayotte	74		127.5 ± 5.7 to 196.1 ± 8.0	Reveillac et al. 2008
	Reunion	9	86–160	120.2 ± 24.7	Robinet et al. 2003
A. bicolor pacifica	Sulawesi	15	158–201	173 ± 20.9	Arai et al. 2001, 2003
	Indonesia	12	165–256	199 ± 27.0	Marui et al. 2001
A. bicolor bicolor	Indonesia	50		106.4 ± 11.1	Budimawan and Lecomte-Finiger 2007
	Indonesia	12	148–202	177 ± 16.4	Arai et al. 1999a
	Reunion	11	68–96	79.8 ± 7.7	Robinet et al. 2003
A. reinhardtii	Australia	176	140–227	181.7 ± 16.5	Shiao et al. 2002
A. mossambica	Reunion	12	96–151	123.6 ± 17.0	Robinet et al. 2003
A. bengalensis labiata	Reunion	2	143–145	144 ± 1.4	Robinet et al. 2003

collections and ocean circulation models indicate durations of about 2 years. A study on Japanese leptocephali has shown that low temperature has an effect on the rate of ring deposition in the otoliths (Fukuda et al. 2009). These results, when integrated into Lagrangian models of larval drift, result in much longer periods of larval transport, being two years and seven months for *A. anguilla* and *A. japonica* respectively (Zenimoto et al. 2011).

Glass eels are negatively buoyant when they arrive in fresh water (Hickman 1981; Williamson 1987; Tsukamoto et al. 2009a). The swimbladder only becomes fully functional after several weeks in fresh water. A complex gastro-intestinal tract develops (e.g., Monein-Langle 1985; Ciccotti et al. 1993; Tesch 2003) from the relatively simple tube of the leptocephalus. Like other marine teleosts, osmoregulation is mostly carried out by the chloride cells in the gills, and to a lesser extent in the digestive system and the kidneys. Glass eels are able to cope with large variations in salinity when they arrive in estuaries, but this ability diminishes with residence in fresh water in response to reduced activity of the chloride cells responsible for ion exchange (Wilson et al. 2004, 2007). Hence fully pigmented glass eels will die if transferred directly to salt water (Crean et al. 2007).

Feeding ceases from the onset of metamorphosis to the middle of the glass eel stage, and the body relies on lipids and glycosaminoglycan (GAG) compounds (Mochioka 2003; Miller 2009) as the sources of energy to fuel the metamorphosis into the glass eel stage and arrive in fresh water (Tesch 2003; Otake 2003). On entry into freshwater, the stomachs of glass eels are typically empty and muscle tissue has marine isotopic signatures (Bardonnet et al. 2005). Active feeding commences within estuarine areas, and feeding rates and associated weight increases are important prerequisites for upstream migration (Bureau du Colombier et al. 2008). The delay before first feeding is dependent on temperature in *A. japonica* (Fukuda et al. 2009), with no feeding observed at temperatures of 5 or 10°C. Tesch (2003) recorded a range of amphipods, copepods, and chironomid larvae, as well as polychaetes, oligochaetes, and various insect larvae, in glass eel stomachs. Detritus in stomachs is thought to be deliberately ingested rather than accidentally ingested during burrowing on ebb tides, and includes estuarine and riverine plankton, periphyton, and a variety of detrital particulate organic matter (Bardonnet et al. 2005). In captivity, glass eels are usually induced to commence feeding on natural foods like *Tubifex* worms (Heinsbroek 1991) or fish roe, before substituting various pastes or propriety foods (e.g., Ingram et al. 2001).

An important survival mechanism in freshwater eels is their ability to obtain some of their oxygen requirements through the skin. When exposed to the air, *A. anguilla* is able to obtain 2/3 of its oxygen needs from cutaneous respiration (Berg and Steen 1966). The international transport of glass eels takes advantage of this ability by packing chilled glass eels in plastic bags inflated with oxygen but containing only enough water to keep the eels moist.

The ability to survive in cool air, together with a highly developed climbing ability due to a high surface-area to mass ratio, means that glass eels are also accomplished climbers, and can leave freshwater to surmount obstacles they encounter during upstream migration.

Seasonality

Temperate species of *Anguilla* have a clearly defined recruitment season (winter/spring, e.g., *A. rostrata* Sullivan et al. 2006; *A. anguilla* Tesch 2003; Arribas et al. 2012; *A. japonica* Kawakami et al. 1999a; *A. australis* and *A. dieffenbachii* Jellyman et al. 1999) (Fig. 4), although small numbers of *A. anguilla* glass eels can be observed all year round (Désaunay et al. 1996a; Arribas et al. 2012). Seasonal data show a clear peak in abundance, but this is subject to yearly variation (Aoyama et al. 2012). In contrast, tropical species recruit all year round (e.g., Jellyman 2003; Arai et al. 2001; Robinet et al. 2003).

The Australian longfin eel, *A. reinhardtii*, has an extensive latitudinal range, and recruits year-round in the tropical part of its range (Beumer and Sloane 1990), but seasonally, in the temperate part (Sloane 1984). It has been suggested (Tsukamoto et al. 2002) that the ancestral anguillids were tropical species, probably found in the Indonesian area, and the spawning grounds were relatively local, requiring short marine migrations in a similar manner to that of current Indonesian species. For example, *A. celebesensis* spawns within Tomini Bay, Sulawesi (Aoyama et al. 2003), and the range in distance from there to the full extent of the species distribution is 80–300 km. Some anguillids dispersed widely from Indonesia to form the temperate eel species, but these have retained tropical spawning grounds, necessitating much more extensive migrations, e.g., 4000–8000 km for *A. anguilla* (Aoyama et al. 2003). A comparison of the life history characteristics of tropical and temperate larvae showed that tropical species grow faster than temperate species and metamorphose earlier at a relatively fixed size, whereas temperate species grow slower, have a body shape better specialised for long-distance migration, and show flexibility in their sizes at metamorphosis and recruitment (Kuroki et al. 2014a).

Localised spawning of tropical species (e.g., Aoyama et al. 2003; Arai 2014) almost certainly results in a smaller geographic range than the more spatially widespread spawning of temperate eels. An exception to this assumption is *A. marmorata*, the most widespread of all the freshwater eel species, which has a range of 180 degrees of longitude from Eastern Africa to Tahiti. This species is recognized as having four genetically different populations (North Pacific, South Pacific, Indian Ocean, Guam region; Minegishi et al. 2008). The geographically separate populations of other tropical species (*A. bicolor*, *A. bengalensis*, *A. celebesensis*, *A. megastoma*), also indicate the likelihood of separate spawning populations (Miller et al. 2009; Watanabe et al. 2014; Schabetsberger et al. 2015), although this has yet to be confirmed.

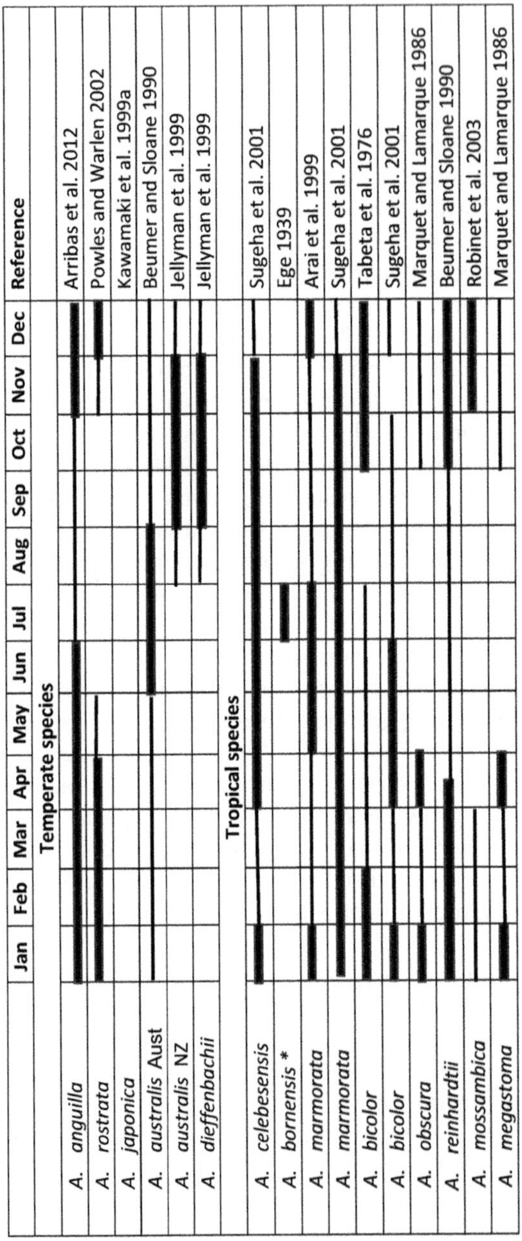

	Jan	Feb	Mar	Apr	May	Jun	Jul	Aug	Sep	Oct	Nov	Dec	Reference
Temperate species													
A. anguilla													Arribas et al. 2012
A. rostrata													Powles and Warlen 2002
A. japonica													Kawamaki et al. 1999a
A. australis Aust													Beumer and Sloane 1990
A. australis NZ													Jellyman et al. 1999
A. dieffenbachii													Jellyman et al. 1999
Tropical species													
A. celebesensis													Sugeha et al. 2001
A. bornensis *													Ege 1939
A. marmorata													Arai et al. 1999
A. marmorata													Sugeha et al. 2001
A. bicolor													Tabeta et al. 1976
A. bicolor													Sugeha et al. 2001
A. obscura													Marquet and Lamarque 1986
A. reinhardtii													Beumer and Sloane 1990
A. mossambica													Robinet et al. 2003
A. megastoma													Marquet and Lamarque 1986

Fig. 4. Arrival times of glass eels in fresh water (from selected studies). Aust = Australia, NZ = New Zealand. * = data deficient.

The lack of seasonal recruitment of tropical species may be due to a number of factors including:

- a year-round growing season for tropical species—these species experience warm temperatures all year whereas winter/spring recruitment in temperate species results in the maximum growing season of spring and summer during their first year in fresh water
- lack of a well-defined season for silver eel migration—the migration periodicity of tropical silver eels is poorly understood (Aoyama 2009; Schabetsberger et al. 2013), but based on the year-round recruitment of glass eels, it is likely that it is also extensive (e.g., Miller et al. 2002; Jellyman 2003)
- short oceanic migrations of tropical eels might mean that adult migrations do not need to be as highly synchronised as the more extensive migrations of temperate species
- multiple spawning populations could result in overlapping recruitment of glass eels originating from the different populations

Knowledge of recruitment seasons is of obvious importance for eel conservation and management, and will enable collection efforts or passage facilities to be targeted to operate during these periods.

Behaviour

Arrival from the sea

The behaviour of glass eels in the inshore marine environment is poorly known.

Glass eels have been found up to 1000 km from the European coast (Tesch 2003); they continue with the same diel vertical migrations as leptocephali (Tesch 1980; Tesch et al. 1986) and use tidal (Creutzberg 1958) and wind driven currents (Westerberg 1998) to progress towards the coast. If the temperatures are warm enough (e.g., exceed 4°C, Dutil 2009), active swimming can be sustained over long distances and can enable glass eels to arrive in coastal areas (Wuenschel and Able 2008). For glass eels contained in directed currents like the Gulf Stream or Kuroshio, active swimming is required to detrain from such currents (*A. japonica* Shinoda et al. 2011; *A. rostrata* Kleckner and McCleave 1985). In contrast, Westerberg (1998) found that the transport of glass eels towards the Baltic is mostly passive, and that wind driven currents are responsible for different distribution patterns between years. Very few studies have focused on the behaviour of glass eels at sea, and while glass eels are magnetosensitive (Nishi and Kawamura 2005) and have an acute sense of smell (Tesch 2003; Sorensen 1986), their orientation and assumed ability to navigate if swept away from adjacent landmasses (Jellyman and Bowen 2009) remain matters of speculation.

The distribution of glass eels at the coast is largely determined by ocean currents, but will also reflect the length of their migration route and their energetic status (Kawakami et al. 1999b). In *A. rostrata*, the glass eels from the northern or southern extremities of their geographic range have a higher RNA–DNA ratio and lower condition factor than glass eels from the centre of the range; it is hypothesised that these features are a reflection of the more physiologically stressful conditions experienced by those glass eels at the edge of the range (Laflamme et al. 2012). The existence of a similar stress, linked with increased metabolic expenditure, has been associated with the seasonal reduction in condition factor (Kawakami et al. 1999b) and hormone levels (Lambert et al. 2003) in glass eels of *A. japonica* and *A. anguilla* respectively.

The high mortality expected at the leptocephalus stage (Bonhommeau et al. 2009), might be the consequence of extensive oceanic migration (Miller et al. 2014), and also the result of a coastal selection process as shown by the differences in allele frequencies at different sites along the coast in *A. rostrata* (Gagnaire et al. 2012) and *A. anguilla* (Ulrik et al. 2014). At the completion of the marine migration, a possible genetic selection of successful oceanic migrants may not have selected individuals best-suited for the freshwater environment. Adaptation to this new environment is probably achieved by a phenotypic plasticity (Edeline 2007) which might be the result of the expression of paralogous genes (Gagnaire et al. 2012) or by the expression of phenotypic traits via a combination of genes (Ulrik et al. 2014).

Estuarine invasion

Glass eels entering an estuary do so mainly at night (Gandolfi et al. 1984; Dou and Tsukamoto 2003; Trancart et al. 2012), usually during the first hours of darkness (Bardonnet et al. 2005). Light directly influences glass eel migration and capture (Elie 1979; Jellyman 1979; Jellyman and Lambert 2003), and with residence in freshwater, glass eels become more light-tolerant (Sorensen and Bianchini 1986; Geffroy et al. 2014).

Their transparency and avoidance of shoaling behaviour would reduce the impact of predation on glass eels at sea. Once they enter freshwater, mortality will be high (Tesch 1980), and survival will be largely dependant on arrival during a favourable environmental window. In addition, the presence of predatory fishes can influence the survival and residence of glass eels—for instance, a study of factors influencing the settlement of glass eels in a French lagoon system (Bevacqua et al. 2011) found that settlement was negatively correlated with abundance of European catfish (*Silurus glanis*).

There are few data on glass eel survival rates, but Graynoth et al. (2008) recorded that the proportion of glass eels of *A. australis* and *A. dieffenbachii* that survived to 400 mm was highly variable in three different streams, ranging from 0.02% to 20%. Juvenile eels are important in the diet of a variety of birds, especially cormorants (Carpentier et al. 2009), as well as otters (Carss et al. 1999; Britton et al. 2006). Recruitment success measured as recruitment to adult eel stocks is in turn affected by the extent of density-dependant mortality (Lobon-Cervia and Iglesias 2008; Graynoth et al. 2008), plus a series of anthropogenic impacts like the installation of dams and weirs, and the extent of harvest of eels at a variety of life history stages.

One of the main mechanisms explaining variations in glass eel abundance and the initial colonisation of estuaries is the use of selective tidal stream transport (STST). STST is used by glass eels to make progress upstream in estuaries during the flood tide while taking shelter in the substratum during the ebb tide (Deelder 1952, 1958; Creutzberg 1961; Jellyman 1979; McCleave and Kleckner 1982; Gascuel 1986). If the substrate is fine and dense, glass eels "cling" to it but it is if cobbly they will enter the interstitial spaces between the cobbles (Creutzberg 1961). Laboratory experiments of habitat preferences showed that glass eels preferred habitats containing structure, especially rocks and cobbles although preferences varied slightly between species (Silberschneider et al. 2004).

Although migration behaviour is triggered by an endogenous rhythm (Wippelhauser and McCleave 1987), it requires external cues related to the tide such as odour or current variation, to be maintained (Wippelhauser and McCleave 1987; Bolliet et al. 2007; Trancart et al. 2012). Field evidence suggests that STST is not always best synchronized with the tide and results in some individuals being flushed downstream; for instance, a study in the Gironde River found that the rate of upstream transport was only 15% of what would have been achieved by making full use of the tidal currents (Beaulaton and Castelnaud 2005).

Although most glass eels migrate at night (Fig. 5), 30% have been found to migrate during the day but they swim at a greater depth; as many as 15% of migrating glass eels have been found in the water column during the ebb tide (P. Lambert Cemagref, France, pers. comm.). During the ebb tide, the directed downstream current forces migrating glass eels to swim along the edge of the estuary where they experience reduced downstream velocity. This behaviour makes them more susceptible to hand net fisheries, especially in long estuaries such as the Gironde and Severn (Cantrelle 1981; Harrison et al. 2014). Glass eels will progressively migrate upstream in estuaries using STST and accumulate in an area where riverine currents reduce the efficiency of upstream transport (Cieri 1999; Edeline et al. 2007). The behaviour of glass eels in response to tide cycles is shown in Fig. 6.

Active upstream migration

In spring, when temperature conditions are favourable, an active migration phase (Creutzberg 1961) enhances the upstream transport achieved by STST. During this phase, glass eels are less prone to be flushed downstream by flood events (Gascuel 1986), they migrate near the surface and swim actively against the current. In historical times, this behaviour resulted in a visible ribbon of glass eels migrating upstream adjacent to the streambank (Tesch 2003). Studies of the swimming ability of glass eels have shown that they are capable of sustained swimming speeds of > 0.5 m.s^{-1} (Langdon and Collins

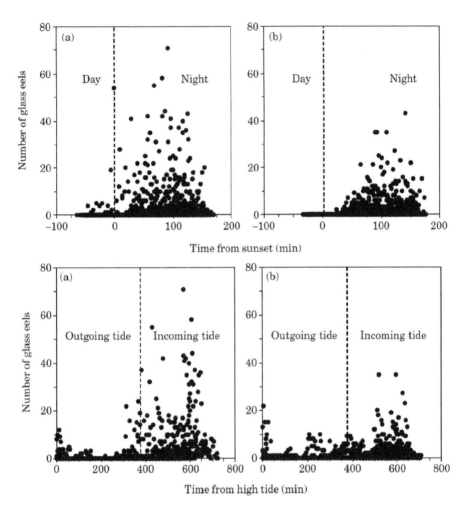

Fig. 5. Catches of glass eels at 5 min intervals in the Grey River, New Zealand in (a) 2000 and (b) 2001 in relation to the time from sunset (upper panel) and time from high tide (lower panel). From Jellyman and Lambert (2003).

2000). During this phase, glass eels will also climb up vertical surfaces to bypass waterfalls or man-made obstacles. This ability to climb vertical surfaces is lost at a size of 100 mm (Legault 1988). The many eel ladders and ramps used internationally to capture juvenile eels during their upstream migration phase (e.g., Feunteun 2012), take advantage of the eels' climbing ability and their ability to respire cutaneously.

The transition to active migration is probably mediated by thyroid hormones (Jegstrup and Rosenkilde 2003; Edeline et al. 2005; Imbert 2008) and is related to temperature—time until active behaviour commences is estimated at 40 days at 8°C but only 10 days at 14°C (Briand 2009). The onset of the active

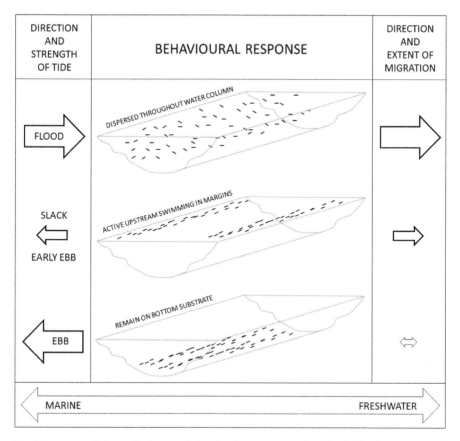

Fig. 6. Summary of glass eel migratory behaviour in upper estuaries. Flood tide; glass eels spread throughout the water column. Slack/early ebb tide; glass eels move to the margins and actively swim upstream. Ebb tide; glass eels remain on or in the bottom substrate. From Harrison et al. (2014).

migration phase also depends on the density of recruits in the estuary which acts as a triggering mechanism (Briand 2009).

Differences in individual activity and aggression can be observed at this stage (Geffroy et al. 2014). Glass eels that feed more are also more active and will grow faster (Bureau du Colombier et al. 2008). The energy content of glass eels explains much of their migratory behaviour (Bureau du Colombier et al. 2007) and those eels with a lower energy content will settle early at sea or in the estuary (Bureau du Colombier et al. 2011; Sullivan et al. 2009). The duration of the active migration phase is short (Briand 2009), as evidenced by the small proportion of active glass eels (10%) found in one estuary (Briand et al. 2006). There is often a delay of several months to one year before the resumption of the migration behaviour at the juvenile yellow eel stage (Sloane 1984; Moriarty 1986; Michaud et al. 1988; Haro and Krueger 1991; Lobón-Cerviá et al. 1995; Jessop et al. 2002). During that period, the size of the individuals (Imbert 2008) and intraspecific competition will play a major role in subsequent migratory behaviour (Lambert 2005; Lambert and Rochard 2007).

This two-stage migration mechanism (Pease et al. 2003) probably increases the fitness of individuals which choose to move upstream to the freshwater zone where the cost of osmoregulation is higher but competition is lower than in the inter-tidal areas where they accumulate using STST (Arai et al. 2006; Edeline 2007). In places where several species compete for resources, active upstream migration facilitates different inter-specific use of available habitats (Jellyman et al. 1999; Shiao et al. 2003).

Environmental factors

The migration of glass eels is affected by two thresholds of temperature. The first is associated with STST. Temperatures falling below 4°C to 6°C will result in cessation of activity and a consequent reduction in fisheries catch has been observed for *A. anguilla* (Deelder 1952; Elie 1979; Lecomte-Finiger and Razouls 1981; Hvidsten 1983; Gascuel 1986; Briand 2009). At sea, *A. rostrata* glass eels migrating upstream in the Gulf of St. Lawrence must do so at temperatures near the lower limit of 4°C, conditions considered marginal for recruitment (Dutil 2009). The second temperature threshold is associated with active swimming against the current and this requires higher temperatures of 9°C to 12°C in *A. anguilla* (Deelder 1958; Lowe 1961; Hvidsten 1983; Tongiorgi et al. 1986; Dekker 1998), *A. rostrata* (Sorensen and Bianchini 1986; McCleave and Wippelhauser 1987; Haro and Krueger 1988; Martin 1995; Jessop 2003; Sullivan et al. 2006; Linton et al. 2007; Overton and Rulifson 2008; Sullivan et al. 2009), *A. japonica* (Han 2011), *A. australis* and *A. dieffenbachii* (Jellyman 1977; Sloane 1984; McKinnon and Gooley 1998; August and Hicks 2008). The temperature threshold for climbing is higher than for swimming in both *A. anguilla* and *A. rostrata* (Linton et al. 2007). Temperature plays a key role in the final transition from glass eel to the juvenile yellow eel stage (Briand et al. 2005; Briand 2009) and an increase in temperature results in more rapid shifts between STST and active behaviour.

Monthly tide cycles are governed by the lunar cycle which makes it difficult to isolate their respective effects (Elie and Rochard 1994). A tidal cycle effect on migration has been demonstrated in many field studies (Jellyman 1979; Sorensen and Bianchini 1986; Ciccotti et al. 1995; Martin 1995; De Casamajor 1998; McKinnon and Gooley 1998; Jellyman et al. 1999; Sugeha et al. 2001; Jellyman and Lambert 2003; August and Hicks 2008; Mouton et al. 2014; Hwang et al. 2014). In many estuaries, the largest catches are made during spring tides (Jellyman 1979; Jellyman and Lambert 2003). At the top of the estuary, the increased tide levels of spring tides progressively concentrate glass eels already present with those newly arrived from the sea. During the shift from spring to neap tide, the glass eels progressively lose the STST tidal signal and this results in them becoming more widely distributed.

The combination of light and turbidity in the estuary influences the depth that glass eels swim at and hence their catchability (De Casamajor et al. 1999, 2000). Glass eels become more sensitive to light as they pigment (Bardonnet et al. 2005) and in some instances are more likely to migrate at night as they progress upstream in the estuary (Arribas et al. 2012). This result is in contrast to what is described in the upper parts of estuaries, where glass eels, while still migrating mostly at night, can also migrate during the day (Jellyman 1979; Sorensen and Bianchini 1986; Tesch 2003), especially when large migrations are underway. An increased level of thyroid hormones (Edeline et al. 2005), is probably responsible for such migratory behaviour during the day.

River flows can have beneficial or adverse effects on glass eel migrations. A negative effect can be observed on catches in the upper reaches of estuaries where large freshwater flows can constrain the rate of upstream movement (Jellyman et al. 2009). A positive effect associated with increased flow attracting glass eels to migrate, has been suggested by several authors (Elie and Rochard 1994; Jessop et al. 2002; Lambert 2005), and increases in discharge are an important factor explaining variations in catches (Zompola et al. 2008; Arribas et al. 2012); also the larger offshore plume of freshwater during floods serves to intercept and attract more glass eels (Crivelli et al. 2008). Thus larger rivers attract more glass eels than smaller rivers (Jellyman et al. 2009).

The attraction of an earthy scent (geosmin) to glass eels in the laboratory has been demonstrated by various authors (Miles 1968; Sorensen 1986; Sola 1995; Sola and Tongiorgi 1996; McCleave and Jellyman 2002) and a preference for water from the habitat occupied by adults of the same species has been demonstrated in *A. australis* (McCleave and Jellyman 2002). However, the latter research was unable to elicit a species-specific response to the varying water types that would provide a means self-sorting of *A. australis* and *A. dieffenbachii* as these two species occur in mixed shoals.

Although not extensively researched, the annual recruitment of glass eels will result in the transfer of large quantities of marine-derived nutrient to freshwater ecosystems. An analysis of the net carbon flux in a small river in northern France (Laffaille et al. 2000) found that the annual C input was only 12% of the total annual output, the latter being due to the emigration of adult eels. As pointed out by the authors, such balances could be substantially affected by the presence of instream barriers.

Worldwide trends and concerns

With current scientific knowledge, it is not possible to determine the biomass of spawning eels of any of the three major species of world commerce, *A. japonica*, *A. anguilla* and *A. rostrata*, that would be required to maintain or enhance the continental eel stocks. However, given the uncertainties of spawning success, and the subsequent larval dispersal by changeable oceanic currents, the most suitable measure of spawner success is the extent of glass eel recruitment.

For *A. anguilla*, the ICES Working Group on eel has developed a series of recruitment indices from different locations in Europe (Moriarty and Dekker 1997), based on catches from a variety of experimental fisheries, traps and commercial fisheries. Some series, especially those associated with commercial fisheries, have shown catch abnormalities resulting from changes in management. The remaining series show generally similar short-term trends as they are largely dependant on local environmental factors, but in the long term, they all show a consistent decreasing trend from 1979 to 2010. There are regional differences between recruitment recorded at North Sea sites and sites for the rest of Europe (Dekker 2000b; ICES 2010), with the decline in the North Sea sites being greater. Ideally, all sampling would involve capture near the mouth of an estuary. However, given the limited availability of such data, information from other sites is also used, and data standardized and pooled over a season can provide consistent historical trends (Sullivan et al. 2009).

The recruitment index of *A. anguilla* has shown a continuous declining trend from the beginning of the 1980's to 2010 (Fig. 7) (ICES 2014). This decline is also observed in the ocean (Hanel et al. 2014). About a decade later, significant reduction in recruitment of *A. rostrata* was noted, especially towards the edges

Fig. 7. Trends in recruitment indices of glass eels for the North Sea sampling sites (NS) and the rest of Europe (EE). Lines are means of estimated (GLM) recruitment for each area, scaled to the 1960–79 average (log scale).

of its geographic range (Casselman 2003). Catches of *A. japonica* glass eels have also dropped from 174 t in 1969 to 14 t in 2001 (Tsukamoto et al. 2009a). Such reductions led to an expression of worldwide concern about the status of *Anguilla* species (Dekker et al. 2003).

These declining trends have influenced the glass eel trade and the eel aquaculture industry as all production, whether from wild fisheries or culture, is dependent upon the use of wild glass eels as seed stock. Japan is the world's largest consumer of eels, with up to 90,000 t imported annually (although this has reduced to < 10,000 t in 2009; Monticini 2014). Japan has a current production of 22,000 t but China is the largest producer of eels, and accounts for ¾ of the total world production (in 2011) of 268,000 t (Monticini 2014). Prior to 1990, virtually all eel farming was done in Southeast Asia and used glass eels of *A. japonica* (Crook and Nakamura 2013).

Declines in recruitment of *A. japonica* lead to many farms, especially in China, switching to *A. anguilla* glass eels in the late 1990's. In December 2010, after a progressive decrease in the amount of glass eel trade, the implementation of the Convention on Trade of Endangered Species (CITES) led EU member states to suspend all exports and imports of *A. anguilla* from and to the EU—this effectively terminated any residual glass eel trade to Asia. The European Eel Management Plan also required the allocation of 60% of the glass eel catch to restocking by 2013.

The decline in *A. japonica* and *A. anguilla* glass eels prompted the development of the market for *A. rostrata* glass eels from the USA and Canada. This market ranged from 0.1 to 10 t per year until 2010 but increased dramatically to over 50 t in the first six months of 2011 alone (Crook and Nakamura 2013). Prices for glass eels reflect their availability and international trade decisions. Prices paid to French fishers in 2008 were €750/kg (~ $US 850/kg) (Briand et al. 2008), and recent prices paid to Maine fishers for *A. rostrata* glass eels are as high as $US 1180/kg (Sneed 2014). Should the U.S. Fish and Wildlife Service list the American eel under the Endangered Species Act, this could potentially prohibit glass eel fishing in all states.

In the early 1970's a small glass eel fishery commenced in New Zealand and up to 10 t were exported to Japan over a 4-year period—these were a mix of ~ 90% *A. australis* and 10% *A. dieffenbachii* (Jellyman 1979). No glass eel exports from New Zealand or Australia are permitted at present, and Asian glass eel buyers are now turning to Africa, especially Madagascar, for supplies of species like *A. mossambica* and also to South East Asia for *A. marmorata* (Crook and Nakamura 2013); at this stage, quantities are relatively small (up to 1.4 t per year), but the demand is growing. There are already concerns that a number of the tropical eel species are vulnerable to over-exploitation as their biology is poorly understood (Crook and Nakamura 2013). The rise in price is also raising concerns about the development of a lucrative black market in Europe.

A number of researchers have speculated on reasons for the marked decline in the abundance of the Northern hemisphere glass eels (e.g., Castonguay et al. 1994a,b; Richkus and Whalen 2000; Feunteun 2002; Wirth and Bernatchez 2003;

Dekker 2003; Knights 2003; Kettle et al. 2008; Tsukamoto et al. 2009a; Baltazar-Soares et al. 2014). Suggested reasons include anthropogenic influences during the freshwater stage (chemical contamination, disease prevalence, overfishing, habitat modifications, turbine mortality), and changes in ocean circulation. Oceanic changes are known to have a direct influence on entrainment of some species into favourable, or otherwise, currents.

For instance, catches of *A. japonica* glass eels tend to be low in *El Nino* years as this is usually associated with a southward shift of the salinity front which, in turn affects the latitude of spawning sites (Kimura and Tsukamoto 2006). A southward shift of 1° in the spawning site may result in greater entrainment of larvae in the Mindanao Current which transports larvae south with consequently reduced numbers entering the Kuroshio Current that normally transports larvae to Japan, China and Taiwan (Zenimoto et al. 2009). Likewise, Tzeng et al. (2012) found significant correlations between long-term catches of *A. japonica* glass eels in Taiwan and a range of climate indices that affect oceanic production and eddy activities.

In the Atlantic, a positive phase of the NAO (North Atlantic Oscillation) appears to have a negative impact on larval survival and migration from the Sargasso Sea (Kettle et al. 2008). Ocean-atmospheric changes observed in the Sargasso Sea, such as a northward shift in the 22.5°C isotherm bounding the spawning area of *A. anguilla,* and a shallowing of the mixed layer depth might explain the decline in recruitment of European and possibly American eels (Friedland et al. 2007). Baltazar-Soares et al. (2014) also found that the major driver of declines has been the variation in atmospherically driven ocean currents in the Sargasso Sea.

Climate change is the main driver of changes in oceanic circulation patterns. Knights (2003) hypothesised that trends in global warming resulted in inhibition of spring thermocline mixing and nutrient circulation in the spawning areas of the European, American and Japanese eels, and this resulted in most of the observed declines in the recruitment of these species. Bonhommeau et al. (2008) also reviewed recruitment patterns of these species and concluded that changes in marine productivity were related to global warming. In their review of factors affecting recruitment variability of *A. japonica* in Taiwan, Tzeng et al. (2012) concluded that "Japanese eel recruitment may be influenced by multi-timescale climate variability" and a southward shift of spawning sites might explain the larger proportion of *A. japonica* glass eels found in the Philippines (Yoshinaga et al. 2014). Results from a Lagrangian passive drift model for *A. anguilla* (Pacariz et al. 2014) showed that post-1970, fewer larvae would have been transported to northern latitudes, an outcome consistent with observations of reduced recruitment to the Baltic (Svärdson 1976); however the model did not support the hypothesis that shifts in gyre circulation were responsible for declines in recruitment of glass eels to Europe, although indirect effects of climate change like food availability, were not considered. Not only do changes in oceanic circulation affect the numbers of glass eels recruiting to freshwater, such changes can also affect the size of glass eels—thus Dekker (2004) found that changes in the size of glass eels recruiting to Den Oever, the Netherlands correlated with NAO.

Recently, the ICES eel recruitment index increased from very low levels of 1%–4% levels in 2010 to 5%–12% in 2014 (ICES 2014). It is too soon to speculate whether this increased recruitment corresponds to the increased levels of silver eel escapement achieved over the preceding two years through the respective management plans of the EU member countries, or whether it represents some serendipitous changes in ocean circulation patterns.

Conservation and management

Changes to oceanic circulation patterns are obviously controlled by factors external to those managed by fishery managers. However, an understanding of the importance of oceanic factors does not preclude the possible simultaneous and interactive impact of continental anthropogenic factors affecting eel recruitment (Friedland et al. 2007), especially drought and extensive dam construction (Kettle et al. 2011). The most effective management action for conserving stocks is to increase the number of silver eels escaping to spawn (e.g., Kuroki et al. 2014b), which in turn implies addressing issues of up- and downstream access, habitat protection and enhancement, water quality and quantity issues, maintaining eel health and well-being, and managing catches of all continental life history stages to sustainable levels.

Managing the harvest of glass eels is an essential component of stock management. While glass eel exploitation rates are believed to be high (80% to 85%; Dekker 2000a; Tsukamoto et al. 2009a), they have seldom been evaluated directly (Tzeng 1984; Jessop 2000; Beaulaton and Briand 2007; Bru et al. 2009). Among those studies, the model developed by Beaulaton and Briand (2007) predicts the proportion of settled glass eels relative to a non-impacted situation with current or pristine recruitment, and can provide an initial assessment of anthropogenic influences and be used to test various management scenarios. Outputs can also be related to the percentage of spawners-per-recruit, with a target of at least 40% virgin biomass being one of the main targets of the European eel management plans (ICES 2001). Theoretically, in rivers where recruitment is very high but much of it is lost through density dependent mortality, a harvest rate greater than that required to achieve the above spawner escapement could be considered. However, the lack of knowledge about the carrying capacity of continental habitats and the depleted state of the stock means it would be unwise to implement this option at present.

Although considerable progress has been made in the laboratory propagation of glass eels (Okamura et al. 2014), it is not yet possible to mass-produce glass eels. Therefore, for the foreseeable future, the world's production of farmed eels will continue to depend upon the harvest and culture of wild glass eels. European eel management plans also require that the majority of glass eels be used to restock waterways. Collectively, such demands on the wild resource mean that continued research, and conservative management must be encouraged to ensure that the presently fragile stocks of the three Northern hemisphere species of *Anguilla*, do not diminish to below recoverable levels. With the exception of *A. marmorata*, the Indo-Pacific species have more geographically-limited distributions than the temperate species (Jellyman 2003), and particular care must be taken to not over-exploit glass eels of these species, especially as there is currently little information on their basic life history parameters.

The international trade and air-freighting of glass eels provides a ready vector for disease dispersal. Already, the Eel Virus European (EVE) which caused mass mortalities of *A. japonica* in culture conditions, has been traced to imports of European glass eels (Haenen et al. 2009). The spread of the swim bladder nematode, *Anguillicoloides* (= *Anguillicola*) *crassus* that has caused high mortalities of European eels, is thought to have originated when Japanese eels were imported into Europe for experimental purposes (Szekely et al. 2009). Because of their limited residence time in freshwater, glass eels have a relatively low disease loading, but *Anguillicoloides* has been recorded from this stage (Nimeth et al. 2004) meaning that relocation of glass eels is a potential means of spread of this pathogen. Furthermore, four species of *Anguillicoloides/ Anguillicola* have been recorded from Pacific and African species of *Anguilla* (Szekely et al. 2009), and the transport of glass eels of these eel species to Asia or the Northern hemisphere could result in the inadvertent introduction of further parasites.

Conclusions

The extent of historical recruitment is legendary, with observations of migrations of *A. anguilla* glass eels "as dense ribbons of fish, hundreds deep and tens of miles long" (Righton and Roberts 2014). In New Zealand, Cairns (1941) recorded a shoal in the Waikato River that measured 4.5 m wide and 3.0 m deep and took 8 hours to pass. Such extensive migrations are not reported today.

The timing of glass eel migrations are predictable, and the eels can then be readily caught in large numbers, often with relatively simple equipment like hand nets. Coupled with high prices for aquaculture, this makes these migrations particularly vulnerable to overfishing. Until glass eels can be produced in hatcheries at commercially acceptable quantities and costs, the extensive aquaculture and restocking industries will be dependant upon the supply of wild glass eels. Firm precautionary actions advocated by Dekker et al. (2003) have been implemented in the EU, and may have resulted in the significantly increased recruitment observed since 2013, although a continuation of this trend would be required to confirm this.

Long-lived and semelparous species like anguillids are especially vulnerable to over-exploitation. Being highly fecund provides some life history insurance, and given that anguillids populate extensive areas of the globe, this is obviously a life-history strategy that has served the species well, but not one that can readily cope with high levels of anthropogenic impacts. There are also concerns that the Anthropogenic Allee effect (Courchamp et al. 2006) might further reduce spawner biomass below the required thresholds for successful spawning to occur (Dekker 2004; Briand et al. 2008). The likelihood that eels spawn at a variety of sites within a larger geographic area like the Sargasso Sea (Miller et al. 2014), would amplify the risk of such an Allee effect occurring, due to the reduced likelihood of individuals encountering other mature eels. Continued conservative exploitation and greater habitat protection are needed in order to ensure adequate protection for all temperate species of *Anguilla*, and further research is required to understand the extent and seasonality of the recruitment of tropical species.

Keywords: recruitment, seed stock, artificial production, otolith microstructure, freshwater invasion, freshwater migrations, seasonality, behaviour, trends, reduced abundance, management

References

Als, T.D., M.M. Hansen, G.E. Maes, M. Castonguay, L. Riemann, K. Aarestrup, P. Munk, H. Sparholt, R. Hanel and L. Bernatchez. 2011. All roads lead to home: panmixia of European eel in the Sargasso Sea. Molec. Ecol. 20: 1333–1346.

Antunes, C. and F.-W. Tesch. 1997. A critical consideration of the metamorphosis zones when identifying daily rings in otoliths of European eel, *Anguilla anguilla*. Ecol. Freshw. Fish 6: 102–107.

Aoyama, J. 2009. Life history and evolution of migration in catadromous eels (genus *Anguilla*). Aqua-BioSci. Monogr. 2: 1–42.

Aoyama, J. and K. Tsukamoto. 1997. Evolution of the freshwater eels. Naturwissenschaften 84: 17–21.

Aoyama, J., S. Wouthuyzen, M.J. Miller, T. Inagaki and K. Tsukamoto. 2003. Short-distance spawning migration of tropical freshwater eels. Biol. Bull. 204: 104–108.

Aoyama, J., A. Shinoda, T. Yoshinaga and K. Tsukamoto. 2012. Late arrival of *Anguilla japonica* glass eels at the Sagami River estuary in two recent consecutive year classes: ecology and socio-economic impacts. Fish. Sci. 78: 1195–1204.

Arai, T. 2014. Evidence of local short-distance spawning migration of tropical freshwater eels, and implications for the evolution of freshwater eel migration. Ecol. Evol. 4: 3812–3819.

Arai, T., J. Aoyama, D. Limbong and K. Tsukamoto. 1999a. Species composition and inshore migration of the tropical eels *Anguilla* spp. recruiting to the estuary of the Poigar River, Sulawesi Island. Mar. Ecol. Prog. Ser. 188: 299–303.

Arai, T., T. Otake, D.J. Jellyman and K. Tsukamoto. 1999b. Differences in the early life history of the Australasian shortfinned eel *Anguilla australis* from Australia and New Zealand, as revealed by otolith microstructure and microchemistry. Mar. Biol. 135: 381–389.

Arai, T., T. Otake and K. Tsukamoto. 2000. Timing of metamorphosis and larval segregation of the Atlantic eels *Anguilla rostrata* and *A. anguilla*, as revealed by otolith microstructure and microchemistry. Mar. Biol. 137: 39–45.

Arai, T., D. Limbong, T. Otake and K. Tsukamoto. 2001. Recruitment mechanisms of tropical eels *Anguilla* spp. and implications for the evolution of oceanic migration in the genus *Anguilla*. Mar. Ecol. Prog. Ser. 216: 253–264.

Arai, T., M. Marui, M.J. Miller and K. Tsukamoto. 2002. Growth history and inshore migration of the tropical eel, *Anguilla marmorata*, in the Pacific. Mar. Biol. 140: 309–316.

Arai, T., D. Limbong and K. Tsukamoto. 2003. Migration and recruitment of tropical glass eels to the mouth of the Poigar River, Sulawesi Island, Indonesia. Amer. Fish. Soc. Symp. 33: 37–46.

Arai, T., A. Kotake and T.K. McCarthy. 2006. Habitat use by the European eel *Anguilla anguilla* in Irish waters. Estuar. Coast. Shelf Sci. 67: 569–578.

Arai, T., N. Chino and A. Kotake. 2009. Occurrence of estuarine and sea eels *Anguilla japonica* and a migrating silver eel *Anguilla anguilla* in the Tokyo Bay area, Japan. Fish. Sci. 75: 1197–1203.

Arribas, C., C. Fernández-Delgado, F.J. Oliva-Paterna and P. Drake. 2012. Oceanic and local environmental conditions as forcing mechanisms of the glass eel recruitment to the southern most European estuary. Estuar. Coast. Shelf Sci. 107: 46–57.

August, S.M. and B.J. Hicks. 2008. Water temperature and upstream migration of glass eels in New Zealand: implications of climate change. N.Z. J. Mar. Freshw. Res. 81: 195–205.

Baltazar-Soares, M., A. Biastoch, C. Harrod, H. Hanel, L. Marohn, E. Prigge, D. Evans, K. Bodles, E. Behrens, C.W. Böning and C. Eizaguirre. 2014. Recruitment collapse and population structure of the European eel shaped by local ocean current dynamics. Curr. Biol. 24: 104–108.

Bardonnet, A. and P. Riera. 2005. Feeding of glass eels (*Anguilla anguilla*) in the course of their estuarine migration: new insights from stable isotope analysis. Estuar. Coast. Shelf Sci. 63: 201–209.

Bardonnet, A., V. Bolliet and V. Belon. 2005. Recruitment abundance estimation: role of glass eel (*Anguilla anguilla*) response to light. J. Exp. Biol. Ecol. 321: 181–190.

Beaulaton, L. and G. Castelnaud. 2005. The efficiency of selective tidal stream transport in glass eels entering the Gironde (France). Bull. Fr. Pêche Piscic. 378-379: 5–21.

Beaulaton, L. and C. Briand. 2007. Effect of management measures on glass eel escapement. ICES J. Mar. Sci. 64: 1402–413.

Berg, T. and J.B. Steen. 1966. Regulation of ventilation in eels exposed to air. Comp. Biochem. Physiol. 18: 511–516.

Beumer, J. and R. Sloane. 1990. Distribution and abundance of glass-eels *Anguilla* spp. in east Australian waters. Int. Rev. Gesamten Hydrobiol. 75: 721–736.

Bevacqua, D., M. Andrello, P. Melia, S. Vincenzi, G.A. De Leo and A.J. Crivelli. 2011. Density-dependent and inter-specific interactions affecting European eel settlement in freshwater habitats. Hydrobiol. 671: 259–265.

Boetius, J. 1976. Elvers (*Anguilla anguilla*) and (*Anguilla rostrata*), from two Danish localities. Size, body weight, developmental stage and number of vertebrae related to time of ascent. Dana 7: 199–220.

Bolliet, V., P. Lambert, J. Rives and A. Bardonnet. 2007. Rythmic swimming activity in *Anguilla anguilla* glass eels: Synchronisation to water current reversal under laboratory conditions. J. Exp. Biol. Ecol. 344: 54–66.

Bonhommeau, S., E. Chassot and E. Rivot. 2008. Fluctuations in European eel (*Anguilla anguilla*) recruitment resulting from environmental changes in the Sargasso Sea. Fish. Ocean. 17: 32–44.

Bonhommeau, S., O. Le Pape, D. Gascuel, B. Blanke, A.-M. Tréguier, N. Grima, Y. Vermard, M. Castonguay and E. Rivot. 2009. Estimates of the mortality and the duration of the trans-Atlantic migration of European eel *Anguilla anguilla* leptocephali using a particle tracking model. J. Fish Biol. 74: 1891–1914.

Bonhommeau, S., M. Castonguay, E. Rivot, R. Sabatie and O. Le Pape. 2010. The duration of migration of the Atlantic *Anguilla* larvae. Fish Fisheries 11: 289–306.

Briand, C. 2009. Dynamique de population et de migration des civelles en estuaire de Vilaine. Population dynamics and migration of glass eels in the Vilaine estuary. Ph.D. Thesis. Agrocampus Ouest.

Briand, C., D. Fatin, E. Ciccotti and P. Lambert. 2005. A stage-structured model to predict the effect of temperature and salinity on glass eel *Anguilla anguilla* pigmentation development. J. Fish Biol. 67: 993–1009.

Briand, C., D. Fatin, G. Fontenelle and E. Feunteun. 2006. Estimating the stock of glass eels in an estuary by mark-recapture experiments using vital dyes. Bull. Fr. Pêche Piscic. 378: 23–46.

Briand, C., S. Bonhommeau, G. Castelnaud and L. Beaulaton. 2008. An appraisal of historical glass eel fisheries and markets: landings, trade routes and future prospect for management. Proc. Inst. Fish. Man. 2007, Wesport, Ireland.

Britton, J.R., J. Pegg, J.S. Shepherd and S. Toms. 2006. Revealing the prey items of the otter *Lutra lutra* in South West England using stomach contents analysis. Folia Zool. 55: 167–174.

Bru, N., P. Prouzet and M. Lejeune. 2009. Daily and seasonal estimates of the recruitment and biomass of glass eels runs (*Anguilla anguilla*) and exploitation rates in the Adour open estuary (Southwestern France). Aquat. Liv. Res. 22: 509–523.

Budimawan and R. Lecomte-Finiger. 2005. Inshore migration of the tropical glass eels *Anguilla marmorata* recruiting to Poso estuary, Sulawesi Island, Indonesia. Vie Mil. Envir. 55: 7–14.

Bureau du Colombier, S., V. Bolliet, P. Lambert and A. Bardonnet. 2007. Energy and migratory behaviour in glass eels (*Anguilla anguilla*). Physiol. Behav. 92: 684–690.

Bureau du Colombier, S., P. Lambert and A. Bardonnet. 2008. Is feeding behaviour related to glass eel propensity to migrate ? Estuar. Coast. Shelf Sci. 30: 323–329.

Bureau du Colombier, S., V. Bolliet, P. Lambert and A. Bardonnet. 2011. Metabolic loss of mass in glass eels at different salinities according to their propensity to migrate. Estuar. Coast. Shelf Sci. 93: 1–6.

Cairns, D. 1941. Life-history of the two species of New Zealand freshwater eel. Part I. Taxonomy, age, growth, migration, and distribution. N.Z. J. Sci. 23: 53–72.

Cairns, D.K., J.C. Shiao, Y. Iizuka, W.-N. Tzeng and C.D. MacPherson. 2004. Movement patterns of American eels in an impounded watercourse, as indicated by otolith microchemistry. N. Amer. J. Fish. Manag. 24: 452–458.

Campana, S.E. and J.D. Neilson. 1985. Microstructure of fish otoliths. Can. J. Fish. Aquat. Sci. 42: 1014–1032.

Cantrelle, I. 1981. Etude de la migration et de la pêche des civelles (*A. anguilla* L. 1758) dans l'estuaire de la Gironde. Ph.D. Thesis, Université Paris VI.

Carpentier, A., L. Marion, J.M. Paillisson, A. Acou and E. Feunteun. 2009. Effects of commercial fishing and predation by cormorants on the *Anguilla anguilla* stock of a shallow eutrophic lake. J. Fish Biol. 74: 2132–2138.

Carss, D.N., D.A. Elston, K.C. Nelson and H. Kruuk. 1999. Spatial and temporal trends in unexploited yellow eel stocks in two shallow lakes and associated streams. J. Fish Biol. 55: 636–654.

Casselman, J.M. 2003. Dynamics of resources of the American eel, *Anguilla rostrata*: declining abundance in the 1990s. pp. 255–274. *In*: K. Aida, K. Tsukamoto and K. Yamauchi (eds.). Eel Biology. Springer, Tokyo.

Castonguay, M., P.V. Hodson, C. Moriarty, K.F. Drinkwater and B.M. Jessop. 1994a. Is there a role of ocean environment in American and European eel decline? Fish. Ocean. 3: 197–203.

Castonguay, M., P.V. Hodson, C.M. Couillard, M.J. Eckersley, J.-D. Dutil and G. Verreault. 1994b. Why is recruitment of the American eel, *Anguilla rostrata*, declining in the St. Lawrence River and Gulf? Can. J. Fish. Aquat. Sci. 51: 479–488.

Cheng, P.W. and W.-N. Tzeng. 1996. Timing of metamorphosis and estuarine arrival across the dispersal range of the Japanese eel (*Anguilla japonica*). Mar. Ecol. Prog. Ser. 131: 87–96.

Chisnall, B.L., D.J. Jellyman, M.L. Bonnett and J.R.E. Sykes. 2002. Spatial and temporal variability in length of glass eels (*Anguilla* spp.) in New Zealand. N.Z. J. Mar. Freshw. Res. 36: 89–104.

Ciccotti, E., E. Macchi, A. Rossi, E. Cataldiand and S. Cataudells. 1993. Glass eel (*Anguilla anguilla*) acclimatation to freshwater and seawater: morphological changes of the digestive tract. J. Appl. Ichth. 9: 74–81.

Ciccotti, E., T. Ricci, M. Scardi, E. Fresi and S. Cataudella. 1995. Intraseasonal characterization of glass eel migration in the River Tiber: space and time dynamics. J. Fish Biol. 47: 248–255.

Cieri, M.D. 1999. Migration, growth, and early life history of the American eel (*Anguilla rostrata*). Ph.D. Thesis, University of Maine.

Cieri, M.D. and J.D. McCleave. 2001. Validation of daily otolith increments in glass-phase American eels *Anguilla rostrata* (Lesueur) during estuarine residency. J. Exp. Mar. Biol. Ecol. 257: 219–227.

Courchamp, F., E. Angulo, P. Rivalan, R.J. Hall, L. Signoret, L. Bull and Y. Meinard. 2006. Rarity value and species extinction: the anthropogenic Allee effect. Plos Biology 4: 2405–2140.

Crean, S.R., J.T.A. Dick, D.W. Evans, R.S. Rosell and R.W. Elwood. 2007. Survival of juvenile European eels (*Anguilla anguilla*), transferred among salinities, and developmental shifts in their salinity preference. J. Zool. 266: 11–14.

Creutzberg, F. 1958. Use of tidal streams by migrating elvers (*Anguilla vulgaris* Turt.). Nature 181: 857–858.

Creutzberg, F. 1961. On the orientation of migrating elvers (*Anguilla vulgaris* Turt.) in a tidal area. Neth. J. Sea Res. 1: 257–338.

Crook, V. and M. Nakamura. 2013. Assessing supply chain and market impacts of a CITES listing on *Anguilla* species. TRAFFIC Bulletin 25: 24–30.

Daverat, F. and J. Tomas. 2006. Tactics and demographic attributes in the European eel *Anguilla anguilla* in the Gironde watershed, SW France. Mar. Ecol. Prog. Ser. 307: 247–257.

De Casamajor, M.N. 1998. Comportement migratoire de la civelle d'anguille (*Anguilla anguilla* L.) dans l'estuaire de l'Adour en fonction de la variabilité des conditions environnementales. Ph.D. thesis, Université de Pau et des pays de l'Adour.

De Casamajor, M.N., N. Bru and P. Prouzet. 1999. Influence de la luminosité nocturne sur le comportement migratoire de la civelle d'anguille (*Anguilla anguilla* L.) dans l'estuaire de l'Adour. Bull. Fr. Pêche Piscic. 355: 327–347.

De Casamajor, M.N., N. Bru and P. Prouzet. 2000. Fluctuation des captures de civelles (*Anguilla anguilla*, L.) et analyse de la variation de leur capturabilité dans l'estuaire de l'Adour. Bull. Fr. Pêche Piscic. 357/358: 387–404.

Deelder, C.L. 1952. On the migration of the elver (*Anguilla vulgaris* Turt.) at sea. J. Cons. Int. Explor. Mer 18: 187–218.

Deelder, C.L. 1958. On the behaviour of elvers (*Anguilla vulgaris* Turt.) migrating from the sea into fresh water. J. Cons. Int. Explor. Mer 24: 135–146.

Dekker, W. 1998. Long-term trend in the glass eels immigrating at Den Oever, The Netherland. Bull. Fr. Pêche Piscic. 349: 199–214.

Dekker, W. 2000a. A procrustean assessment of the European eel stock. ICES J. Mar. Sci. 57: 938–947.

Dekker, W. 2000b. The fractal geometry of the European eel stock. ICES J. Mar. Sci. 57: 109–121.

Dekker, W. 2003. Status of the European eel stocks and fisheries. pp. 237–254. *In*: K. Aida, K. Tsukamoto and K. Yamauchi (eds.). Eel Biology. Springer, Tokyo.

Dekker, W. 2004. Slipping through our hands. Population dynamics of the European eel. Ph.D. Thesis. University of Amsterdam.

Dekker, W. and L. Beaulaton. In press. Faire mieux que la nature ? The history of eel restocking in Europe. Environ. History.

Dekker, W., J.M. Casselman, D.K. Cairns, K. Tsukamoto, D.J. Jellyman and H. Lickers. 2003. Worldwide decline of eel resources necessitates immediate action. Quebec declaration of concern. Fisheries 28: 28–30.

Désaunay, Y. and D. Guérault. 1997. Seasonal and long-term changes in biometrics of eel larvae: a possible relationship between recruitment variation and North Atlantic ecosystem productivity. J. Fish Biol. 51(A): 317–339.

Désaunay, Y., D. Guérault and R. Lecomte-Finiger. 1996a. Variation of the oceanic larval migration of *Anguilla anguilla* (L.) glass eels from a two year study in the Vilaine estuary (France). Arch. Pol. Fish. 4: 195–210.

Désaunay, Y., R. Lecomte-Finiger and D. Guérault. 1996b. Mean age and migration patterns of *Anguilla anguilla* (L.) glass eels from three French estuaries. Arch. Pol. Fish. 4: 187–194.

Dou, S. and K. Tsukamoto. 2003. Observations on the nocturnal activity and feeding behavior of *Anguilla japonica* glass eels under laboratory conditions. Environ. Biol. Fishes 67: 389–395.

Dutil, J.D. 2009. *Anguilla rostrata* glass eel migration and recruitment in the estuary and Gulf of St. Lawrence. J. Fish Biol. 74: 1970–1984.

Edeline, E. 2007. Adaptive phenotypic plasticity of eel diadromy. Mar. Ecol. Prog. Ser. 341: 229–232.

Edeline, E., S. Dufour, C. Briand, D. Fatin and P. Elie. 2004. Thyroid status is related to migratory behavior in *Anguilla anguilla* glass eels. Mar. Ecol. Prog. Ser. 282: 261–270.

Edeline, E., A. Bardonnet, V. Bolliet, S. Dufour and P. Elie. 2005. Endocrine control of *Anguilla anguilla* glass eel dispersal: effect of thyroid hormones on locomotor activity and rheotactic behavior. Horm. Behav. 48: 53–63.

Edeline, E., P. Lambert, C. Rigaud and P. Elie. 2006. Effect of body condition and water temperature on *Anguilla anguilla* glass eel migratory behavior. J. Exp. Mar. Biol. Ecol. 331: 217–225.

Edeline, E., L. Beaulaton, R. Le Barh and P. Elie. 2007. Dispersal in metamorphosing juvenile eel *Anguilla anguilla*. Mar. Ecol. Prog. Ser. 344: 213–218.

Ege, V. 1939. A revision of the genus *Anguilla* Shaw. A systematic, phylogenetic and geographical study. Dana Rep. 3: 1–256.

Elie, P. 1979. Contribution à l'étude des montées de civelles d'*Anguilla anguilla* L. (poisson téléostéen anguilliforme) dans l'estuaire de la Loire : pêche, écologie, écophysiologie et élevage. Unpublished Ph.D. thesis, Université de Rennes.

Elie, P. and E. Rochard. 1994. Migration des civelles d'anguilles (*Anguilla anguilla* L.) dans les estuaires, modalités du phénomène et caractéristiques des individus. Bull. Fr. Pêche Piscic. 335: 81–98.

Elie, P., R. Lecomte-Finiger, I. Cantrelle and N. Charlon. 1982. Définition des limites des différents stades pigmentaires durant la phase civelle d'*Anguilla anguilla* L. Vie et Milieu 32: 149–157.

Feunteun, E. 2002. Management and restoration of European eel population (*Anguilla anguilla*): an impossible bargain. Ecol. Eng. 18: 575–591.

Feunteun, E. 2012. Le rêve de l'anguille. Une sentinelle en danger, Buchet/Chastel Ecologie, Paris.

Feunteun, E. and T. Robinet. 2014. Freshwater eels and people in France. pp. 75–89. *In*: K. Tsukamoto and M. Kuroki (eds.). Eels and Humans. Springer, Tokyo, Japan.

Friedland, K.D., M.J. Miller and B. Knights. 2007. Oceanic changes in the Sargasso Sea and declines in recruitment of the European eel. ICES J. Mar. Sci. 64: 519–530.

Fukuda, N., M. Kuroki, A. Shinoda, Y. Yamada, A. Okamura, J. Aoyama and K. Tsukamoto. 2009. Influence of water temperature and feeding regime on otolith growth in *Anguilla japonica* glass eels and elvers: does otolith growth cease at low temperatures? J. Fish Biol. 74: 1915–1933.

Fukuda, N., M.J. Miller, J. Aoyama, A. Shinoda and K. Tsukamoto. 2013. Evaluation of the pigmentation stages and body proportions from the glass eel to yellow eel in *Anguilla japonica*. Fisheries Sci. 79: 425–438.

Gagnaire, P.-A., E. Normandeau, C. Côté, M.M. Hansen and L. Bernatchez. 2012. The genetic consequences of spatially varying selection in the panmictic American eel (*Anguilla rostrata*). Genetics 190: 725–736.

Gandolfi, G., M. Pesaro and P. Tongiorgo. 1984. Environmental factors affecting the ascent of elvers, *Anguilla anguilla* (L.), into the Arno River. Oebalia 10: 17–35.

Gascuel, D. 1986. Flow carried and swimming migration of the glass eel (*Anguilla anguilla*) in the tidal area of a small estuary on the French Atlantic coast. Helgoläender Meeresun. 40: 321–326.

Geffroy, B., B. Sadoul and A. Bardonnet. 2014. Behavioural syndrome in juvenile eels and its ecological implications. Behaviour 152: 147–166.

Graynoth, E., R.I.C.C. Francis and D.J. Jellyman. 2008. Factors influencing juvenile eel (*Anguilla* spp.) survival in lowland New Zealand streams. N.Z. J. Mar. Freshw. Res. 42: 153–172.

Haenen, O., V. van Ginneken, M. Engelsma and G. Van den Thillart. 2009. Impact of eel viruses on recruitment of European eel. pp. 387–400. *In*: G. Van den Thillart, S. Dufour and J.C. Rankin (eds.). Spawning Migration of the European Eel. Springer, Netherlands.

Han, Y.-S. 2011. Temperature-dependent recruitment delay of the Japanese glass eel *Anguilla japonica* in East Asia. Mar. Biol. 158: 2349–2358.

Hanel, R., D. Stepputtis, S. Bonhommeau, M. Castonguay, M. Schaber, K. Wysujack, M. Vobach and M.J. Miller. 2014. Low larval abundance in the Sargasso Sea: new evidence about reduced recruitment of the Atlantic eels. Naturwissenschaften: DOI 10:1007/s00114-014-1243-6.

Haro, A.J. and W.H. Krueger. 1988. Pigmentation, size, and migration of elvers (*Anguilla rostrata* (Lesueur)) in a coastal Rhode Island stream. Can. J. Zool. 66: 2528–2533.

Haro, A. and W.H. Krueger. 1991. Pigmentation, otolith rings, and upstream migration of juvenile American eels (*Anguilla rostrata*) in a coastal Rhode Island stream. Can. J. Zool. 69: 812–814.

Harrison, A.J., A.M. Walker, A.C. Pinder, C. Briand and M.W. Aprahamian. 2014. A review of glass eel migratory behaviour, sampling techniques and abundance estimates in estuaries: implications for assessing recruitment, local production and exploitation. Rev. Fish Biol. Fish. 24: 967–983.

Heinsbroek, L.T.N. 1991. A review of eel culture in Japan and Europe. Aquaculture Fish. Manag. 22: 57–72.

Hickman, R.A. 1981. Densities and swimbladder development of juvenile American eels (*Anguilla rostrata*) (Lesueur), as related to energetics of migration. J. Fish Biol. 18: 507–517.

Hulet, W.H. 1978. Structure and functional development of the eel leptocephalus *Ariosoma balearicum* (de la roche, 1809). Philos. T. R. Soc. Lond. 282: 107–138.

Hulet, W.H. and R. Robins. 1989. The evolutionary significance of the leptocephalus larvae. pp. 669–679. *In*: E.A. Böhlke (ed.). Fishes of the Western North Atlantic, part nine, volume two. Sears Foundation for Marine Research, New Haven.

Hunt, W. 2007. The Victorian Elver Wars. Cheltenham, Reardon Publishing. 52p.

Hvidsten, N.A. 1983. Ascent of elvers (*Anguilla anguilla*) in the stream Imsa Norway. Rep. Inst. Freshw. Res. 62: 71–74.

Hwang, S.D., T.W. Lee, I.S. Choi and S.W. Hwang. 2014. Environmental factors affecting the daily catch Levels of *Anguilla japonica* glass eels in the Geum River estuary, South Korea. J. Coast. Res. 30: 954–960.

ICES. 2001. Report of the ICES/EIFAC Working Group on Eels. St. Andrews, Canada. ICES Document CM 2001/ACFM: 03.

ICES. 2010. The report of the 2010 Session of the Joint ICES/EIFAC Working Group on Eels, Hamburg, Germany. ICES Document CM 2009/ACOM: 18.

ICES. 2014. The report of the Joint EIFAAC/ICES/GFCM Working Group on Eel. Rome, Italy. ICES Document CM 2014/ACOM: 18.

Iglesias, T., J. Lobon Cervia, S. Costa Dias and C. Antunes. 2010. Variation of life traits of glass eels of *Anguilla anguilla* (L.) during the colonization of Rios Naln and Minho estuaries (northwestern Iberian Peninsula). Hydrobiol. 651: 213–223.

Imbert, H. 2008. Relationships between locomotor behavior, morphometric characters and thyroid hormone levels give evidence of stage-dependent mechanisms in European eel upstream migration. Horm. Behav. 53: 69–81.

Ingram, B.A., G.J. Gooley, S.S. DeSilva, B.J. Larkin and R.A. Collins. 2001. Preliminary observations on the tank and pond culture of the glass eels of the Australian shortfin eel, *Anguilla australis* Richardson. Aquaculture Res. 32: 833–848.

Ishikawa, S., K. Tsukamoto and M. Nishida. 2004. Genetic evidence for multiple geographic populations of the giant mottled eel *Anguilla marmorata* in the Pacific and Indian oceans. Ichthyol. Res. 51: 343–353.

Jegstrup, I.M. and P. Rosenkilde. 2003. Regulation of post-larval development in the European eel: thyroid hormone level, progress of pigmentation and changes in behaviour. J. Fish Biol. 63: 168–175.

Jellyman, D.J. 1977. Invasion of a New Zealand freshwater stream by glass-eels of two *Anguilla* spp. N.Z. J. Mar. Freshw. Res. 11: 193–209.

Jellyman, D.J. 1979. Upstream migration of glass-eels (*Anguilla* spp.) in the Waikato River. N.Z. J. Mar. Freshw. Res. 13: 13–22.

Jellyman, D.J. 2003. The distribution and biology of the South Pacific species of *Anguilla*. pp. 275–292. *In*: K. Aida, K. Tsukamoto and K. Yamauchi (eds.). Eel Biology. Springer, Tokyo.

Jellyman, D.J. and P.W. Lambert. 2003. Factors affecting recruitment of glass eels into the Grey River, New Zealand. J. Fish Biol. 63: 1067–1079.

Jellyman, D.J. and M.M. Bowen. 2009. Modeling larval migration routes and spawning areas of anguillid eels of Australia and New Zealand. Am. Fish. Soc. Symp. 69: 255–274.

Jellyman, D.J., B.L. Chisnall, M.L. Bonnett, M.L. and J.R.E. Sykes. 1999. Seasonal arrival patterns of juvenile freshwater eels (*Anguilla* spp.) in New Zealand. N.Z. J. Mar. Freshw. Res. 33: 249–262.

Jellyman, D.J., D.J. Booker and E. Watene. 2009. Recruitment of *Anguilla* spp. glass eels in the Waikato River, New Zealand. Evidence of declining migrations? J. Fish Biol. 74: 2014–2033.

Jessop, B.M. 1998. Geographic and seasonal variation in biological characteristics of American eel elvers in the Bay of Fundy area and on the Atlantic coast of Nova Scotia. Can. J. Zool. 76: 2172–2185.

Jessop, B.M. 2000. Estimates of population size and instream mortality rate of American eel elvers in a Nova Scotia river. Trans. Am. Fish. Soc. 129: 514–526.

Jessop, B.M. 2003. Annual variability in the effects of water temperature, discharge, and tidal stage on the migration of American eel elvers from estuary to river. Am. Fish. Soc. Symp. 33: 3–16.

Jessop, B.M., J.C. Shiao, Y. Iizukaand and W.-N. Tzeng. 2002. Migratory behaviour and habitat use by American eels *Anguilla rostrata* as revealed by otolith microchemistry. Mar. Ecol. Prog. Ser. 233: 217–229.

Kawakami, Y., N. Mochioka and A. Nakazono. 1999a. Immigration patterns of glass eels *Anguilla japonica* entering river in Northern Kyushu, Japan. Bull. Mar. Sci. 64: 315–327.

Kawakami, Y., N. Mochioka, R. Kimura and A. Nakazono. 1999b. Seasonal changes of the RNA/DNA ratio, size and lipid content and migration adaptability of Japanese glass-eels, *Anguilla japonica*, collected in northern Kyushu, Japan. J. Exp. Mar. Biol. Ecol. 238: 1–19.

Kettle, A.J., D.C.E. Bakker and K. Haines. 2008. Impact of the North Atlantic Oscillation on the trans-Atlantic migrations of the European eel (*Anguilla anguilla*). J. Geophys. Res. 113: G03004.

Kettle, A.J., L.A. Vollestad and J. Wibig. 2011. Where once the eel and the elephant were together: decline of the European eel because of changing hydrology in southwest Europe and northwest Africa? Fish Fisheries 12: 380–411.

Kimura, S. and K. Tsukamoto. 2006. The salinity front in the North Equatorial Current: A landmark for the spawning migration of the Japanese eel (*Anguilla japonica*) related to the stock recruitment. Deep-Sea Res. Pt. II. 53: 315–325.

Kleckner, R.C. and J.D. McCleave. 1985. Spatial and temporal distribution of American eel larvae in relation to North Atlantic Ocean current systems. Dana 4: 67–92.

Knights, B. 2003. A review of the possible impacts of long-term oceanic and climate changes and fishing mortality on recruitment of anguillid eels of the Northern Hemisphere. Sci. Total Environ. 310: 237–244.

Kotake, A., T. Arai, M. Ohji, S. Yamane, N. Miyazaki and K. Tsukamoto. 2004. Application of otolith microchemistry to estimate the migratory history of Japanese eel *Anguilla japonica* on the Sanriku Coast of Japan. J. Appl. Ichth. 20: 150–153.

Kuroki, M., M.J. Miller and K. Tsukamoto. 2014a. Diversity of early life-history traits in freshwater eels and the evolution of their oceanic migrations. Can. J. Zool. 92: 749–770.

Kuroki, M., M.J.P. van Oijen and K. Tsukamoto. 2014b. Eels and the Japanese: an inseparable, longstanding relationship. pp. 91–108. *In*: K. Tsukamoto and M. Kuroki (eds.). Eels and Humans. Springer, Tokyo, Japan.

Laffaille, P., E. Feunteun, A. Acou and J.C. Lefeuve. 2000. Role of European eel (*Anguilla anguilla* L.) in the transfer of organic matter between marine and freshwater systems. Verh. Internat. Verein. Limnol. 27: 616–619.

Laflamme, S., C. Côté, P.-A. Gagnaire, M. Castonguay and L. Bernatchez. 2012. RNA/DNA ratios in American glass eels (*Anguilla rostrata*): evidence for latitudinal variation in physiological status and constraints to oceanic migration? Ecol. Evol. 2: 875–884.

Lambert, P. 2005. Exploration multiscalaire des paradigmes de la dynamique de la population d'anguilles européennes à l'aide d'outils de simulation. Ph.D. Thesis. Université Bordeaux.

Lambert, P. and E. Rochard. 2007. Identification of the inland population dynamics of the European eel using pattern oriented modelling. Ecol. Model. 206: 166–178.

Lambert, P., M. Sbaihi, E. Rochard, J. Marchelidon, S. Dufour and P. Elie. 2003. Variabilités morphologique et du taux d'hormone de croissance des civelles d'anguilles européennes dans l'estuaire de la Gironde au cours de la saison 1997–1998. Bull. Fr. Pêche Piscic. 368: 69–84.

Lambert, P., B. Durozoi and L. Beaulaton. 2007. Validation des hypothèses comportementales utilisées dans la quantification du flux d'anguille (*Anguilla anguilla*) dans l'Isle. Technical report, CEMAGREF Bordeaux.

Langdon, S.A. and A.L. Collins. 2000. Quantification of the maximal swimming performance of Australasian glass eels, *Anguilla australis* and *Anguilla reinhardtii*, using a hydraulic flume swimming chamber. N.Z. J. Mar. Freshw. Res. 34: 629–636.

Lecomte-Finiger, R. and C. Razouls. 1981. Influence des facteurs hydrologiques et météorologiques sur la migration anadrome des civelles dans le golfe du Lion. Cah. Lab. Montereau 12: 13–16.

Lecomte-Finiger, R., C. Maunier and M. Khafif. 2004. Les larves leptocephales, ces méconnues. Cybium 28: 83–95.

Legault, A. 1988. Le franchissement des barrages par l'escalade de l'anguille. Etude en Sèvre Niortaise. Bull. Fr. Pêche Piscic. 308: 1–10.

Linton, E.D., B. Jonsson and D. Noakes. 2007. Effect of water temperature on the swimming and climbing behaviour of glass eels, *Anguilla* spp. Environ. Biol. Fishes 78: 189–192.

Lobón-Cerviá, J. and T. Iglesias. 2008. Long-term numerical changes and regulation in a river stock of European eel *Anguilla anguilla*. Freshw. Biol. 53: 1832–1844.

Lobón-Cerviá, J., C.G. Utrilla and P.A. Rincón. 1995. Variations in the population dynamics of the European eel *Anguilla anguilla* (L.) along the course of a Cantabrian river. Ecol. Freshw. Fish 4: 17–27.

Lowe, H. 1961. Factors influencing the run of elvers in the River Bann, Northern Ireland. J. Cons. Int. Pour Explor. Mer XVII: 299–315.

Martin, M.H. 1995. The effects of temperature, river flow, and tidal cycles on the onset of glass eel and elver migration into fresh water in the American eel. J. Fish Biol. 46: 891–902.

Martin, J., F. Daverat, C. Pecheyran, T.D. Als, E. Feunteun and E. Reveillac. 2010. An otolith microchemistry study of possible relationships between the origins of leptocephali of European eels in the Sargasso Sea and the continental destinations and relative migration success of glass eels. Ecol. Freshw. Fish 19: 627–637.

Marquet, G. and P. Lamarque. 1986. Acquisitions recente sur la biologie des anguillies de Tahiti et de Moorea (Polynesie Francaise): *A. marmorata, A. megastoma, A. obscura*. Vie Milieu 36: 311–315.

Marui, M., T. Arai, M.J. Miller, D.J. Jellyman and K. Tsukamoto. 2001. Comparison of early life history between New Zealand temperate eels and Pacific tropical eels revealed by otolith microstructure and microchemistry. Mar. Ecol. Prog. Ser. 213: 273–284.

McCarthy, T.K. 2014. Eels and people in Ireland: from mythology to international eel stock conservation. pp. 13–40. *In*: K. Tsukamoto and M. Kuroki (eds.). Eels and Humans. Springer, Tokyo, Japan.

McCleave, J.D. 2008. Contrasts between spawning times of *Anguilla* species estimated from larval sampling at sea and from otolith analysis of recruiting glass eels. Mar. Biol. 155: 249–262.

McCleave, J.D. and R.C. Kleckner. 1982. Selective tidal stream transport in the estuarine migration of glass eels of the American eel (*Anguilla rostrata*). J. Cons. Cons. Int. Pour Explor. Mer 40: 262–271.

McCleave, J.D. and G. Wippelhauser. 1987. Behavioral aspects of selective tidal stream transport in juvenile American eel. Am. Fish. Soc. Symp. 1: 138–150.

McCleave, J.D. and D.J. Jellyman. 2002. Discrimination of New Zealand stream waters by glass eels of *Anguilla australis* and *Anguilla dieffenbachii*. J. Fish Biol. 61: 785–800.

McKinnon, L.J. and G.J. Gooley. 1998. Key environmental criteria associated with the invasion of *Anguilla australis* glass eels into estuaries of southern Australia. Bull. Fr. Pêche Piscic. 349: 117–128.

Michaud, M., J.D. Dutil and J.J. Dodson. 1988. Determination of the age of young American eels, *Anguilla rostrata*, in fresh water, based on otolith surface area and microstructure. J. Fish Biol. 32: 179–189.

Miles, S.G. 1968. Rheotaxis of elvers of the American eel (*Anguilla rostrata*) in the laboratory to water from different streams in Nova Scotia. J. Fish. Res. Board Can. 25: 1591–1602.

Miller, M.J. 2009. Ecology of anguilliform leptocephali: remarkable transparent fish larvae of the ocean surface layer. Aqua-BioSci. Monogr. 2: 1–94.

Miller, M.J., N. Mochioka, T. Otake and K. Tsukamoto. 2002. Evidence of a spawning area of *Anguilla marmorata* in the western North Pacific. Mar. Biol. 140: 809–814.

Miller, M.J., J. Aoyama and K. Tsukamoto. 2009. New perspectives on the early life history of tropical anguillid eels: implications for resource management. Am. Fish. Soc. Symp. 58: 71–84.

Miller, M.J., S. Bonhommeau, P. Munk, M. Castonguay, R. Hanel and J.D. McCleave. 2014. A century of research on the larval distributions of the Atlantic eels: a re-examination of the data. Biol. Rev.: 000-000. DOI: 10.1111/brv.12144.

Miller, M.J., J. Dubosc, E. Vourey, K. Tsukamoto and V. Allain. 2015. Low occurrence rates of ubiquitously present leptocephalus larvae in the stomach contents of predatory fish. ICES J. Mar. Sci. 72(3).

Minegishi, Y., J. Aoyama and K. Tsukamoto. 2008. Multiple population structure of the giant mottled eel *Anguilla marmorata*. Molec. Ecol. 17: 3109–3122.

Mochioka, N. 2003. Leptocephali. pp. 51–60. *In*: K. Aida, K. Tsukamoto and K. Yamauchi (eds.). Eel Biology. Springer, Tokyo.

Monein-Langle, D. 1985. Morphologie et physiologie digestive de la civelle d'*Anguilla anguilla* (Linnaeus, 1758) en phase préalimentaire dans les conditions naturelles et en régime thermique particulier. Ph.D. Thesis. Université de Perpignan.

Monticini, P. 2014. Eel (*Anguilla* spp.): production and trade according to Washington Convention Legislation. FAO GLOBEFISH Research Programme 114: 77p.

Moriarty, C. 1986. Riverine migration of young eels *Anguilla anguilla* (L.). Fish. Res. 4: 43–58.

Moriarty, C. and W. Dekker. 1997. Management of the European eel. Fish. Bull. 15: 1–110.

Mouton, A.M., S. Huysecom, D. Buysse, M. Stevens, T. Van den Neucker and J. Coeck. 2014. Optimisation of adjusted barrier management to improve glass eel migration at an estuarine barrier. J. Coast. Cons. 18: 111–120.

Nimeth, K., P. Zwerger, J. Würtz, W. Salvenmoser and B. Pelster. 2000. Infection of the glass-eel swimbladder with the nematode *Anguillicola crassus*. Parasitology 121: 75–83.

Nishi, T. and G. Kawamura. 2005. *Anguilla japonica* is already magnetosensitive at the glass eel phase. J. Fish Biol. 67: 1213–1224.

Okamura, A., N. Horie, N. Mikawa, Y. Yamada and K. Tsukamoto. 2014. Recent advances in artificial production of glass eels for conservation of anguillid eel populations. Ecol. Freshw. Fish 23: 95–110.

Otake, T. 2003. Metamorphosis. pp. 61–74. *In*: K. Aida, K. Tsukamoto and K. Yamauchi (eds.). Eel Biology. Springer, Tokyo.

Overton, A.S. and R.A. Rulifson. 2008. Annual variability in upstream migration of glass eels in a southern USA coastal watershed. Environ. Biol. Fishes 84: 29–37.

Pacariz, S., H. Westerberg and G. Björk. 2014. Climate change and passive transport of European eel larvae. Ecol. Freshw. Fish 23: 86–94.

Pannella, G. 1971. Fish otoliths: daily growth layers and periodical patterns. Science 173: 1124–1127.

Pease, B.C., V. Silberschneider and T. Walford. 2003. Upstream migration by glass eels of two *Anguilla* species in the Hacking River, New South Wales, Australia. Am. Fish. Soc. Symp. 33: 47–61.

Powles, P.M. and S.M. Warlen. 2002. Recruitment season, size, and age of young American eels (*Anguilla rostrata*) entering an estuary near Beaufort, North Carolina. Fishery Bull. 100: 299–306.

Reveillac, E., E. Feunteun, P. Berrebi, P.-A. Gagnaire, R. Lecomte-Finiger, P. Bosc and T. Robinet. 2008. *Anguilla marmorata* larval migration plasticity as reveled by otolith microstructural analysis. Can. J. Fish. Aquat. Sci. 65: 2127–2137.

Richkus, W.A. and K. Whalen. 2000. Evidence for a decline in the abundance of the American eel, *Anguilla rostrata* (LeSueur), in North America since the early 1980s. Dana 83–97.

Righton, D. and M. Roberts. 2014. Eels and people in the United Kingdom. pp. 1–12. *In*: K. Tsukamoto and M. Kuroki (eds.). Eels and Humans. Springer, Tokyo, Japan.

Ringuet, S., F. Muto and C. Raymakers. 2002. Eels: their harvest and trade in Europe and Asia. TRAFFIC Bulletin 19: 2–27.

Robinet, T., R. Lecomte-Finiger, K. Escoubeyrou and E. Feunteun. 2003. Tropical eels *Anguilla* spp. recruiting to Reunion Island in the Indian Ocean: taxonomy, patterns of recruitment and early life histories. Mar. Ecol. Prog. Ser. 259: 263–272.

Schabetsberger, R., F. Økland, K. Aarestrup, D. Kalfatak, U. Sichrowsky, M. Tambets, G. Dall'Olmo, R. Kaiser and P.I. Miller. 2013. Oceanic migration behaviour of tropical Pacific eels from Vanuatu. Mar. Ecol. Prog. Ser. 475: 177–190.

Schabetsberger, R., F. Økland, D. Kalfatak, U. Sichrowsky, M. Tambets, K. Aarestrup, C. Gubili, J. Sarginson, B. Boufana, R. Jehle, G. Dall'Olmo, M.J. Miller, A. Scheck, R. Kaiser and G. Quartly. 2015. Genetic and migratory evidence for sympatric spawning of tropical Pacific eels from Vanuatu. Mar. Ecol. Prog. Ser. 521: 171–187.

Schmidt, J. 1925. On the distribution of the fresh-water eels (*Anguilla*) throughout the world. - II. Indo-Pacific region. Kong. Danske Vidensk. 8: 329–382.

Shiao, J.C., W.-N. Tzeng, A. Collins and D.J. Jellyman. 2001. Dispersal pattern of glass eel stage of *Anguilla australis* revealed by otolith growth increments. Mar. Ecol. Progr. Ser. 219: 241–250.

Shiao, J.C., W.-N. Tzeng, A. Collins and Y. Iizuka. 2002. Role of marine larval duration and growth rate of glass eels in determining the distribution of *Anguilla reinhardtii* and *A. australis* on Australian eastern coasts. Mar. Freshw. Res. 53: 687–695.

Shiao, J.C., Y. Iizuka, C.W. Chang and W.-N. Tzeng. 2003. Disparities in habitat use and migratory behavior between tropical eel *Anguilla marmorata* and temperate eel *A. japonica* in four Taiwanese rivers. Mar. Ecol. Progr. Ser. 261: 233–242.

Shiao, J.C., L. Lozys, Y. Iizuka and W.-N. Tzeng. 2006. Migratory patterns and contribution of stocking to the population of European eel in Lithuanian waters as indicated by otolith Sr: Ca ratios. J. Fish Biol. 69: 749–769.

Shinoda, A., J. Aoyama, M.J. Miller, T. Otake, N. Mochioka, S. Watanabe, Y. Minegishi, M. Kuroki, T. Yoshinaga, K. Yokouchi, N. Fukuda, R. Sudo, S. Hagihara, K. Zenimoto, Y. Suzuki, M. Oya, T. Inagaki, S. Kimura, A. Fukui, T.W. Lee and K. Tsukamoto. 2011. Evaluation of the larval distribution and migration of the Japanese eel in the western North Pacific. Rev. Fish Biol. Fisheries 21: 591–611.

Silberschneider, V., B.C. Pease and D.J. Booth. 2004. Estuarine habitat preferences of *Anguilla australis* and *A. reinhardtii* glass eels as inferred from laboratory experiments. Environ. Biol. Fish. 71: 395–402.

Sinha, V.R.P. and J.W. Jones. 1975. The European Freshwater Eel. Liverpool University, Liverpool.

Sloane, R.D. 1984. Invasion and upstream migration by glass-eels of *Anguilla australis australis* Richardson and *A. reinhardtii* Steindachner in Tasmanian freshwater streams. Aust. J. Mar. Freshw. Res. 35: 47–50.

Sneed, A. 2014. Glass Eel Gold Rush Casts Maine Fishermen against Scientists. *Scientific American* August 2014. http://www.scientificamerican.com/article/glass-eel-gold-rush-casts-maine-fishermen-against-scientists/.

Sola, C. 1995. Chemoattraction of upstream migrating glass eels *Anguilla anguilla* to earthy and green odorants. Environ. Biol. Fish. 43: 179–185.

Sola, C. and P. Tongiorgi. 1996. The effect of salinity on the chemotaxis of glass eels, *Anguilla anguilla*, to organic earthy and green odorants. Environ. Biol. Fishes 47: 213–218.

Sorensen, P.W. 1986. Origins of the freshwater attractant(s) of migrating elvers of the American eel *Anguilla rostrata*. Environ. Biol. Fishes 17: 185–200.

Sorensen, P.W. and M.L. Bianchini. 1986. Environmental correlates of the freshwater migration of elvers of the American eel in a Rhode Island brook. Trans. Am. Fish. Soc. 115: 258–268.

Strubberg, A. 1913. The metamorphosis of elvers as influenced by outward conditions. Med. Komm. Havunders. Ser. Fisheries 4: 1–11.

Sugeha, H.Y., T. Arai, M.J. Miller, D. Limbong and K. Tsukamoto. 2001. Inshore migration of the tropical eels *Anguilla* spp. recruiting to the Poigar River estuary on north Sulawesi island. Mar. Ecol. Prog. Ser. 221: 233–243.

Sullivan, M.C., K.W. Able, J.A. Hare and H.J. Walsh. 2006. *Anguilla rostrata* glass eel ingress into two, US east coast estuaries: patterns, processes and implications for adult abundance. J. Fish Biol. 69: 1081–1101.

Sullivan, M.C., M.J. Wuenschel and K.W. Able. 2009. Inter and intra-estuary variability in ingress, condition and settlement of the American eel *Anguilla rostrata*: implications for estimating and understanding recruitment. J. Fish Biol. 74: 1949–1969.

Svärdson, G. 1976. The decline of the Baltic eel population. Rep. Inst. Freshw. Res., Drottingholm 55: 136–143.

Szekely, C., A. Palstra, K. Molnar and G. Van den Thillart. 2009. Impact of the swim-bladder parasite on the health and performance of European eels. pp. 201–226. *In*: G. Van den Thillart, S. Dufour and J.C. Rankin (eds.). Spawning Migration of the European Eel. Springer, Netherlands.

Tabeta, O. and N. Mochioka. 2003. The glass eel. pp. 75–87. *In*: K. Aida, K. Tsukamoto and K. Yamauchi (eds.). Eel Biology. Springer, Tokyo.

Tabeta, O., T. Tanimoto, T. Takai, I. Matsui and T. Imamura. 1976. Seasonal occurrence of Anguillid elvers in Cagayan River, Luzon Island, the Philippines. Bull. Jap. Soc. Sci. Fisheries 42: 421–426.

Tanaka, H., H. Kagawa, H. Ohta, T. Unuma and K. Nomura. 2003. The first production of glass eel in captivity: fish reproductive physiology facilitates great progress in aquaculture. Fish Physiol. Biochem. 28: 493–497.

Tesch, F.-W. 1980. Occurrence of eel *Anguilla anguilla* larvae west of the European continental shelf, 1971–1977. Environ. Biol. Fishes 5: 185–190.

Tesch, F.-W. 2003. The Eel. Blackwell Science Ltd., Oxford. 408p.

Tesch, F.-W., U. Niermann and A. Plaga. 1986. Differences in development stage and stock density of larval *Anguilla anguilla* off the west coast of Europe. Vie Milieu 36: 255–260.

Thibault, I., J.J. Dodson, F. Caron, W.-N. Tzeng, Y. Iizuka and J.C. Shiao. 2007. Facultative catadromy in American eels: testing the conditional strategy hypothesis. Mar. Ecol. Prog. Ser. 344: 219–229.

Tongiorgi, P., L. Tosi and M. Balsamo. 1986. Thermal preferences in upstream migrating glass-eels of *Anguilla anguilla* (L.). J. Fish Biol. 28: 501–510.

Trancart, T., P. Lambert, E. Rochard, F. Daverat, J. Coustillas and C. Roqueplo. 2012. Alternative flood tide transport tactics in catadromous species: *Anguilla anguilla, Liza ramada* and *Platichthys flesus*. Estuar. Coast. Shelf Sci. 99: 191–198.

Tsukamoto, K., J. Aoyama and M.J. Miller. 2002. Migration, speciation, and the evolution of diadromy in anguillid eels. Can. J. Fish. Aquat. Sci. 59: 1989–1998.

Tsukamoto, K., J. Aoyama and M.J. Miller. 2009a. Present status of the Japanese eel: resources and recent research. Am. Fish. Soc. Symp. 58 : 21–35.

Tsukamoto, K., Y. Yamada, A. Okamura, T. Kaneko, H. Tanaka, M.J. Miller, N. Horie, N. Mikawa, T. Utoh and S. Tanaka. 2009b. Positive buoyancy in eel leptocephali: an adaptation for life in the ocean surface layer. Mar. Biol. 156: 835–846.

Tucker, D.W. 1959. A new solution to the Atlantic eel problem. Nature 183: 495–501.

Tzeng, W.-N. 1984. Dispersal and upstream migration of marked anguillid eel, *Anguilla japonica*, elvers in the estuary of the Shuang River, Taiwan. Bull. Jap. Soc. Fish. Ocean. 45: 10–19.

Tzeng, W.-N., J.C. Shiao, Y. Yamada and H.P. Oka. 2003. Life history patterns of Japanese eel *Anguilla japonica* in Mikawa Bay, Japan. Amer. Fish. Soc. Symp. 33: 285–293.

Tzeng, W.-N., Y.-H. Tseng, Y.-S. Han, C.-C. Hsu, C.-W. Chang, C.C. Hsu, E. Di Lorenzo and C.H. Hsieh. 2012. Evaluation of multi-scale climate effects on annual recruitment levels of the Japanese eel, *Anguilla japonica*, to Taiwan. PLoS ONE 7(2): e30805. doi:10.1371/journal. pone.0030805.

Ulrik, M.G., J.M. Pujolar, A.-L. Ferchaud, M.W. Jacobsen, T.D. Als, P.A. Gagnaire, J. Frydenberg, P.K. Bøcher, B. Jónsson, L. Bernatchez and M.M. Hansen. 2014. Do North Atlantic eels show parallel patterns of spatially varying selection? BMC Evol. Biol. 14(1): 138.

Umezawa, A. and K. Tsukamoto. 1991. Factors influencing otolith increment formation in Japanese eel (*Anguilla japonica*), T. and S., elvers. J. Fish Biol. 39: 211–223.

Wang, C.H. and W.-N. Tzeng. 2000. The timing of metamorphosis and growth rates of American and European eel leptocephali: A mechanism of larval segregative migration. Fish. Res. 46: 191–205.

Watanabe, S., M.J. Miller, J. Aoyama and K. Tsukamoto. 2014. Evaluation of the population structure of *Anguilla bicolor* and *A. bengalensis* using total number of vertebrae and consideration of the subspecies concept for the genus *Anguilla*. Ecol. Freshw. Fish 23: 77–85.

Westerberg, H. 1998. Short note—oceanographic aspects of the recruitment of eels to the Baltic Sea. Bull. Fr. Pêche Piscic. 349: 177–185.

Williamson, G.R. 1987. Vertical drifting position of glass eel, *Anguilla rostrata*, off Newfoundland. J. Fish Biol. 31: 587–588.

Wilson, J.M., J.C. Antunes, P.D. Bouca and J. Coimbra. 2004. Osmoregulatory plasticity of the glass eel of *Anguilla anguilla*: freshwater entry and changes in branchial ion-transport protein expression. Can. J. Fish. Aquat. Sci. 61: 432–442.

Wilson, J.M., P. Reis-Santos, A.V. Fonseca, J.C. Antunes, P.D. Bouça and J. Coimbra. 2007. Seasonal changes in ionoregulatory variables of the glass eel *Anguilla anguilla* following estuarine entry: comparison with resident elvers. J. Fish Biol. 70: 1239–1253.

Wippelhauser, G. and J.D. McCleave. 1987. Precision of behavior of migrating juvenile American eels (*Anguilla rostrata*) utilizing selective tidal stream transport. J. Cons. Perm. Int. Pour Explor. Mer 44: 80–89.

Wirth, T. and L. Bernatchez. 2003. Decline of North Atlantic eels: a fatal synergy? Proc. Roy. Soc. B. (London) 270: 681–688.

Wuenschel, M.J. and K.W. Able. 2008. Swimming ability of eels (*Anguilla rostrata, Conger oceanicus*) at estuarine ingress: contrasting patterns of cross-shelf transport? Mar. Biol. 154: 775–786.

Yoshinaga, T., J. Aoyama, A. Shinoda, S. Watanabe, R.V. Azanza and K. Tsukamoto. 2014. Occurrence and biological characteristics of glass eels of the Japanese eel *Anguilla japonica* at the Cagayan River of Luzon Island, Philippines in 2009. Zool. Stud. 53: 65–70.

Zenimoto, K., T. Kitagawa, S. Miyazaki, Y. Sasai, H. Sasaki and S. Kimura. 2009. The effects of seasonal and interannual variability of oceanic structure in the western Pacific North Equatorial Current on larval transport of the Japanese eel *Anguilla japonica*. J. Fish Biol. 74: 1878–1890.

Zenimoto, K., Y. Sasai, H. Sasaki and S. Kimura. 2011. Estimation of larval duration in *Anguilla* spp., based on cohort analysis, otolith microstructure, and Lagrangian simulations. Mar. Ecol. Prog. Ser. 438: 219–228.

Zompola, S., G. Katselis, C. Koutsikopoulos and Y. Cladas. 2008. Temporal patterns of glass eel migration (*Anguilla anguilla* L., 1758) in relation to environmental factors in the Western Greek inland waters. Estuar. Coast. Shelf Sci. 80: 330–338.

Juvenile Eels: Upstream Migration and Habitat Use

Donald J. Jellyman[1],* and *Takaomi Arai*[2]

Introduction

The life-history of freshwater eels is characterised by extensive marine migrations, initially as larvae and finally as maturing adults. However, many eels also undergo significant migrations within freshwater when as juvenile fish, they gradually move upstream to find and colonise new habitats. This upstream migration of small pigmented eels (elvers) is best known for temperate species where it takes place during summer. At natural and man-made obstacles like waterfalls and dams respectively, large numbers of these juvenile eels often accumulate during their endeavours to surmount such obstacles and penetrate further upstream. These accumulations provide important monitoring sites for determining an index of juvenile eel recruitment; they also provide harvesting sites to seed upstream habitats or other catchments.

Inshore coastal zones and estuaries, the arrival areas for glass eels, are more productive than riverine habitats in temperate regions, and hence there must be advantages which cause small eels to vacate such areas and migrate

[1] National Institute of Water and Atmosphere, PO Box 8602, Christchurch 8053, New Zealand.
[2] Environmental and Life Sciences Programme, Faculty of Science, Universiti Brunei Darussalam, Jalan Tungku Link, Gadong, BE 1410, Brunei Darussalam.
  E-mail: takaomi.arai@ubd.edu.bn
* Corresponding author: don.jellyman@niwa.co.nz

extensively in fresh water. Indeed, the summer upstream migration of juvenile eels serves several purposes:

- It enables progressive recolonisation of habitats left vacant by other juvenile eels that have migrated further upstream, and also of habitats previously occupied by silver eels that have migrated to sea.
- It allows small eels to locate their preferred habitat type and to utilise different seasonal habitats (e.g., summer feeding, winter shelter).
- It distributes eels over an extensive longitudinal gradient.
- It provides for development of female eels as eels at high density in lower river sites tend to develop as males whereas eels at low density in upstream sites tend to develop as females.

A further advantage of migration *en masse* could be reduced predation, as predator satiation could occur rapidly when a seasonal migration is underway. Shoaling in fishes is also one of the mechanisms for reducing predation (e.g., Krause et al. 2002) as predators' sensory systems become overwhelmed by the sheer numbers and behaviour of the would-be prey. Of course, the same migratory behaviour could also expose elvers to increased predation when accumulations of elvers occur at natural (waterfalls) or unnatural (dams and weirs) obstacles—such places provide enhanced feeding opportunities for predators, and a variety of fish, birds and rodents take advantage of such seasonal aggregations to predate on elvers (e.g., Jellyman 1977).

Species and seasons of upstream migration

Species

Freshwater migrations have been recorded for a range of temperate species, e.g., *A. anguilla*: Tesch (2003), White and Knights (1997a), Naismith and Knights (1988), Moriarty (1986), Hvidsten (1985), *A. rostrata*: Haro and Krueger (1991), *A. australis*: Jellyman (1977), Beentjes et al. (1997), Sloane (1984). *A. dieffenbachii*: Jellyman (1977), Beentjes et al. (1997); there are no records of this migration occurring in *A. japonica* although this might be largely a consequence of very extensive harvest at the glass eel stage. Reports for tropical species are less common and are often confined to generalised references like Schmidt (1925) who reported an extensive migration of elvers in New Guinea, and noted that this was "the first recorded observation in the tropics of an ascent of eel young, such as is well known from temperate countries". There are records of *A. mossambica* elver migrations from South Africa (Skead 1959; Jubb 1970; Jubb 1961; Wasserman et al. 2012), and Bruton et al. (1987) noted that "young eels employ a series of migrations to move into the upstream reaches of rivers"—presumably this behaviour occurred in all the four species discussed by these authors (*A. bicolor bicolor, A. marmorata, A. bengalensis labiata,* and *A. mossambica*), although it may be more pronounced in the latter two species as these penetrate further inland, and elvers of these species have been reported as migrating hundreds of kilometers upstream (Tamatamah 2012). Elvers of *A. bicolor* have been recorded congregating below a dam in India during October–January, following heavy rains (Nair 1973).

Where two species coexist, elvers of both species migrate together. Thus mixed shoals of *A. australis* and *A. dieffenbachii* occur in the lower Waikato River, New Zealand. Although the composition of glass eel samples in this river is dominated by *A. australis* (90%, Jellyman 1979), historically the proportion of this species in elver samples recorded at Karapiro Dam 130 km upstream have reduced to ~ 60% (Jellyman 1977) because many elvers of this species take up residence in the lower river and lakes as they migrate upstream. The proportion of *A. dieffenbachii* at Karapiro Dam has averaged 10% over recent seasons (2008/09–2011/12; Martin et al. 2013), which may reflect a reduction in the recruitment of this species—similar to the reduced representation of this species in glass eel catches in this river (Jellyman et al. 2009). Although *A. reinhardtii* occurs in Tasmania, this species was virtually absent from elvers caught at the Trevallyn Dam, probably because it has limited upstream distribution in Tasmania (Sloane 1984), although it penetrates well inland in mainland Australia (Pusey et al. 2004).

Seasons

These secondary migrations occur mainly during summer months (e.g., Jellyman 1977; Naismith and Knights 1988). In a seasonal study at the outlet of a small New Zealand lake, Jellyman and Ryan (1983) found distinct seasonal patterns in elver recruitment, although peak months varied between years, and some recruitment was recorded during every month of the year. Records of elver arrivals at four major hydro dams in New Zealand indicated that while elvers arrive from mid-November to mid-March, about half the total seasonal catch is recorded in January (Martin et al. 2013). In Tasmania, Sloane (1984) reported elvers of *A. australis* arriving at a power station 55 km from the sea, during late spring and through summer, i.e., also mid-November to mid-March, with peak catches in January and early February. Migrations of *A. anguilla* elvers in the River Thames, England, take place over a 7-month period (April to October) although most eels move during May and June (Naismith and Knights 1988), a pattern generally similar to that observed in the Avon and Severn Rivers by White and Knights (1997b).

Although these upstream migrations usually occur during the warmer summer months, some elvers have been recorded accompanying glass eels during their arrival in fresh water (e.g., Jellyman 1977; Dutil et al. 1989; White and Knights 1997b; Jessop et al. 2002). Jellyman (1977) assumed that such eels had been resident in the lower reaches of estuaries or adjacent coastal areas, otolith microchemistry of *A. rostrata* elvers showed that they exhibited a variety of life history patterns including coastal or estuarine residence for a year or more (Jessop et al. 2002).

Sizes and ages

Because elvers migrate upstream over successive years, migrations can be comprised of a mix of species and sizes, and the average size increases with increasing distance upstream (Naismith and Knights 1988; Jellyman 1977; Haro and Krueger 1991; Lobón-Cervia et al. 1995). Naismith and Knights (1988) noted a similar increase in the age cohorts represented with the distance upstream, with elvers leaving the Thames estuary mainly being 1–3 years in freshwater (length < 14 cm), compared with most elvers at a site 15 km upstream being 4–8 years in freshwater (length range 20–30 cm). Jellyman and Ryan (1983) found 6 cohorts represented among recruits in a small lowland lake in New Zealand; although ages 0+ and 1+ predominated, the relative glass eel class strengths varied considerably between seasons. Elvers collected from Karapiro Dam, New Zealand also comprised a mix of 3 and 4 age classes (Jellyman 1977; Beentjes et al. 1997). In Tasmania, Sloane (1984) found that 0–3 year old elvers of *A. australis* dominated migrations, but eels up to 10 years old were also present. At the elver ladder on the Moses-Saunders Dam (approximately 400 km upstream of the St. Lawrence River estuary), the mean age in 1975 was 6 years (modal age 5 years), but eels up to 11 years old were also present (Marcogliese and Casselman 2009).

Regular upstream migrations may persist for many years of an eel's life, enabling substantial recolonisation of upstream habitats. Migratory juvenile eels up to 31 cm were recorded at an inland hydro dam by Jellyman (1977). In a small catchment in northwest France, Laffaille et al. (2005a) found that yellow eels > 20 cm adopt a sedentary lifestyle, and can remain within the same general location for many years. Although many glass eels choose to stay within the tidal reaches of rivers for extended periods and sometimes take up permanent residence there, most glass eels use selective tidal stream transport (STST) to migrate within estuaries and tidal reaches of rivers (see Chapter 7 of this book), and accumulate at the upper limit of tidal activity (e.g., Edeline et al. 2007). From here they generally disperse upstream, possibly as a result of density-dependant processes linked to food availability, and a change from "pelagic" migration to "benthic" settlement behaviour (Edeline et al. 2007). The switch from the largely passive mode of STST to active swimming against the current is associated with increased thyroid hormone levels (Edeline et al. 2004). Indeed, migratory elvers of *A. rostrata* had thyroxine levels twice that of sedentary elvers (Castonguay et al. 1990). Not all individuals have high enough thyroid activity to facilitate this switch (Edeline et al. 2007), and choose to settle at the interface of tidal influence and riverine habitat, causing a "traffic jam" (Edeline et al. 2007). This accumulation of small eels is compounded by the subsequent arrival of new glass eel recruits, and the ensuing competition for food and space results in a density-dependent dispersal (Feunteun et al. 2003; Edeline et al. 2007).

Duration of migration season

The longer the migration season, the further upstream elvers can penetrate; Naismith and Knights (1988) found that the average duration of the migratory season in the Thames River was only 47 days and during this time window, the actual distance that the elvers were able to move upstream could be considerably restricted by the presence of dams and weirs. Baras et al. (1996) recorded a rather similar migration periodicity of 44 ± 7 days in the River Meuse, Belgium. At five hydro stations in New Zealand, the duration of the migration season ranged from 7.4 to 14.8 weeks (52–104 days), with an average of 12 weeks (Martin et al. 2013). Likewise, from 1977–81, Sloane (1984) recorded that the elvers arrived at dams in Tasmania over a 14-week period. Duration of the migration period can vary with size of eels, with elvers > 15 cm tending to migrate earlier in the season than those < 10 cm (Moriarty 1986).

Swimming and climbing ability

Upstream-migrating juvenile eels show strong positive rheotaxis, but they have limited sustained swimming ability against velocities > 0.2–0.5 m.s^{-1} (Sorensen 1951; Mitchell 1989; Knights and White 1998). Swimming ability increases with size and elvers > 10 cm long can swim against currents of 1.5–2.0 m.s^{-1} (Sorensen 1951; Tesch 2003; Knights and White 1998). The climbing ability of small eels is legendary, and eels up to 12 cm can climb damp vertical surfaces, adhering by surface tension (Skead 1959; Jellyman 1977) (Fig. 1). The reduction in climbing ability associated with increasing size emphasises the need to ensure that upstream migration is as unimpeded as possible so that small eels can penetrate into small steep tributaries before they become too large for vertical climbing. During climbing, they are able to respire through cutaneous respiration, although elevated temperatures during daylight hours can be fatal (e.g., Beentjes et al. 1997) as they cause desiccation and ultimately death if elvers do not have access to shade and flowing water.

Estimates of the speed of upstream migration of elvers in European Rivers generally range between 10 to 45 km.y^{-1} (Moriarty 1986; Aprahamian 1988; Mann and Blackburn 1991); from recaptures of marked fish, Baras et al. (1996) also recorded an annual rate of elver migration of 45 km.y^{-1}. However, there is considerable variation in migration speeds and Feunteun et al. (2003) recorded some eels covering 200 km.y^{-1}, while others achieved only 50 km.y^{-1}, even in the absence of major obstacles. In the Waikato River (New Zealand) hydro lakes, Beentjes et al. (1997) found that the elvers released in several of the lakes continued to migrate upstream and were found at the base of the next upstream dam a few days later—these movements corresponded to a swimming speed of 2.5 km.d^{-1}. Suppression of such migratory behaviour

Fig. 1. Elvers climbing a damp vertical wall at Wairua Falls power station, New Zealand. The rope at the left of the picture is a mussel (*Mytilus edulis, Perna canaliculus*) spat collection rope, and provides additional purchase for the juvenile eels. Photo. Courtesy J. Boubée, NIWA, Hamilton, New Zealand.

would be a useful management tool to encourage the settlement of elvers transferred into a body of water, but there are no known mechanisms to achieve this. Naismith and Knights (1988) noted that elvers migrating early in the season showed the strongest migratory behaviour and travelled the greatest distances upstream, so retention of elvers for several weeks prior to transfer might reduce their migratory urge. The speed of migration increases with increasing water temperature (White and Knights 1997b).

Behaviour and migration triggers

Density-dependent processes

Density-dependent processes are widely considered to be important triggers leading to the commencement of upstream migrations, and they also affect the mortality rates of juvenile eels. Thus Lobón-Cervia et al. (1995) found that both mortality and migration rates correlated with eel numbers, suggesting such a density-dependant relationship. In a study of the movement of juvenile eels in the River Shannon, Ireland, Moriarty (1986) noted that the largest migrations of elvers occurred in the year following the greatest migration of glass eels to the river, suggesting that migration may be influenced by the population pressure in the lower reaches of the river. Intraspecific competition and agonistic behaviour increases with increasing population density and biomass (White and Knights 1997a), and also with increased water temperature (Nyman 1972), and such features could serve as important precursors to the onset of migration. Feunteun et al. (2003) also suggested that high densities of juvenile eels in their first year in freshwater were responsible for this cohort being "pushed" further upstream, but while such density-dependent behaviour was evident *within* size classes, it was not evident *between* size classes.

As with many aspects of eel life-history, the tendency for upstream migration varies considerably between individuals (Naismith and Knights 1988; White and Knights 1997a). Thus Naismith and Knights (1988) found that some of the marked elvers took up to two years to reappear in a trap when released only a few hundred metres downstream. Feunteun et al. (2003) classified juvenile eels as either "pioneers" or "founders", depending on whether they moved extensive or limited distances upstream; after the first or second year in freshwater, these authors suggested that elvers were either "nomadic" or "home range dwellers". However, they noted that "These four different behaviours are probably not mutually exclusive, and occur alternatively throughout each individual eel's life, according to age and to variations of environmental conditions".

An interesting consequence of reduced recruitment noted in a 21-year data set for Rio Esva, Spain, was that there was a compensatory increase in the survival of young cohorts, with the result that the number of eels achieving silvering was similar for both the strongest and weakest cohorts (Lobon-Cervia and Iglesias 2008).

Environmental triggers

As eels are negatively phototrophic and mainly active during darkness (e.g., Glova and Jellyman 2000), elver migrations are very largely nocturnal (e.g., Dutil et al. 1989; Jellyman and Ryan 1983). There is some evidence of crepuscular activity at dusk and dawn (Jellyman and Ryan 1983), a finding similar to the activity patterns of the larger tagged eels (Jellyman and Sykes 2003).

Many researchers have concluded that water temperature is an important factor governing both the strength and seasonal duration of upstream migrations. For instance, Naismith and Knights (1988) found that water temperatures of 10–15°C were required to stimulate upstream migration, but the end of the most intensive period of migration did not appear to be temperature-related—this agreed with Moriarty (1986) who suggested that temperature was involved in the commencement of migration but not in the cessation. Other authors have noted that the minimum threshold temperature for commencement of elver migration is usually between 10°C and 17°C (Deelder 1984; Hvidsten 1985). Likewise White and Knights (1997a) studying elver eel migrations in the Severn and Avon Rivers, England, found that water temperature was the key stimulus for migration, with a threshold temperature between 14–16°C and maximum catches were made at 18 to 20°C. When reviewing the timing of elver migrations at New Zealand hydro dams, Martin et al. (2013) concluded that migrations variously started or stopped at water temperatures of 17–18°C, although this threshold effect did not hold true at all sites or through all years. The influence of water temperature on migrations of a range of *Anguilla* species has also been noted by a number of authors including Jellyman 1977; Jellyman and Ryan 1983; Sloane 1984; Vollestad and Jonsson 1988; Schmidt et al. 2009.

Flow also affects migrations, and a weak positive association between elver catches and flow was noted by White and Knights (1997a), while Jellyman and Ryan (1983) found a strong positive (but lagged) relationship between passage of a flood and subsequent catches of elvers entering a trap at the outlet of a small lake in New Zealand; high autocorrelations were found between elvers caught on any particular day with elvers caught on preceding days indicating strong synergistic behaviour, caused by an escalating response to increases in flow. Conversely, Sloane (1984) recorded a negative relationship between elver catches and flows at a Tasmanian power station outfall, as increased flows and their associated velocities restricted the ability of elvers to migrate upstream and not because of a reduced attraction during high flows. Sloane (1984) also considered day length and water temperature (> 10°C) to be important factors in the initiation of elver migrations. A combination of dark nights and high flows (freshets) were associated with *A. rostrata* elvers using an eel ladder in a tributary of the Hudson River, New York (Schmidt et al. 2009). Eels have a very highly developed sense of smell (Tesch 2003); the odour of other eels is an important feature in trap capture efficiency, and Briand et al. (2002) found that directing the outflow water from a holding bin onto the access ladder increased the total catch by a factor of 1.4.

Habitat use and distribution

Diversity of habitats used

Habitat use by freshwater eels varies with species (interspecific differences, habitat segregation), size (ontogenetic shifts), time of day (diel shifts) and season. Eels show specific habitat choice (habitat selection) which is sometimes most evident in smaller eels (Johnson and Nack 2013), as larger eels occupy a greater diversity of habitats and have been described as habitat generalists (Bain et al. 1988; Glova et al. 1998).

Many researchers have commented upon the high degree of plasticity apparent in many aspects of the ecology and behaviour of freshwater eels (e.g., Daverat et al. 2006; Thibault et al. 2007) that provides a remarkable "bet-hedging" strategy for anguillids as a group (Daverat et al. 2006). Much of the success of anguillids in terms of their global distribution and the diversity of habitats they are able to occupy, is due to these features (Helfman et al. 1987; Righton et al. 2012).

In a study of *A. rostrata* in a coastal watershed, Lamson et al. (2006) noted a range of life-history strategies from purely saltwater residence to purely freshwater residence, and found that shifting between the habitats increased with age. Thibault et al. (2007) tested the hypothesis that the choice of a particular life-history tactic could be controlled by a conditional strategy with individual size, age and/or sex of eels being determining factors—although the data collected did not support this hypothesis, the authors found that eels that remained in estuarine areas grew more rapidly than eels that migrated to freshwater.

Habitat shifts

The Sr:Ca ratio in the otoliths of fishes differs according to the time they spend in fresh water and sea water respectively (Campana 1999). Early studies of the otoliths of *Anguilla japonica* showed that the Sr:Ca level was strongly correlated with the salinity of the water and was little affected by other factors such as water temperature, food and physiological factors (Tzeng 1996; Kawakami et al. 1998). Thus, Sr:Ca ratios could determine the amount of time individual eels spent in freshwater, estuarine or marine environments and how often they moved between these different habitats. Surprisingly, studies showed that some yellow and silver eels never migrate into fresh water, but spend their entire life in the ocean (Tsukamoto et al. 1998). Thus, the classification of anguillid eels as being catadromous species with a freshwater growth stage needs revision, as their movement into fresh water is not obligatory, and their life history should be defined as a facultative catadromy (Tsukamoto and Arai 2001), with ocean and estuarine residents as ecophenotypes.

Studies of a range of *Anguilla* species have found otolith signatures intermediate to those of marine and freshwater, indicating estuarine residence, often with switches between environments of different salinity, e.g., *A. anguilla* (Tsukamoto et al. 1998; Tzeng et al. 2000; Arai et al. 2006; Daverat et al. 2006), *A. japonica* (Tsukamoto et al. 1998; Tsukamoto and Arai 2001; Arai et al. 2003a,b; Kotake et al. 2003, 2005; Chino et al. 2008; Chino and Arai 2009; Yokouchi et al. 2012), *A. rostrata* (Jessop et al. 2002, 2004, 2006, 2008; Lamson et al. 2006), *A. australis* and *A. dieffenbachii* (Arai et al. 2004), *A. marmorata* (Chino and Arai 2010a; Arai et al. 2013), *A. bicolor bicolor* (Chino and Arai 2010b,c) and *A. bicolor pacifica* (Arai et al. 2013). Figure 2 shows typical otolith transects of saltwater residence, switch-type (freshwater to saltwater) and freshwater residence.

In the lower reaches of rivers and in estuaries, a predominance of eels that move once or several times between habitats of different salinity (River-Estuary switch types, RE) has been reported for many eel species and regions (Arai and Chino 2012) and appears to be the most common habitat shift pattern in anguillid eels (Daverat et al. 2006; Arai and Chino 2012).

Density of juvenile eels is a major driver of habitat shifts. For example, *A. japonica* in the Hamana Lake system of Japan that had ratios < 2.5 at recruitment became predominantly river-residents (Yokouchi et al. 2012), while eels with ratios > 2.5 at recruitment became either river residents or RE switch types. However, eels with the ratios > 6 showed no evidence of ever entering fresh water. These results suggest that for *A. japonica*, it might be important to arrive in fresh water early in season to be able to successfully establish river residence, as many later arrivals tend to shift back to the estuary (Yokouchi et al. 2012). An advantage of estuarine residence is that it usually results in more rapid growth than does freshwater residence (e.g., *A. rostrata*: Helfman et al. 1984, 1987; Morrison and Secor 2003; *A. anguilla*: Acou et al. 2003; Melia et al. 2006; *A. reinhardtii*: Walsh et al. 2006).

Habitat segregation

In coexisting populations of more than one species of *Anguilla*, there is almost invariable evidence of spatial segregation and differences in habitat preferences. Thus in northeast Australia, *A. australis* is essentially a still water species with a lowland distribution, whereas *A. reinhardtii* prefers more flowing water (Pusey et al. 2004) and penetrates further inland. Likewise, *A. australis* in New Zealand is largely a species of lowland lakes and low gradient rivers, while *A. dieffenbachii* has a preference for stony and higher gradient rivers and penetrates further inland to high country lakes (McDowall 1990; McDowall and Taylor 2000). Differences in the longitudinal distribution of *A. australis* and *A. dieffenbachii* and in their daytime habitat usage were also found by Glova et al. (1998). Comparison of the diets of these two species in a small stream showed considerable overlaps in prey, and there was no evidence of temporal or trophic segregation, leading the authors (Sagar et al. 2005) to suggest that habitat separation is the main mechanism to reduce interspecific competition between these co-occurring species.

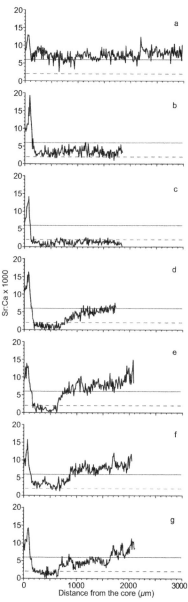

Fig. 2. Plots of the otolith Sr:Ca ratios along transect lines from the core to the edge of the otolith for seven representative specimens of a range of *Anguilla* species. Eels exhibiting a constant type lived in either marine water (a), brackish water (b) or freshwater (c) throughout their lives. Eels exhibiting a switch type moved among habitats. The varieties included eels that moved from freshwater to brackish water (d), freshwater to marine water (e), brackish water to marine water (f), multiple movements among freshwater, brackish water, and marine water (g). The solid line in each panel indicates marine water life period ($\geq 6.0 \times 10^{-3}$ in Sr:Ca ratios), and the dotted line in each panel indicates freshwater life period ($< 2.5 \times 10^{-3}$ in Sr:Ca ratios). From Arai and Chino 2012. Reproduced with permission.

Longitudinal differences in distribution were also evident in a study of *A. marmorata* and *A. japonica* in four Taiwanese rivers, with *A. japonica* being more abundant in the lower reaches and *A. marmorata* in the upper reaches (Shiao et al. 2003). Chino and Arai (2010b,c) recorded *A. bicolor bicolor* in either brackish waters or seawater areas, a finding similar to Briones et al. (2007) who found this species in the downstream area of a river in the Philippines. Habitat choice in *A. marmorata* is varied, and in a study of this species in allopatry, three different life history patterns were evident that indicated freshwater residence, continuous brackish water residence, or a combination of the two (Chino and Arai 2010a). In sympatry with other species, *A. marmorata* seems to be mainly resident in freshwater (Shiao et al. 2003; Briones et al. 2007), but in a large estuary in Vietnam, *A. marmorata* was primarily found in estuarine areas along with *A. bicolor pacifica* (Arai et al. 2013). This study concluded that both *A. marmorata* and *A. bicolor pacifica* were able to utilise a full range of saline-influenced habitats, similar to that in a number of temperate eel species. Given the greater productivity of fresh water than sea water in tropical areas, and the reverse in temperate areas (Gross et al. 1988), it might be expected that tropical freshwater eel species would show a greater likelihood of growing in a freshwater environment and an associated reduced likelihood of dwelling in saline areas. Arai et al. (2013) did not find supporting evidence for this hypothesis, and noted that both tropical and temperate species display the ability to exist in a wide range of habitats of varying salinities and distances inland. However, observations by Daverat et al. (2006) that temperate eels living at high latitudes (i.e., where ocean productivity exceed freshwater productivity) had a greater probability of remaining in the lower reaches of waterways than did eels living at lower latitudes, provide some confirmation of this hypothesis.

Ontogenetic habitat shifts

Ontogenetic shifts in habitat preferences have been noted by a number of researchers, e.g., Glova et al. 1998; Laffaille et al. 2003; Jellyman et al. 2003. In an extensive study of habitat use by *A. australis* and *A. dieffenbachii* in four physically contrasting waterways in New Zealand, Jellyman et al. (2003) found interspecific differences between similar size groups of both species, and size-related preferences within species; thus juvenile eels (< 10 cm) of both species preferred slow (< 0.5 m.s^{-1}) and shallow (< 0.2 m deep) reaches, while eels > 50 cm of both species preferred deeper (> 0.3 m) but slow flowing water, and were strongly associated with cover (undercut banks, aquatic plants, instream debris). This study also found changes in habitat preferences for *A. australis* > 20 cm and *A. dieffenbachii* > 30 cm, probably associated with the shift from living with the substrates to living in open water. A subsequent study by Graynoth et al. (2008) confirmed the presence of a "habitat bottleneck" that affected the survival of eels when they made shifts from living *within* the substrates to living *above* the substrates in the limited areas of suitable cover (banks and debris clusters).

Confirmation of the strong association of juvenile eels (> 10 cm) and cover was obtained from laboratory activity trials of *A. australis* and *A. dieffenbachii* (Glova and Jellyman 2000), where such eels spent substantially more time within the substrates than did larger eels (10–19.9 cm, 20–29.9 cm). Diet studies of *A. australis* (Sagar and Glova 1998) showed that eels > 20 cm fed exclusively during the night, but for smaller eels the pattern was less distinct with some feeding occurring during the day, from which the authors concluded that these smaller eels were mainly feeding within the substrate rather than above the substrate.

Size and age-related shifts in habitat use have been shown to occur around an age of three in *A. anguilla* (Daverat et al. 2006; Thibault et al. 2007) and also at the same age in *A. japonica* and *A. rostrata* (Daverat et al. 2006), while Feunteun et al. (2003) states that habitat preferences change markedly in *A. anguilla* at about 30 cm, the size at which this species starts to feed regularly on fish. Laffaille et al. (2003) found small *A. anguilla* in shallow habitats that had abundant aquatic vegetation, whereas large eels preferred greater depth and less vegetation; these authors also concluded that European eels change their behaviour and microhabitat characteristics at a size of 30 cm. Most shifts in *A. japonica* occurred within their first year in fresh water (Yokouchi et al. 2012). Similarly, Thibault et al. (2007) found some shifts in *A. rostrata* in the St. Jean River, eastern Canada, which took place at ages 0 and 1, although the majority occurred at ages 2–4, similar to ages 3–5 found by Jessop et al. (2008) in the East River, Nova Scotia.

A study of *A. rostrata* in a tributary of the Hudson River, New York, found that all size groups of eels showed diel differences in habitat use, with larger substrate and more cover used during the day than at night (Johnson and Nack 2013). Movement between daytime sheltering habitat and night-time foraging habitat, is often limited to < 100 m (e.g., Chisnall and Kalish 1993; Beumer 1979; Oliveira 1997; Jellyman and Sykes 2003), which suggests a narrow home range (Feunteun et al. 2003). Likewise in lakes, the movement of tagged fish is usually limited (e.g., Jellyman et al. 1996; Beentjes and Jellyman 2003), although over intervals of years, the movement can be extensive.

In a shallow lake in the North Island of New Zealand, Chisnall (1996) showed that *A. australis* < 35 cm in length occurred almost exclusively around the lake margins, with a preference for emergent reed habitat, the most complex of the five habitat types surveyed. A similar distribution was noted by Jellyman and Chisnall (1999) in a large shallow coastal South Island lake, where small *A. australis* (< 30 cm) were also associated with inshore areas, and depths of 0.6–1.2 m; eels larger than this showed no preference for particular depths or distances offshore.

Other factors influencing habitat choice

Habitat choice by juvenile eels is partly influenced by the presence of conspecifics. This influence may be either positive (e.g., odour attraction, social interaction) or negative (e.g., avoidance of larger predatory eels). In experiments to test cover preferences of juvenile *A. australis* and *A. dieffenbachii* (< 30 cm), Glova (1999) concluded that both size and species interactions influenced the choice of habitats used, with small (< 10 cm) and medium (10–19.9 cm) eels of both species preferring cobbles and macrophytes more or less equally, whereas larger eels (20–29.9 cm) showed a clear preference for macrophytes. In a further experiment, Glova (2001) concluded that the presence of potentially predatory *A. dieffenbachii* (42–87.5 cm) restricted the cover preferences of smaller *A. australis* and *A. dieffenbachii*, a result similar to that found by Chisnall et al. (2003) after the selective removal of large *A. dieffenbachii* in a natural stream. In trials to test the effects of eel density on the cover preference of small *A. australis*, Glova (2002) found that at densities ranging from 4 to 83 eels.m^{-2}, small eels preferred to crowd together within cobbles and to a lesser extent within macrophytes, rather than spreading themselves more uniformly among other cover types. Seemingly, small eels especially have a high tolerance for sharing available cover, and body contact may be an important part of their social organisation (Glova 2002). The use of cover makes small eels less vulnerable to predation, and small eels readily occupied small brush piles placed on barren areas of lowland lakes (Jellyman and Chisnall 1999). Cannibalism is not uncommon in eels (e.g., Jellyman 1989), and piscivory has been recorded at lengths > 40 cm (Jellyman 1989). Lack of suitable cover for small eels (2–5 g) stocked in waterways throughout Denmark, was suggested as one of the main reasons for the high rate of their disappearance (Pedersen 2009).

Collectively, a range of factors have been associated with habitat choice in eels including physical features (water velocity, substrate type, depth, cover, shade), and various social/behavioural and biotic features (e.g., food availability, the presence of other eels). Ultimately though, the choice of habitats by small eels is likely to be made on the basis of multiple, rather than singular, favourable factors. Thus, a study of eel abundance and habitat utilisation in five rivers in Maryland, USA, found that the main predictors of eel density (of a range of sizes) were velocity and depth diversity, distance inland, density of non-eel fishes, and distance to a semi-passable or impassable barrier (Wiley et al. 2004).

Seasonal shifts in habitat

Seasonal shifts in habitat have been reported in temperate *Anguilla* species. For example, Johnson and Nack (2013) recorded seasonal habitat shifts in *A. rostrata*, with eels in autumn occupying slower flowing areas with finer substrates than the ones occupied during summer. During winter,

A. dieffenbachii in the South Island of New Zealand occupy deeper waters with finer sediments than the ones occupied during spring and summer (Broad et al. 2001), probably because such winter habitats provide better opportunities to bury within the substrates. Localised seasonal movements by yellow eels between summer feeding habitats and overwintering habitats have been rarely reported in the literature, but are commonly known by the eel fishers, especially in lakes and lagoons (Feunteun et al. 2003). For example, the Mi'kmaq first nation people of western Newfoundland, Canada, recognised three main periods for eel harvesting—the first being in May when eels are moving from inland waters to the sea, the second is during the main summer months of June to August when gaffing of eels takes place at the coast during low tides, and the third is in late fall and early winter when holes are cut in the ice of inland waters to spear eels at depths of ~ 1.5 m (Parrill and Vodden 2014). In *A. rostrata*, burrows of 1–1.5 m are one of the habitats used in winter (Tomie et al. 2013).

Management and fish passes

Understanding the specific habitat requirements of juvenile eels has important implications for habitat restoration and protection activities (Johnson and Nack 2013). A number of studies that have considered the possible reasons for decline of temperate *Anguilla* species have included habitat quality and quantity (e.g., Richkus and Whalen 2000; Feunteun 2002; Kettle et al. 2011). Of course, given seasonal and size-related changes in habitat use, access to- and connectivity of habitats are essential components of any such restoration activities. Improved longitudinal connectivity could result in the whole of the Loire River catchment becoming accessible to eels (Laffaille et al. 2009), compared with the present situation where eels were present at only 50% of the 123 sites sampled, with tributary barriers being a major reason for restricted access. The importance of lateral connectivity of tributaries was also highlighted by Lasne et al. (2008), especially for eels < 15 cm.

   While anguillid eels are a very adaptable and ubiquitous species, "one size doesn't fit all" as far as habitat is concerned, and habitat enhancement and restoration techniques need to be cognisant of the species and size of the eel being targeted, as well as differences in diel and seasonal habitat preferences. In habitat use trials, juvenile *A. australis* preferred natural cover to artificial cover (Glova 2002), although in the absence of natural cover, pipes will serve as an alternative. The strong association of eels for cover is utilised in some traditional and experimental capture methods, especially in Japan where hollowed-out bamboo or plastic pipes, bundles of twigs, and rock mounds built at low tide, are variously used as capture methods (Kuroki et al. 2014).

   In a reclaimed marsh, Laffaille et al. (2004) found that eel densities were related to the width of ditches, depth of silt, and the density of emergent plants-knowing such relationships provides specific guidelines for further habitat enhancement, including recommendations by these authors that regular dredging should be managed to maintain a mosaic of permanent habitats. Given the variety of habitat shifts that eels undergo, habitat heterogeneity is an important principle in restoration projects.

A consideration of habitat scale is also important, and though micro-habitat preferences can be established from the abundance of eels relative to various habitat parameters, broad-scale meso-habitat features should not be ignored. Thus, when comparing the relative importance of landscape-scale variables (e.g., distance inland, channel slope) and local-scale physical habitat variables (e.g., water velocity, cover) on size-distribution patterns in New Zealand eels, Booker and Graynoth (2013) found that local-scale variables explained more of the patterns of *A. dieffenbachii* whereas landscape-scale variables were more influential for *A. australis*. Similarly, when considering distribution patterns of American eels in streams in Virginia, USA, Smogor et al. (1995) cautioned that small-scale patterns like locally high densities of juvenile eels at downstream sites, probably reflect density-dependent processes that rarely influence eel distribution patterns, whereas distribution patterns reflect larger-scale processes like distance inland.

Barriers to migration

Many large catchments have multiple barriers for eel movement. For instance, the River Thames, England, has over 500 barriers affecting up- and downstream movement (Gollock et al. 2011).

When reviewing the need for decision making and cost-benefit analysis for the construction of elver passes, Knights and White (1998) suggested that information was needed on:

- the number and types of obstructions within the catchment
- estimation of the relative significance of each obstruction in terms of its passability, e.g., steepness and height
- The extent and quality of habitat that could be opened up by a fishway
- The location and effectiveness of existing fishways
- The number and sizes of eels (and possibly other species) that might benefit from improved/renewed access

Eels of all sizes can climb slopes that have surface regularities or vegetation (e.g., Knights and White 1998), and either natural or artificial materials are commonly used to construct ladders that enable elver passage. The requirements for such ladders are relatively simple, being (after Knights and White 1998):

- a flow of water sufficient to attract elvers to the ladder entrance
- suitable design and placement of the entrance and exit
- suitable water velocities and the provision of some form of material to assist climbing

There is extensive literature on suitable upstream passage facilities for elvers (e.g., Bell 1986; Legault 1992; Knights and White 1998; Tesch 2003; Gough et al. 2012). However, as noted by Tesch (2003), even though normal fish passes will allow the upstream passage of elvers, specifically designed elver ladders should be installed at barriers, as they will be more effective and economic than normal passes. The benefits of restoring passage for juvenile eels have been demonstrated in a number of studies including Laffaille et al. (2005b).

Although high pipe and "bottle-brush" passes have been constructed (e.g., 68 m at Patea Dam, New Zealand), most passes do not exceed 10 m in height. The Patea Dam pass had problems with elevated water temperatures, especially during the day as many elvers were not able to traverse the whole length overnight; also, the pass was found to be selective for only smaller *A. australis* and few larger *A. dieffenbachii* were able to negotiate it (Boubée et al. 2003). As a result of such experiences, all juvenile eel passage facilities in New Zealand are "catch-and-carry" whereby elvers ascend a short ramp or ladder to a holding box and are then manually transported upstream. Advantages with this method are that it enables an estimate of elver numbers to be made (estimated from total weight of elvers), can respond rapidly to peak migrations (especially if traps are monitored by observers or remotely operated video cameras), mortality is very low (provided predators are excluded through netting and ladder constrictions, and ladders are monitored regularly during the peak migration periods), and releases can be made into various waterways on a planned basis. Also, this avoids the increased predation that can occur at the outlet of elver passes, when predators learn of an easy meal. Installation costs of ramps are relatively low, and retro-fitting is common, especially when observations indicate the areas where elvers naturally accumulate. Multiple ramps can be installed should elvers accumulate at more than one location. Figure 3 shows a range of small ramp-traps used in New Zealand too collect elvers.

New Zealand does not allow export of glass eels, and apart from some historic records and experimental catches (e.g., Jellyman et al. 2002; Jellyman and Lambert 2003; Jellyman et al. 2009) there are no data on glass eel recruitment. As a surrogate, the numbers of elvers arriving at various hydro dams are monitored each year, as part of the catch-and-carry requirements for power companies to transfer juvenile eels upstream. A consistency in fishing effort and gear is required for valid inter-annual catch comparisons to be made, and the presence of multiple year classes is a confounding factor. Nevertheless, trends in elver time series covering up to 20 years are now an important part of the annual review of the status of eel stocks (Martin et al. 2013, Fig. 4). Gollock et al. (2011) reviewed elver catches from three sites in the River Thames between 2005 and 2009, and concluded that catches were generally low, but there was a lack of data on eel recruitment for the United Kingdom as a whole. The use of fish passes as monitoring sites for elver

Fig. 3. A variety of floating traps and ramps used in New Zealand to capture elvers at hydro dams. a and b = floating ramps (lowered to operate); c–e = fixed ramps. [Collectively such traps catch between 4 and 8 million elvers annually.]

migrations was advocated by Baras et al. (1996) who recorded elver numbers, migration periodicity, and the rate of movement from two Denil fish passes on the River Meuse, Belgium.

The importance of monitoring the numbers and sizes of elvers using particular fishways is evident from the experience of North American researchers. The St. Lawrence River and catchment is the largest source of silver female eels of *A. rostrata*, and the best index of recruitment for this species

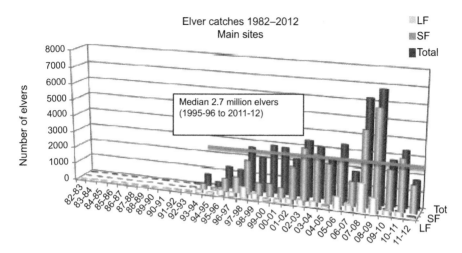

Fig. 4. Seasonal catches of elvers at the main monitored sites in New Zealand. LF = *A. dieffenbachii*, SF = *A. australis*. From Martin et al. 2013, reproduced with permission.

comes from the number of juvenile eels ascending the ladder at the Moses Saunders hydroelectric dam (Casselman 2003). Castonguay et al. (1994) noted an 81-fold decline in recruitment from 1985–1992. From a peak 1982–1983, recruitment declined substantially through the 1980's and 1990's, and the few eels that ascended the ladder in the 1990's were also larger and older than recruits in the 1970's and 1980's (Casselman 2003; Figs. 5 and 6). Recruitment virtually ceased since the late 1990's (Marcogliese and Casselman 2009) and as a result, trial stocking of the upper St. Lawrence River and Lake Ontario with 3.8 m glass eels took place from 2006–2009 (Pratt and Threader 2011).

Transfers and stocking

Much of the wild eel fishery in Australia is dependent upon the stocking of lowland lakes with juvenile eels. Many of these elvers are sourced from Tasmania where they are harvested as they congregate in late spring and summer below stream barriers, especially below two hydro dams. The largest migration takes place at Trevallyn Dam in northern Tasmania where between 3–5 m elvers take part each year (Sloane 1984). Many of these elvers are captured and transported to mainland Australia for stocking of suitable waterways (see Chapter 14 of this book). In a review of stocking strategies for *A. anguilla*, Knights and White (1997) noted that stocking with glass eels and elvers is widely carried out in Europe, and that stocking of lakes has been preferred to rivers, and suggested that typical survival from glass eel or elver to harvestable- sized eels would be 20–30%. Stocking of Lough Neagh, Northern Ireland, has taken place from 1932–1947 and from 1960 onwards (Parsons et al. 1977), but natural recruitment to the River Bann has declined to the

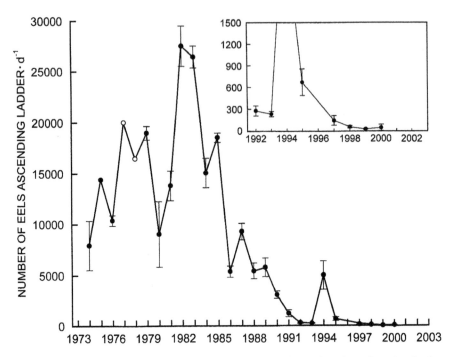

Fig. 5. The number of elvers ascending the eel ladder per day at the Moses-Saunders hydro-electric dam, St. Lawrence River, Canada. From Casselman 2003, reproduced with permission.

extent that supplementary stocks of glass eels have been imported (Knights and White 1997). Poor recruitment to the Baltic Sea has led to stocking since 1894, but hydro dams now exacerbate passage issues (Knights and White 1997), and the extent that stocked eels contribute to silver eels remains uncertain (e.g., Limburg et al. 2003; Prigge et al. 2013).

The Waikato River (North Island, New Zealand) has been extensively developed for hydro generation via the construction of 8 dams; historically, few eels were able to negotiate the upper reaches as access was limited by numerous rapids and waterfalls. No fish-passage facilities were installed at any of the dams, with the result that only small numbers of elvers were able to negotiate the lower two dams and populate the hydro reservoirs. However, the successful implementation of a catch-and-carry scheme has meant that all lakes are now stocked proportionally with the 1–2 m elvers collected annually at Karapiro Dam, the lowermost dam (Martin et al. 2013), which has resulted in the creation of new and productive fisheries. From a series of up to 19 capture sites in New Zealand (some of which are fished irregularly), annual totals of 4–8 m elvers are caught and transferred above hydro dams or falls (Martin et al. 2013), of which ~ 84% are *A. australis* and the remainder *A. dieffenbachii*. Ironically, the presence of yellow eels in these new habitats as a result of this

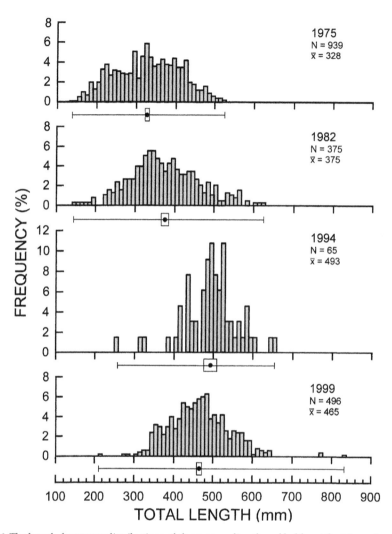

Fig. 6. The length-frequency distributions of elvers ascending the eel ladder at the Moses-Saunders hydro-electric dam, St. Lawrence River, Canada. The horizontal bars show mean, 95% confidence limits, and range in length. From Casselman 2003, reproduced with permission.

stocking, has led to some subsequent turbine mortality of silver eels—no exclusion devices are present at turbine intakes because no eels were present when the dams were constructed.

Transfer of larger *A. dieffenbachii* (mean weight 173 g; Beentjes and Jellyman 2003) to a high country lake in the same (Clutha) catchment, resulted in high survival and enhanced growth, with the latter being sufficient to act as a discernible inflection point in otoliths and validate otolith ageing in a sample of 79 eels recaptured after 10 years at large (Beentjes and Jellyman 2015).

To encourage recovery of European eel stocks, the European Council regulations (2007) require that 35% of glass eels caught annually be used for the restocking of European waterways. In a test of the effectiveness of stocking as a conservation tool, Desprez et al. (2013) found no difference in survival of glass eels or juvenile eels (20–30 cm), and concluded that such stocking in warm regions was a potentially effective management tool as production of silver eels commenced three years after stocking.

Conclusions

The summer upstream migration of juvenile eels is a very important process for colonisation of habitats and to achieve longitudinal distribution of yellow eels in rivers. Unfortunately, such upstream passages have often been compromised by the installation of dams and weirs, and such impacts are thought to have contributed towards reduced recruitment of Northern hemisphere temperate eel species. With increased concern about the well-being of eel stocks internationally, there is often an associated requirement to address up- and downstream eel passage, and the benefits of eel ladders or catch-and-carry practices are widely recognized. Annual counts of elvers at such facilities provide important time-series of recruitment strength, while the advent of otolith microchemistry has been able to provide detailed information about the variable life-history patterns that characterise freshwater eels worldwide. It is very likely that such variability and associated risk-spreading, is one of the main reasons that eels have become widespread and ubiquitous, leading to the claim of Helfman et al. (1987) that "American eels also occur naturally in perhaps the broadest diversity of habitats of any fish species in the world".

Keywords: Elvers, *Anguilla*, upstream migration, triggers, migration seasons, swimming speeds, otolith microchemistry, density-dependent processes, habitat use, distribution, elver passes, stocking, monitoring

References

Acou, A., F. Lefebvre, P. Contournet, G. Poizat, A. Panfili and A.J. Crivelli. 2003. Silvering of female eels (*Anguilla anguilla*) in two sub-populations of the Rhone Delta. Bull. Fr. Peche. Piscic. 368: 55–68.

Aprahamian, M.W. 1988. Age structure of eel, *Anguilla anguilla* (L.), populations in the River Severn, England, and the River Dee, Wales. Aquac. Fish. Manage. 19: 365–376.

Arai, T. and N. Chino. 2012. Diverse migration strategy between freshwater and seawater habitats in the freshwater eels genus *Anguilla*. J. Fish Biol. 81: 442–455.

Arai, T., A. Kotake, M. Ohji, N. Miyazaki and K. Tsukamoto. 2003a. Migratory history and habitat use of Japanese eel *Anguilla japonica* in the Sanriku Coast of Japan. Fish. Sci. 69: 813–818.

Arai, T., A. Kotake, M. Ohji, M.J. Miller, K. Tsukamoto and N. Miyazaki. 2003b. Occurrence of sea eels of *Anguilla japonica* along the Sanriku Coast of Japan. Ichthyol. Res. 50: 78–81.

Arai, T., A. Kotake, P.M. Lokman, M.J. Miller and K. Tsukamoto. 2004. Evidence of different habitat use by New Zealand freshwater eels, *Anguilla australis* and *A. dieffenbachii*, as revealed by otolith microchemistry. Mar. Ecol. Prog. Ser. 266: 213–225.

Arai, T., A. Kotake and T.K. McCarthy. 2006. Habitat use by the European eel *Anguilla anguilla* in Irish waters. Estuar. Coast. Shelf Sci. 67: 569–578.

Arai, T., N. Chino and D.A. Le. 2013. Migration and habitat use of the tropical eels *Anguilla marmorata* and *A. bicolor pacifica* in Vietnam. Aquat. Ecol. 47: 57–65.

Bain, M.B., J.T. Finn and H.E. Booke. 1988. Stream flow regulation and fish community structure. Ecol. 69: 382–392.

Baras, E., J.C. Philippart and B. Salmon. 1996. Estimation of migrant yellow eel stock in large rivers through the survey of fish passes: a preliminary investigation in the River Meuse (Belgium). pp. 82–92. *In*: I.G. Cowx (ed.). Stock Assessment in Inland Fisheries. Fishing News Books, Oxford, England.

Beentjes, M.P. and D.J. Jellyman. 2003. Enhanced growth of longfin eels, *Anguilla dieffenbachii*, transplanted into Lake Hawea, a high country lake in South Island, New Zealand. N.Z. J. Mar. Freshw. Res. 37: 1–11.

Beentjes, M.P. and D.J. Jellyman. 2015. Growth patterns and age validation from otolith ring deposition in New Zealand longfin eels *Anguilla dieffenbachii* recaptured after 10 years at large. J. Fish Biol. 86: 924–939.

Beentjes, M.P., B.L. Chisnall, J.A.T. Boubée and D.J. Jellyman. 1997. Enhancement of the New Zealand eel fishery by elver transfers. N.Z. Fish. Tech. Rep. 45: 44pp.

Bell, M.C. 1986. Fisheries handbook of engineering requirements and biological criteria. U.S. Army Corps of Engineers, Portland, Oregon.

Beumer, J.P. 1979. Feeding and movement of *Anguilla australis* and *A. reinhardtii* in Macleods Morass, Victoria, Australia. J. Fish Biol. 14: 573–592.

Booker, D.J. and E. Graynoth. 2013. Relative influence of local and landscape-scale features on the density and habitat preferences of longfin and shortfin eels. N.Z. J. Mar. Freshw. Res. 47: 1–20.

Boubée, J. and D. Jellyman. 2009. Facilitating the upstream and downstream passage of indigenous fish at large dams. pp. 57–63. *In*: E. McSaveney (ed.). Dams—Operating in a Regulated Environment. IPENZ Proceedings of Technical Groups, 35/1.

Boubée, J., B. Chisnall, E. Watene, E. Williams and D. Roper. 2003. Enhancement and management of eel fisheries affected by hydroelectric dams in New Zealand. Amer. Fish. Soc. Symp. 33: 191–205.

Briand, C., D. Fatin and A. Legault. 2002. Role of eel odour on the efficiency of an eel, *Anguilla anguilla*, ladder and trap. Environ. Biol. Fishes 65: 473–477.

Briones, A.A., A.V. Yambot, J.C. Shiao, Y. Iizuka and W.-N. Tzeng. 2007. Migratory pattern and habitat use of tropical eels *Anguilla* spp. (Teleostei: Anguilliformes: Anguillidae) in the Philippines, as revealed by otolith microchemistry. Raff. Bull. Zool. 14: 141–149.

Broad, T.L., C.R. Townsend, G. Closs and D.J. Jellyman. 2001. Microhabitat use by longfin eels in New Zealand streams with contrasting riparian vegetation. J. Fish Biol. 59: 1385–1400.

Bruton, M.N., A.H. Bok and M.T.T. Davies. 1987. Life history styles of diadromous fishes in inland waters of Southern Africa. Amer. Fish. Soc. Symp. 1: 104–121.

Campana, S.E. 1999. Chemistry and composition of fish otolith: pathways, mechanisms and applications. Mar. Ecol. Prog. Ser. 188: 263–297.

Casselman, J.M. 2003. Dynamics of resources of the American eel, *Anguilla rostrata*: Declining abundance in the 1990s. pp. 191–213. *In*: K. Aida, K. Tsukamoto and K. Yamauchi (eds.). Eel Biology. Springer, Tokyo.

Castonguay, M., J.-D. Dutil, C. Audet and R. Miller. 1990. Locomotor activity and concentration of thyroid hormones in migratory and sedentary juvenile American eels. Trans. Am. Fish. Soc. 119: 946–956.

Castonguay, M., P.V. Hodson, C.M. Couillard, M.J. Eckersley, J.-D. Dutil and G. Verreault. 1994. Why is recruitment of the American eel, *Anguilla rostrata*, declining in the St. Lawrence River and Gulf? Can. J. Fish. Aquat. Sci. 51: 479–488.

Chino, N. and T. Arai. 2009. Relative contribution of migratory type on the reproduction of migrating silver eels, *Anguilla japonica*, collected off Shikoku Island, Japan. Mar. Biol. 156: 661–668.

Chino, N. and T. Arai. 2010a. Migratory history of the giant mottled eel (*Anguilla marmorata*) in the Bonin Islands of Japan. Ecol. Freshw. Fish 19: 19–25.

Chino, N. and T. Arai. 2010b. Occurrence of marine resident tropical eel *Anguilla bicolor bicolor* in Indonesia. Mar. Biol. 157: 1075–1081.

Chino, N. and T. Arai. 2010c. Habitat use and habitat transitions in the tropical eel, *Anguilla bicolor bicolor*. Environ. Biol. Fish. 89: 571–578.

Chino, N., T. Yoshinaga, A. Hirai and T. Arai. 2008. Life history patterns of silver eels *Anguilla japonica* collected in the Sanriku Coast of Japan. Coast. Mar. Sci. 32: 54–56.

Chisnall, B.L. 1996. Habitat associations of juvenile shortfinned eels (*Anguilla australis*) in shallow Lake Waahi, New Zealand. N.Z. J. Mar. Freshw. Res. 30: 233–237.

Chisnall, B.L. and J.M. Kalish. 1993. Age validation and movement of freshwater eels (*Anguilla dieffenbachii* and *A. australis*) in a New Zealand pastoral stream. N.Z. J Mar. Freshw. Res. 27: 333–338.

Chisnall, B.L., M.L. Martin and B.J. Hicks. 2003. Effect of harvest on size, abundance, and production of freshwater eels *Anguilla australis* and *A. dieffenbachii* in a New Zealand stream. Amer. Fish. Soc. Symp. 33: 177–189.

Daverat, F., K.E. Limburg, I. Thibault, J.C. Shiao, J.J. Dodson, F.O. Caron, W.-N. Tzeng, Y. Iizuka and H. Wickstrom. 2006. Phenotypic plasticity of habitat use by three temperate eel species, *Anguilla anguilla*, *A. japonica* and *A. rostrata*. Mar. Ecol. Prog. Ser. 308: 231–241.

Deelder, C.L. 1984. Synopsis of biological data on the eel *Anguilla anguilla* (Linnaeus, 1758). FAO Fisheries Synopsis 80: 73pp.

Desprez, M., A.J. Crivelli, I. Lebel, G. Massez and O. Gimenez. 2013. Demographic assessment of a stocking experiment in European Eels. Ecol. Freshw. Fish 22: 412–420.

Dutil, J.-D., M. Michaud and A. Giroux. 1989. Seasonal and diel patterns of stream invasion by American eels (*Anguilla rostrata*) in the northern Gulf of St. Lawrence. Can. J. Zool. 67: 182–188.

Edeline, E., S. Dufour, C. Briand, D. Fatin and P. Elie. 2004. Thyroid status is related to migratory behavior in *Anguilla anguilla* glass eels. Mar. Ecol. Prog. Ser. 282: 261–270.

Edeline, E., L. Beaulaton, R. Le Barh and P. Elie. 2007. Dispersal in metamorphosing juvenile eel *Anguilla anguilla*. Mar. Ecol. Prog. Ser. 344: 213–218.

Feunteun, E. 2002. Management and restoration of European eel population (*Anguilla anguilla*): an impossible bargain. Ecol. Eng. 18: 575–591.

Feunteun, E., P. Laffaille, T. Robinet, C. Briand, A. Baisez, J.-M. Olivier and A. Acou. 2003. A review of upstream migration and movements in inland waters by anguillid eels: toward a general theory. pp. 191–213. *In*: K. Aida, K. Tsukamoto and K. Yamauchi (eds.). Eel Biology. Springer, Tokyo.

Glova, G.J. 1999. Cover preference tests of juvenile shortfinned eels (*Anguilla australis*) and longfinned eels (*A. dieffenbachii*) in replicate channels. N.Z. J. Mar. Freshw. Res. 33: 193–204.

Glova, G.J. 2001. Effects of the presence of subadult longfinned eels (*Anguilla dieffenbachii*) on cover preferences of juvenile eels (*Anguilla* spp.) in replicate channels. N.Z. J. Mar. Freshw. Res. 35: 221–233.

Glova, G.J. 2002. Density effects on juvenile shortfinned eel (*Anguilla australis*) cover preferences in replicate channels. N.Z. J. Mar. Freshw. Res. 36: 483–490.

Glova, G.J. and D.J. Jellyman. 2000. Diel activity patterns of juvenile eels (*Anguilla* spp.) in a laboratory flow tank. Ecol. Freshw. Fish 9: 210–218.

Glova, G.J., D.J. Jellyman and M.L. Bonnett. 1998. Factors associated with the distribution and habitat of eels (*Anguilla* spp.) in three New Zealand streams. N.Z. J. Mar. Freshw. Res. 32: 283–297.

Gollock, M., D. Curnick and A. Debney. 2011. Recent recruitment trends of juvenile eels in tributaries of the River Thames. Hydrobiol. 672: 33–37.

Gough, P., P. Philipsen, P.P. Schollema and H. Wanningen. 2012. From sea to source: international guidance for the restoration of fish migration highways. Regional Water Authority Hunze en Aa's, Veendam, the Netherlands. 300p.

Graynoth, E., R.I.C.C. Francis and D.J. Jellyman. 2008. Factors influencing juvenile eel (*Anguilla* spp.) survival in lowland New Zealand streams. N.Z. J. Mar. Freshw. Res. 42: 153–172.

Gross, M.R., R.M. Coleman and R.M. McDowall. 1988. Aquatic productivity and the evolution of diadromous fish migration. Science 239: 1291–1293.

Haro, A.J. and W.H. Krueger. 1991. Pigmentation, otolith rings and upstream migration of juvenile American eels (*Anguilla rostrata*) in a coastal Rhode Island stream. Can. J. of Zool. 69: 812–814.

Helfman, G.S., E.L. Bozeman and E.B. Brothers. 1984. Size, age, and sex of American eels in Georgia River. Trans. Am. Fish. Soc. 113: 132–141.

Helfman, G.S., D.J. Facey, J.L.S. Hales and J.E.L. Bozeman. 1987. Reproductive ecology of the American eel. Amer. Fish. Soc. Symp. 1: 42–56.

Hvidsten, N.A. 1985. Ascent of elvers (*Anguilla anguilla* L.) in the stream Imsa, Norway. Report of the Institute of Freshwater Research, National Swedish Board of Fisheries, Norway.

Jellyman, D.J. 1977. Summer upstream migration of juvenile freshwater eels in New Zealand. N.Z. J. Mar. Freshw. Res. 11: 61–71.

Jellyman, D.J. 1979. Upstream migration of glass-eels (*Anguilla* spp.) in the Waikato River. N.Z. J. Mar. Freshw. Res. 13: 13–22.

Jellyman, D.J. 1989. Diet of two species of freshwater eel (*Anguilla* spp.) in Lake Pounui, New Zealand. N.Z. J. Mar. Freshw. Res. 23: 1–10.

Jellyman, D.J. and C.M. Ryan. 1983. Seasonal migration of elvers (*Anguilla* spp.) into Lake Pounui, New Zealand, 1974–1978. N.Z. J. Mar. Freshw. Res. 17: 1–15.

Jellyman, D.J. and B.L. Chisnall. 1999. Habitat preferences of shortfinned eels (*Anguilla australis*), in two New Zealand lowland lakes. N.Z. J. Mar. Freshw. Res. 33: 233–248.

Jellyman, D.J. and P.W. Lambert. 2003. Factors affecting recruitment of glass eels into the Grey River, New Zealand. J. Fish. Biol. 63: 1067–1079.

Jellyman, D.J. and J.R.E. Sykes. 2003. Seasonal and diel changes in habitat use and movements of adult freshwater eels (*Anguilla* spp.) in two New Zealand streams. Environ. Biol. Fish 66: 143–154.

Jellyman, D.J., G.J. Glova and P.R. Todd. 1996. Movements of shortfinned eels, *Anguilla australis*, in Lake Ellesmere, New Zealand: results from mark-recapture studies and sonic tracking. N.Z. J. Mar. Freshw. Res. 30: 371–381.

Jellyman, D.J., B.L. Chisnall, J.R.E. Sykes and M.L. Bonnett. 2002. Variability in spatial and temporal abundance of glass eels (*Anguilla* spp.) in New Zealand waterways. N.Z. J. Mar. Freshw. Res. 36: 511–517.

Jellyman, D.J., M.L. Bonnett, J.R.E. Sykes and P. Johnstone. 2003. Contrasting use of daytime habitat by two species of freshwater eel (*Anguilla* spp.) in New Zealand rivers. Am. Fish. Soc. Symp. 33: 63–78.

Jellyman, D.J., D.J. Booker and E. Watene. 2009. Recruitment of *Anguilla* spp. glass eels in the Waikato River, New Zealand. Evidence of declining migrations? J. Fish Biol. 74: 2014–2033.

Jessop, B.M., J.C. Shiao, Y. Iizuka and W.-N. Tzeng. 2002. Migratory behaviour and habitat use by American eels *Anguilla rostrata* as revealed by otolith microchemistry. Mar. Ecol. Prog. Ser. 233: 217–229.

Jessop, B.M., J.C. Shiao, Y. Iizuka and W.-N. Tzeng. 2004. Variation in the annual growth, by sex and migration history, of silver American eels *Anguilla rostrata*. Mar. Ecol. Prog. Ser. 272: 231–244.

Jessop, B.M., J.C. Shiao, Y. Iizuka and W.-N. Tzeng. 2006. Migration of juvenile American eels *Anguilla rostrata* between freshwater and estuary, as revealed by otolith microchemistry. Mar. Ecol. Prog. Ser. 310: 219–233.

Jessop, B.M., D.K. Cairns, I. Thibault and W.-N. Tzeng. 2008. Life history of American eel *Anguilla rostrata*: new insights from otolith microchemistry. Aquat. Biol. 1: 205–216.

Johnson, J.H. and C.C. Nack. 2013. Habitat use of American eel (*Anguilla rostrata*) in a tributary of the Hudson River, New York. J. Appl. Ichthyol. 29: 1073–1079.

Jubb, R.A. 1961. The freshwater eels (*Anguilla* spp.) of Southern Africa. An introduction to their identification and biology. Annals Cape Provin. Mus. 1: 15–48.

Jubb, R.A. 1970. Freshwater eels: *Anguilla* species. The Eastern Cape Naturalist, Wildlife Society— EP Branch, Port Elizabeth. 31p.

Kawakami, Y., N. Mochioka, K. Morishita, H. Toh and A. Nakazono. 1998. Determination of the freshwater mark in otoliths of Japanese eel elvers using microstructure and Sr/Ca ratios. Environ. Biol. Fish. 53: 421–427.

Kettle, A.J., L.A. Vollestad and J. Wibig. 2011. Where once the eel and the elephant were together: decline of the European eel because of changing hydrology in southwest Europe and northwest Africa? Fish Fisheries 12: 380–411.

Knights, B. and E. White. 1997. An appraisal of stocking strategies for the European eel, *Anguilla anguilla*. pp. 121–140. *In*: I.G. Cowx (ed.). Stocking and Introduction of Fish. Fishing News Books, Oxford.

Knights, B. and E.M. White. 1998. Enhancing immigration and recruitment of eels: the use of passes and associated trapping systems. Fish. Manage. Ecol. 5: 459–471.

Kotake, A., T. Arai, T. Ozawa, S. Nojima, M.J. Miller and K. Tsukamoto. 2003. Variation in migratory history of Japanese eels, *Anguilla japonica*, collected in coastal waters of the Amakusa Islands, Japan, inferred from otolith Sr/Ca ratios. Mar. Biol. 142: 849–854.

Kotake, A., A. Okamura, Y. Yamada, T. Utoh, T. Arai, M.J. Miller, H.P. Oka and K. Tsukamoto. 2005. Seasonal variation in migratory history of the Japanese eel, *Anguilla japonica*, in Mikawa Bay, Japan. Mar. Ecol. Prog. Ser. 293: 213–221.

Krause, J., E.M.A. Hensor and G.D. Ruxton. 2002. Fish as prey. pp. 284–297. *In*: P.J.B. Hart and J.D. Reynolds (eds.). Handbook of Fish Biology and Fisheries. Vol. 1. Blackwell Publishing, Oxford, United Kingdom.

Kuroki, M., M.J.P. van Oijen and K. Tsukamoto. 2014. Eels and the Japanese: an inseparable, longstanding relationship. pp. 91–108. *In*: K. Tsukamoto and M. Kuroki (eds.). Eels and Humans. Springer, Tokyo, Japan.

Laffaille, P., E. Feunteun, A. Baisez, T. Robinet, A. Acou, A. Legault and S. Lek. 2003. Spatial organisation of European eel (*Anguilla anguilla* L.) in a small catchment. Ecol. Freshw. Fish 12: 254–264.

Laffaille, P., A. Baisez, C. Rigaud and E. Feunteun. 2004. Habitat preferences of different European eel size classes in a reclaimed marsh: A contribution to species and ecosystem conservation. Wetlands 24: 642–651.

Laffaille, P., A. Acou and J. Guillouet. 2005a. The yellow European eel (*Anguilla anguilla* L.) may adopt a sedentary lifestyle in inland freshwaters. Ecol. Freshw. Fish 14: 191–196.

Laffaille, P., A. Acou, J. Guillouet and A. Legault. 2005b. Temporal changes in European eel, *Anguilla anguilla*, stocks in a small catchment after installation of fish passes. Fish. Manage. Ecol. 12: 123–129.

Laffaille, P., E. Lasne and A. Baisez. 2009. Effects of improving longitudinal connectivity on colonisation and distribution of European eel in the Loire catchment, France. Ecol. Freshw. Fish 18: 610–619.

Lamson, H.M., J.C. Shiao, Y. Iizuka, W.-N. Tzeng and D. Cairns. 2006. Movement patterns of American eels (*Anguilla rostrata*) between salt- and freshwater in a coastal watershed, based on otolith microchemistry. Mar. Biol. 149: 1567–1576.

Lasne, E., A. Acou, A. Vila-Gispert and P. Laffaille. 2008. European eel distribution and body condition in a river floodplain: effect of longitudinal and lateral connectivity. Ecol. Freshw. Fish 17: 567–576.

Legault, A. 1992. Etude de quelques facteurs de sélectivité de passes à anguilles. Bull. Français Pêche Piscic. 325: 83–91.

Limburg, K.L., H. Wickstrom, H. Svedang, M. Elfman and P. Kristiansson. 2003. Do stocked freshwater eels migrate? Evidence from the Baltic suggests "yes". Am. Fish. Soc. Symp. 33: 275–284.

Lobon-Cervia, J. and T. Iglesias. 2008. Long-term numerical changes and regulation in a river stock of European eel *Anguilla anguilla*. Freshw. Biol. 53: 1832–1844.

Lobon-Cervia, J., C.G. Utrilla and P.A. Rincon. 1995. Variations in the population dynamics of the European eel *Anguilla anguilla* (L.) along the course of a Cantabrian river. Ecol. Freshw. Fish 4: 17–27.

Mann, R.H.K. and J.H. Blackburn. 1991. The biology of the eel *Anguilla anguilla* (L.) in an English chalk stream and the interactions with juvenile trout *Salmo trutta* L. and salmon *Salmo salar* L. Hydrobiol. 218: 65–76.

Marcogliese, L.A. and J.M. Casselman. 2009. Long-term trends in size and abundance of juvenile American eels ascending the upper St. Lawrence River. Am. Fish. Soc. Symp. 33: 191–205.

Martin, M., J. Boubée and E. Bowman. 2013. Recruitment of freshwater elvers 1995–2012. New Zealand Fisheries Assessment Report 2013/50. Ministry for Primary Industries, Wellington. 111p.

McDowall, R.M. 1990. New Zealand freshwater fishes: a natural history and guide, Heinemann-Reed, Auckland.

McDowall, R.M. and M.J. Taylor. 2000. Environmental indicators of habitat quality in a migratory freshwater fish fauna. Environ. Manage. 25: 357–374.

Melia, P., D. Bevacqua, A.J. Crivelli, G.A. De Leo, J. Panfili and M. Gatto. 2006. Age and growth of *Anguilla anguilla* in the Camargue lagoons. J. Fish Biol. 68: 876–890.

Mitchell, C.P. 1989. Swimming performances of some native freshwater fishes. N.Z. J. Mar. Freshw. Res. 23: 181–187.

Moriarty, C. 1986. Riverine migration of young eels *Anguilla anguilla* (L.). Fish. Res. 4: 43–58.

Morrison, W.E. and D.H. Secor. 2003. Demographic attributes of yellow-phase American eels (*Anguilla rostrata*) in the Hudson River estuary. Can. J. Fish. Aquat. Sci. 60: 1487–1501.

Nair, R.V. 1973. On the export potential of elvers and cultured eels from India. Ind. J. Fish. 20: 608–616.

Naismith, I.A. and B. Knights. 1988. Migrations of elvers and juvenile European eels, *Anguilla anguilla* L., in the River Thames. J. Fish. Biol. 33: 161–175.

Nyman, L. 1972. Some effects of temperature on eel (*Anguilla*) behaviour. Report—Institute of Freshwater Research Drottningholm 52: 90–102.

Oliveira, K. 1997. Movements and growth rates of yellow-phase American eels in the Annaquatucket River, Rhode Island. Trans. Am. Fish. Soc. 126: 638–646.

Parrill, E. and K. Vodden. 2014. "We always did fish the eels"—Qalipu Mi'kmaq ecological impacts in the American eel fisheries of western Newfoundland. pp. 8–15. *In*: P. McConney, R.P. Medeiros and M. Pena (eds.). Enhancing Stewardship in Small-Scale Fisheries. Practices and Perspectives. CERMES Technical Report No. 73. Special edition.

Parsons, J., K.U. Vickers and Y. Warden. 1977. Relationship between elver recruitment and changes in the sex ratio of silver eels *Anguilla anguilla* L. migrating from Lough Neagh, Northern Ireland. J. Fish. Biol. 10: 211–229.

Pedersen, M.I. 2009. Does stocking of Danish lowland streams with elvers increase European eel populations? Am. Fish. Soc. Symp. 58: 149–156.

Pratt, T.C. and R.W. Threader. 2011. Preliminary evaluation of a large-scale American eel conservation stocking experiment. N. Am. J. Fish. Manage. 31: 619–628.

Prigge, E., L. Marohn and R. Hanel. 2013. Tracking the migratory success of stocked European eels *Anguilla anguilla* in the Baltic Sea. J. Fish. Biol. 82: 686–699.

Pusey, B., M. Kennard and A. Arthington. 2004. Freshwater Fishes of North-Eastern Australia, CSIRO Publishing.

Richkus, W.A. and K. Whalen. 2000. Evidence for a decline in the abundance of the American eel, *Anguilla rostrata* (LeSueur), in North America since the early 1980s. Dana 12: 83–97.

Righton, D., K. Aarestrup, D. Jellyman, P. Sébert, G. van den Thillart and K. Tsukamoto. 2012. The *Anguilla* spp. migration problem: 40 million years of evolution and two millennia of speculation. J. Fish. Biol. 81: 365–386.

Sagar, P.M. and G.J. Glova. 1998. Diel feeding and prey selection of three size classes of shortfinned eel (*Anguilla australis*) in New Zealand. Mar. Freshw. Res. 49: 421–428.

Sagar, P.M., E. Graynoth and G.J. Glova. 2005. Prey selection and dietary overlap of shortfinned (*Anguilla australis*) and longfinned (*A. dieffenbachii*) eels during summer in the Horokiwi Stream, New Zealand. N.Z. J. Mar. Freshw. Res. 39: 931–939.

Schmidt, J. 1925. On the distribution of the fresh-water eels (*Anguilla*) throughout the world. II. Indo-Pacific region. Kong. Danske Vidensk. 8: 329–382.

Schmidt, R.E., C.M. O'Reilly and D. Miller. 2009. Observations of American eels using an upland passage facility and effects of passage on the population structure. N. Am. J. Fish. Manage. 29: 715–720.

Shiao, J.C., Y. Iizuka, C.W. Chang and W.-N. Tzeng. 2003. Disparities in habitat use and migratory behavior between tropical eel *Anguilla marmorata* and temperate eel *A. japonica* in four Taiwanese rivers. Mar. Ecol. Prog. Ser. 261: 233–242.

Skead, C.J. 1959. The climbing of juvenile eels. Piscator 46: 74–86.

Sloane, R.D. 1984. Upstream migration by young pigmented freshwater eels (*Anguilla australis australis* Richardson) in Tasmania. Austr. J. Mar. Freshw. Res. 35: 61–73.

Sorensen, I. 1951. An investigation of some factors affecting the upstream migration of the eel. Report Institute of Freshwater Research Drottningholm 32: 126–132.

Smogor, R.A., P.L. Angermeier and C.K. Gaylord. 1995. Distribution and abundance of American eels in Virginia streams: tests of null models across spatial scales. Trans. Am. Fish. Soc. 124: 789–803.

Tamatamah, R. 2012. Current status of the native catadromous eel species *Anguilla bengalensis labiata* and *Anguilla mossambica* in Tanzania. pp. 128–129. *In*: P. Gough, P. Philipsen, P.P. Schollema and H. Wanningen (eds.). From Sea to Source. International Guidance for the Restoration of Fish Migration Highways. Regional Water Authority, Hunze en Aa's, Veendam, The Netherlands.

Tesch, F.-W. 2003. The Eel. 3rd edit. Blackwell Science Ltd., Oxford.

Thibault, I., J.J. Dodson, F. Caron, W.-N. Tzeng, Y. Iizuka and J.C. Shiao. 2007. Facultative catadromy in American eels: testing the conditional strategy hypothesis. Mar. Ecol. Prog. Ser. 344: 219–229.

Tomie, J.P.N., D.K. Cairns and S.C. Courtenay. 2013. How American eels *Anguilla rostrata* construct and respire in burrows. Aquat. Biol. 19: 287–296.

Tsukamoto, K. and T. Arai. 2001. Facultative catadromy of the eel *Anguilla japonica* between freshwater and seawater habitats. Mar. Ecol. Prog. Ser. 220: 265–276.

Tsukamoto, K., I. Nakai and F. -W. Tesch. 1998. Do all freshwater eels migrate? Nature 396: 635–636.

Tzeng, W.-N. 1996. Effects of salinity and ontogenic movements on strontium:calcium ratios in the otoliths of the Japanese eel, *Anguilla japonica* Temminck and Schlegel. J. Exp. Mar. Biol. Ecol. 199: 111–122.

Tzeng, W.-N., C.H. Wang, H. Wickström and M. Reizenstein. 2000. Occurrence of the semi-catadromous European eel *Anguilla anguilla* in the Baltic Sea. Mar. Biol. 137: 93–98.

Vollestad, L.A. and B. Jonsson. 1988. A 13-year study of the population dynamics and growth of the European eel, *Anguilla anguilla*, in a Norwegian river: evidence for density-dependent mortality, and development of a model for predicting yield. J. An. Ecol. 57: 983–997.

Walsh, C.T., B.C. Pease, S.D. Hoyle and D.J. Booth. 2006. Variability in growth of longfinned eels among coastal catchments of south-eastern Australia. J. Fish Biol. 68: 1693–1706.

Wasserman, R.J., L.L. Pereira-da-Conceicoa, N.A. Strydom and O.L.F. Weyl. 2012. Diet of *Anguilla mossambica* (Teleosti, Anguillidae) elvers in the Sundays River, Eastern Cape South Africa. African J. Aquat. Sci. First, 1–3, DOI: 10.2989/16085914.2012.692320.

White, E.M. and B. Knights. 1997a. Environmental factors affecting migration of the European eel in the Rivers Severn and Avon, England. J. Fish. Biol. 50: 1104–1116.

White, E.M. and B. Knights. 1997b. Dynamics of upstream migration of the European eel, *Anguilla anguilla* (L.), in the Rivers Severn and Avon, England, with special reference to the effects of man-made barriers. Fish. Manage. Ecol. 4: 311–324.

Wiley, D.J., R.P. Morgan, R.H. Hilderbrand, R.L. Raesly and D.L. Shumway. 2004. Relations between physical habitat and American eel abundance in five river basins in Maryland. Trans. Am. Fish. Soc. 133: 515–526.

Yokouchi, K., N. Fukuda, M.J. Miller, J. Aoyama, F. Daverat and K. Tsukamoto. 2012. Influences of early habitat use on the migratory plasticity and demography of Japanese eels in central Japan. Estuar. Coast. Shelf Sci. 107: 132–140.

Feeding Ecology

Hendrik Dörner[a],* and *Søren Berg*[b]

Introduction

"If a fisheries biologist needs to describe something which is both new and simple, go to the tropics, catch some freshwater eels and describe their diet". This phrase epitomises the knowledge distribution of the feeding ecology of anguillid eels today[1] as this knowledge is indeed distributed very unevenly among the species. The feeding ecology of some tropical species of eel still remains almost completely unknown while those temperate species which are of highest commercial (in terms of both fishing as well as aquaculture) or cultural importance are those that have been described best. The four most well investigated species appear to be *Anguilla anguilla*, *Anguilla rostrata*, *Anguilla dieffenbachii*, and *Anguilla australis* and of these, *A. anguilla* is by far the best described. Since the 1930s, commercial and cultural aspects have without doubt been the most important drivers of interest in gaining knowledge on the feeding ecology of the different life-stages of freshwater eels (Godfrey 1957; Tesch 2003). Nevertheless, the current knowledge-base largely relates to the yellow eel stage of anguillid eels and the feeding ecology of the larval stage (leptocephali) is still poorly understood (Terahara et al. 2011; Miller et al. 2013). The severe decline in freshwater eel populations over the last 30 years or more

[a] European Commission, Joint Research Centre JRC, Institute for the Protection and Security of the Citizen IPSC, Maritime Affairs Unit, Via Enrico Fermi 2749, 21027 Ispra (VA), Italy.
[b] DTU Aqua, National Institute of Aquatic Resources, Technical University of Denmark, Vejlsøvej 39, DK-8600 Silkeborg, Denmark.
  E-mail: sbe@aqua.dtu.dk
* Corresponding author: hendrik.doerner@jrc.ec.europa.eu

[1] The authors are aware that in real life, this task might not be so simple and will very likely also be quite expensive.

(Moriarty and Dekker 1997; Dekker 2004) provided an additional important driver for gaining knowledge on their feeding ecology, which over the past 25 years, has led to an improvement in the knowledge-base on leptocephali feeding (Miller 2009; Miller et al. 2013).

The present chapter firstly describes the food and feeding behavior during three of the four developmental stages in the life cycle of eels, leptocephalus, elver, and yellow eel stages but because it is generally accepted that mature or silver eels do not feed (Tesch 2003; Chow et al. 2010), this stage in the life cycle is not described here. To reflect the diversity in feeding ecology during the yellow eel stage, the present chapter also contains sections on food composition in relation to environment/habitat, temperature effects, feeding habits and body size. Subsequent sections describe interactions of eels with other species followed by a section summarizing current knowledge for anguillid species for which information covering the entire post-elver feeding ecology is limited. Finally, the chapter concludes with a section summarizing the patterns common to the feeding ecology of anguillid eel species.

Leptocephali feeding

Several publications relate to feeding of leptocephali of marine eel species from the orders of Elopiformes and Notacanthiformes (e.g., Hulet 1978; Mochioka and Iwamizu 1996; Govoni 2010). Considering the pelagic behavior and morphological similarity of leptocephali around the globe (Miller 2009) it is reasonable to assume and it is also commonly accepted, that the feeding behavior and diet composition of leptocephali of all species of eel is likely to be similar. Consequently, the information presented here, is not restricted to the leptocephali of anguillid eels.

The food and feeding habits of anguillid larvae or leptocephali have long been the subject of speculation and discussion. One theory states that they are pseudo-parasites (Moser 1981). This theory arose in part, due to the fact that despite several investigations, no visible traces of food had been found in the gut of leptocephali (Kratch and Tesch 1981). One of the first serious attempts to explain the nutrition of anguillid leptocephali was made by Hulet (1978) who, based on an anatomical study of the leptocephali of the congrid eel *Ariosoma balearicum*, suggested that leptocephali in general get their nutrition from either particulate (POM) or dissolved (DOM) organic matter. Pfeiler (1986) expanded this theory and argued that evidence pointed to the explanation that stage I leptocephali completely or partly rely on epithelial uptake of DOM. One of the arguments presented in support of such a theory was that the epithelium of leptocephali is only one to three cell layers thick and showed traces of micro-villi (Hulet 1978), making the uptake of DOM theoretically possible. He also argued that in theory, the surface to volume ratio of the leaf-shaped larvae is appropriate to facilitate this type of nutrient assimilation.

Since the early 1990s, more information on the feeding ecology of leptocephali has emerged and the interest in the artificial propagation of eels for aquaculture purposes and the decline in the populations of many species were probably major drivers responsible for the increase in knowledge (Dekker 2004; Tanaka et al. 2003). Observations from laboratory experiments in Japan showed leptocephali of the marine eels *Muraenesox cinereus* and *Conger myriaster* as actively biting off and ingesting pieces of dyed squid paste, which then could be seen in their guts (Mochioka et al. 1993, Fig. 1). Tanaka et al. (1995) also observed *A. japonica* ingesting rotifers in a laboratory experiment. These results clearly indicated nutritional uptake through the intestine. Other observations of small particles in the guts of leptocephali caught in the

Fig. 1. Photograph of *Muraenesox cinereus* leptocephali. The one on the left had ingested red dyed squid paste in a laboratory experiment *ca.* 10 minutes before the picture was taken. White arrow shows the anus. Left hand upper corner: red squid paste attached to the aquarium wall. Reproduced from Mochioka et al. (1993), with kind permission from Springer (License No. 3640681448058).

wild further indicated that leptocephali actively ingest food particles (Otake et al. 1993; Govoni 2010). In addition, in a laboratory experiment the larvae of *A. japonica* were unable to survive and grow in DOM-enriched water (Liao and Chang 2001). A dietary study of the larvae of eight marine species of eel caught off the coast of Japan indicated that their diets comprised of two major food items; larvacean (Subphylum Tunicata, Class Appendicularia) houses and their fecal pellets (Mochioka and Iwamizu 1996, Fig. 2). The larvaceans

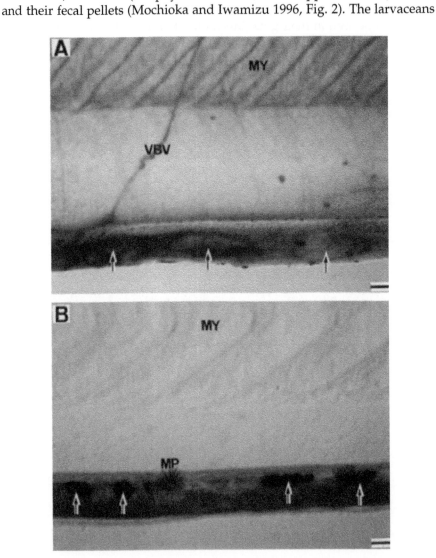

Fig. 2. Anguilloid larvae. (A) Larvacean houses (arrows) in the intestine of *Gymnothorax* sp. leptocephalus. Scale bar = 0.15 mm. (B) Larvacean fecal pellets (arrows) in the intestine of *Conger myriaster* leptocephalus. Scale bar = 0.15 mm (MP melanophore; MY myomere; VBV vertical blood vessel). Reproduced from Mochioka and Iwamizu (1996), with kind permission from Springer (License No. 3640690298773).

themselves were not observed, but it is not clear whether the leptocephali specifically target empty houses or if the larvacean escapes when the house has been seized by a leptocephalus (Miller 2009). Moehioka and Iwamizu (1996) suggested that leptocephali were capable of puncturing and ingesting the contents of larvacean houses that are too big to be swallowed, as they found fecal pellets from such larvaceans in the guts of leptocephali. Ciliates have been reported from the guts of leptocephali of *A. japonica* (Otake et al. 1993) and Govoni (2010) found diatoms, armored dinoflagellates, tintinninds and rotifers in the guts of worm eel leptocephali (*Myrophis* spp., family Ophichthidae), from the northern Gulf of Mexico.

The most recent papers on the diet of anguillid larvae have analyzed the qualitative composition of the diet using methods other than traditional visual observation; two using DNA techniques on *A. anguilla* (Riemann et al. 2010) and *A. japonica* (Terahara et al. 2011), and two papers which examine the trophic position of *A. japonica* by means of stable isotope analysis (Mijazaki et al. 2011; Miller et al. 2013). These papers give the most detailed picture to date on the diet of anguillid leptocephali. The results of Riemann et al. (2010) point to a variety of zooplankton as the most probable diet of *A. anguilla* leptocephali. DNA from one to 17 (mean: three) prey species were identified in different specimens examined, while *ca.* 1/3 of the DNA detected could not be assigned to any known species. Gelatinous zooplankton was found to be the most important component of the diet. It was not possible to decide if larvacean houses were part of the diet as they consist of polysaccharides and consequently cannot be detected by the DNA barcoding techniques used. Terahara et al. (2011) did not find similar results and, based on their finding of relatively short DNA fragments, suggested that POM (or marine snow) is the most likely diet of *A. japonica* leptocephali. Also in this study, a large fraction of the DNA found could not be assigned to any known species (Terahara et al. 2011). In contrast to Riemann et al. (2010), the studies on stable isotope analysis concluded that the leptocephali of *A. japonica* are unlikely to feed extensively on zooplankton or secondary consumers and favors POM as their main food source (Mijazaki et al. 2011; Miller et al. 2013). These results led the authors to question the validity of the results obtained for *A. anguilla*. In conclusion, further work is required, before the food and feeding behavior of anguillid leptocephali can be adequately described and understood, e.g., if larvacean houses or POM are the main food objects, then how important for nutrition are the bacteria found on their surface (Miller et al. 2013)? While the use of shark egg yolk powder (and other ingredients) has been used successfully as the main component in the artificial propagation of *A. japonica* leptocephali, the main constituents of their natural diet remain elusive (Tanaka et al. 2003).

Elver feeding

Compared to both the leptocephalus and especially the yellow eel stages, the elver (glass eel) stage is short-lived and merely constitutes a transitional phase between leptocephali and pigmented eels. Nevertheless, the glass eel stage is vital as this is when they begin their migration into freshwater (Tesch 2003). Thus, the nutritional needs to support the energy demand of a migrating fish must be met. However, knowledge of the natural food and feeding behavior of anguillid elvers is also scarce, even scarcer than that for anguillid leptocephali. It is generally accepted that during metamorphosis from leptocephalus to elver, while the gut is being rearranged and the anus moves forward, anguillid larvae do not eat (Tesch 2003; Miller 2009), but in the early elver stages of *A. anguilla* feeding recommences (Tesch 2003). This is before the elvers have become truly benthic which expectedly is why Bardonnet and Riera (2005) still found evidence of marine origin of a major fraction of *A. anguilla* diet in elvers caught in the estuary of the Adour River in South-Western France. Upon reaching the coastal zone, the elvers begin feeding more on benthic resources. Tesch (2003) reports that *A. anguilla* elvers from the North and Baltic Sea coasts feed on *Mysis*, *Idothea*, copepods and chironomid larvae, and to lesser extent on polychaetes, oligochaetes, amphipods, *Asellus* and various insect larvae. Stable isotope analysis of *A. anguilla* elvers from the estuary of the Adour River indicates that besides marine food the outflowing POM might play a role in the nutrition of elvers in that area (Bardonnet and Riera 2005). When comparing the results reported by Tesch (2003) and Bardonnet and Riera (2005), the observed differences in diet might be due to the fact that the Adour River is located much closer to the Atlantic deep sea than the coasts of the North and Baltic Seas. Thus elvers entering the Adour estuary are less developed than elvers on the more distant coasts of the North and Baltic Seas (Tesch 2003).

When the elvers commence their migration to freshwater, they adopt a benthic habit and as shown by Wasserman et al. (2005), such behavior is reflected in their diet. Wasserman et al. (2005) examined the diet of 45.5–116 mm (TL) *A. mossambica* from the lower reaches of the Sundays River, South Africa. They found the contents of elver stomachs to be dominated by insects (especially chironomids and simulids) while copepod and fish remains were also present but very rare. The observed diet reflected the numerical abundance of the prey categories in the river, indicating that like young yellow eels, the elvers become generalists (Wasserman et al. 2005) and have comparable diets.

Yellow eel feeding in different habitats

Post-elver anguillid or yellow eels have been described as omnivorous and their food composition has been assessed in diverse habitats including coastal and inland lakes, brackish estuaries and lagoons, and from tiny streams to large rivers (Cairns 1950; Tesch 2003). A wide range of items are used as prey by eels and their prey spectrum spans from insect larvae and crustaceans to mussels and fish (Godfrey 1957; Moriarty 1972, 1973; Deelder 1984). Post-elver eels are generalists and feed on benthic macroinvertebrate taxa depending on abundance and availability in their particular habitat (Table 1). They are also able to utilize drifting or inundated food items such as insects and terrestrial annelids (Tesch 2003). This remarkably wide dietary spectrum may also explain why eels are so successful in inhabiting a wide variety of different habitats such as estuaries, rivers, streams, ponds and lakes (Tesch 2003).

The variation in diet between species and habitats indicates a prevalence of comparatively immobile prey (e.g., benthos), providing such prey occurs in sufficiently large numbers (Table 1, Moore and Moore 1976). This observation is further confirmed by a variety of studies, e.g., Dörner et al. (2009) or Burnet (1952), showing that fish were a major prey only under conditions where the invertebrate fauna was not largely available. The diets of *A. anguilla*, *A. rostrata*, *A. dieffenbachii*, *A. australis* are also remarkably different when comparing running waters with lakes (Table 1). Eels in streams and rivers predominantly feed on Trichoptera and Ephemeroptera whereas for eels in lakes choronomids, molluscs and fish appear to be of higher importance. Furthermore, comparing the freshwater diet of eels to the diet of those in brackish coastal and estuary systems (Table 1) also reveals major differences with crustaceans, fish, and amphipods being the predominant food.

Yellow eel feeding and temperature effects

Anguillid eels are poikilothermic species and, according to Tesch (2003), are well adapted to rather warm water conditions showing optimal growth at 23°C to 26°C. Several studies have investigated the influence of temperature on eel feeding activity and have identified temperatures between 5°C to 6°C as lower and 25°C to 30°C as upper boundaries (for *A. anguilla*: Rasmussen and Therkildsen 1979; Gamito et al. 2003; Bouchereau et al. 2009; for *A. australis*: Woods 1964). Generally, feeding is most intense during the spring and summer periods, a pattern which has been shown in numerous studies on *A. anguilla* (Frost 1946; Sinha and Jones 1967; Deelder 1984; De Nie 1987; Bouchereau et al. 2006; Cullen and McCarthy 2007). A similar pattern has been described for *A. rostrata* (Ogden 1970; Smith and Saunders 1955), and for *A. australis* (Beumer 1979; Ryan 1984) and *A. dieffenbachii* (Cairns 1942; Burnet 1952). Seasonal differences in stomach fullness indices and feeding habits, however, were mostly attributed to availability of the different prey types (Cairns 1950; Ryan 1986; De Nie 1987).

Table 1. Diet of *A. anguilla, A. rostrata, A. dieffenbachii,* and *A. australis* in different habitats. Prey organisms are listed in order of importance.

A. anguilla

Lake	Coastal/Estuary	River/Stream
Lake Tjeukemeer, The Netherlands (De Nie 1987)	Tagus estuary, Portugal (Costa et al. 1992)	Lower River Shannon, Ireland (Cullen and McCarthy 2007)
Chironomids most important (percentage of occurrence) food for eels < 25 cm TL. Young-of-the-year smelt and perch and bivalves most important food for eels > 25 cm total length (n analysed stomachs = 1,086). 29 taxonomical groups of invertebrates and 7 teleost species found	Amphipods and shore crab in general most important. In more saline muddy bottom areas polychaetes, bivalves, and shrimp more important than amphipods. Gastropods, mysids, arachnids, and insects incidental prey, with low numeric and weight values below 10%	Hirudinea; Cladocera; Isopoda*; Amphipoda; Gastropoda*; Ephemeroptera**; Trichoptera**; Odonata; Hemiptera; Diptera*; Coleoptera; Fish (*: frequently eaten. ** Main food items in terms of % stomach fullness)
Lake GroßerVätersee, Germany (Dörner et al. 2009)	Estuary of River Severn, UK (Moore and Moore 1976)	12 Rivers, Wales, UK (Sinha 1969)
Composition of diet biomass was 40% percids, 33% cyprinids, 17% crayfish, 8% insect larvae, and 2% Gastropoda	Most important prey species *Crangon vulgaris* and *Neomysis integer* (in terms of prey biomass). Other species eaten: *Praunus flexuosus* and *Gammarus salinus*	Most important prey (percentage of occurrence): Trichoptera (41.7), Crustacea (13.2), fish remains (13.2), Diptera (11.8), fish (11.1)

A. rostrata

Lake	Coastal/Estuary	River/Stream
Lake Ontario–upper St. Lawrence River corridor, USA/Canada (Fitzsimons et al. 2013)	Stanley River estuary complex, Southern Gulf of St. Lawrence, Canada (Schein et al. 2013)	8 streams in New Jersey, USA (Ogden 1970)
Mainly crayfish (49%) and fish (48%) (round goby *Neogobius melanostomus*). Combined amount of alewives *Alosa pseudoharengus*, slimy sculpin *Cottus cognatus*, and rainbow smelt, *Osmerus mordax* was 2%. Results based	Prey biomass: 55.7% fish, 18.1% clams, 10% amphipods, 2.6% shrimp, 2.1% Ulva, and 0.2% snails	Ephemeroptera, Megaloptera, and Trichoptera most important prey (frequence of occurrence) for smaller eels (< 40 cm total length); fish and Crustacea were more frequently found in larger specimens

	(> 40 cm). Prey fish were mainly bottom-dwelling and sluggish species. No trout as prey. One small mammal
	Hannacroix Creek, tributary of the Hudson River, USA (Waldt et al. 2013)
	Most important diet was a variety of many types of aquatic insects. Ephemerellid mayfly nymphs were the major prey consumed with up to 55.8% of prey biomass

A. dieffenbachia

Lake	Coastal/Estuary	River/Stream
Lake Rotoiti, Nelson Lakes National Park, New Zealand (Jellyman 1996)	Ahuriri estuary, Napier, New Zealand (Kilner and Akroyd 1978)	Two streams, New Zealand (Hopkins 1965)
Fish (bullies *Gobiomorphus breviceps* and brown trout *Salmo trutta*), invertebrates (snails, insect larvae (dragonfly)). Fish most important by prey biomass, invertebrates most important by number	Crustaceans (mudflat crabs, amphipods, shrimps, and isopods) most important food (percentage of occurrence)	Most important prey (percentage of occurrence) mayfly larvae, also Mollusca and Trichoptera larvae, no fish, plant material
		Stream, New Zealand (Sagar et al. 2005)
		Gastropods most important food component (percentage of occurrence) of large and medium size classes; larger eels used more taxa than smaller eels; significant overlap in diet of different size classes

on estimates from a mixing model that used the stable isotope signatures of eels and their potential prey

Table 1. contd....

Table 1. contd.

A. australis

Lake	Coastal/Estuary	River/Stream
Macleods Morass, Victoria, Australia (Beumer 1979)	Ahuriri estuary, Napier, New Zealand (Kilner and Akroyd 1978)	Two streams, New Zealand (Hopkins 1965)
Most important prey groups (percentage of occurrence): teleosts, insects. Also: amphibians, molluscs, lepidopterans, hymenopterans	Most important food items were Crustacea 19.1% (mudflat crabs 14.9%) and fish 6.9% (percentage of occurrence)	Most important prey (percentage of occurrence) mayfly larvae, also Mollusca and Trichoptera larvae, no fish, plant material
Lake Ellesmere, Canterbury, New Zealand (Ryan 1986)		Glentui River, Canterbury, New Zealand (Cadwallader 1975)
Most important prey (prey biomass): mollusk Potamopyrgus antipodarum, the isopod Austridotea annectens, the mysid Tenago mysischiltoni, the amphipod Paracalliope fluviatilis, the midge larva Chironomus zealandicus, and the teleosts Retropinna retropinna, Galaxias maculatus, and Gobiomorphus cotidianus		Most important prey (percentage of occurrence): Megaloptera, Plecoptera, Ephemeroptera, Trichoptera, Coleoptera, Gastropoda, Potamopyrgus, Nematomorpha, Oligochaeta; terrestrial Arthropoda, plant material

Yellow eel feeding habit and body size

European eel—A. anguilla

The European eel *A. anguilla* has been described as a nocturnal opportunistic feeder in many studies and there is a close correlation between the availability of benthic food and diet composition (Frost 1946; Tesch 2003; Bouchereau et al. 2009). The size of European eels investigated in the literature studied in preparation for this book chapter ranged from 10 cm (Cullen and McCarthy 2007) to 109 cm total length (Golani et al. 1988). It is apparent that *A. anguilla* broadens the spectrum of potential diet taxa with increasing body size as the eels become increasingly capable of dealing with bigger and also harder food items. Consequently, some studies describe large[2] eels as being more inclined towards fish and/or crayfish as food than smaller eels (Ezzat and El-Seraffy 1977; Deelder 1984; Dörner and Benndorf 2003). However, the composition of the diet of eels of all sizes is positively correlated to the availability of benthic invertebrates, especially chironomids in lakes (Frost 1946; Kangur et al. 1999; Dörner et al. 2009).

American eel—A. rostrata

The size of American eels *A. rostrata* investigated in the literature studied in preparation for this book chapter ranged from 6 cm (Godfrey 1957) to 128 cm total length (Fitzsimons et al. 2013). As in case of the European eel *A. anguilla*, *A. rostrata* has been described as a nocturnal opportunistic feeder preying on the most abundant prey types. American eels < 40 cm body size appear to predominantly feed non-selectively on many benthic macroinvertebrate taxa. On rare occurrences specimens of *A. rostrata* > 15 cm total length, were found to prey on fish fry (*Salmo salar*) but eel predation on the total mortality of salmon fry appeared insignificant (Godfrey 1957). In general, fish and crustaceans are not important as food for small (< 40 cm total length) *A. rostrata*. For larger specimens, fish and crustaceans are more frequently found in the diet and prey fish are mainly bottom-dwelling species (Ogden 1970).

A. australis and A. dieffenbachii

In line with what has already been observed for the European and American eels, *A. australis* and *A. dieffenbachii* have also been identified as opportunistic night-active generalists (Cairns 1950; Ryan 1984). Different feeding and/or habitat preferences may occur where the two species co-exist (Jellyman 1989; Sagar et al. 2005). The size of *A. australis* and *A. dieffenbachii* investigated in the literature studied in preparation for this book chapter ranged from 8 cm (Sagar

---

[2] A body size of *ca.* 40 cm total length has frequently been specified as 'turning point' with eels < 40 cm labeled as small and eels ≥ 40 cm labeled as large.

and Glova 1998) to 90 cm (Kilner and Akroyd 1978) and 12 cm (Cadwallader 1975) to 137 cm total length (Burnet 1952), respectively. Individuals up to 40 cm total length have been shown to predominantly feed on benthic invertebrates (Hopkins 1965; Cadwallader 1975), whereas larger individuals increasingly included various fish and crustacean species in their diet depending on the abundance and availability of potential food sources (Cairns 1950). Large *A. dieffenbachii* > 75 cm total length have been shown to be significant piscivores (Cairns 1942).

Yellow eel interactions with other species

While studying a vast amount of literature in preparation of the present book chapter on the feeding ecology of eels we noticed studies indicating that eels have the potential to influence the structure of invertebrate macrozoobenthos and benthic fish communities in running waters (Stranko et al. 2014 for *A. rostrata*) and to provide a link between benthic and pelagic food webs in lakes (Dörner et al. 2009 for *A. anguilla*). Stranko et al. (2014) investigated wadeable streams in Maryland, USA, by conducting a data meta-analysis to compare 95 sites where American eel access was excluded by dams with 74 sites containing eels. They found benthic fish densities to be significantly lower and benthic macroinvertebrate densities significantly higher at eel-containing sites. However, densities of non-benthic fish, top predators other than eels, and crayfish were not different between the two eel occurrence categories. The authors concluded that American eel may directly influence benthic fish and indirectly influence benthic macroinvertebrate densities and thus potentially play an important role in structuring the food web dynamics in streams.

Stomach contents and stable isotope analysis of the feeding habits of the European eel *A. anguilla* in lakes with differing abundances of macrozoobenthos and small potential prey fish also revealed distinct differences in diet (Dörner et al. 2009). Piscivory among *A. anguilla* was generally determined by the density of macrozoobenthos which in turn, influenced energy flow through the food chains of lake ecosystems (Dörner et al. 2009). More specifically, in systems where macrozoobenthos density is low, *A. anguilla* are predominantly piscivorous (Radke and Eckmann 1996; Dörner et al. 2009) thus coupling benthic and pelagic food webs. In systems with higher macrozoobenthos density, however, this benthic-pelagic food web is decoupled, as the high macrozoobenthos density reduces piscivory in *A. anguilla* (Kangur et al. 1999; Dörner et al. 2009). Despite the fact that abundance of prey fishes is often much higher in such systems, the eels feed on macrozoobenthos which can probably be explained by the lower energy costs associated with foraging for macroinvertebrates as compared to pursuing fish (Helfman and Winkelman 1991).

Numerous studies on the diet of eels have been undertaken because eels were a major concern for recreational fishery managers as they were believed to have strong negative impacts on wild and introduced salmonid stocks through predation (Cairns 1942). The notion that salmonid stocks would benefit through the elimination of eels (Burnet 1968) was founded in the 1930s and was still prevalent in the 1950s (Godfrey 1957). Even as recent as the late 1980s the 'eel negatively impact trout' concept led to a study on the diet of the African mottled eel *Anguilla labiata* in the upper Kairezi River, Zimbabwe (Butler and Marshall 1996). However, none of the studies that have investigated eel—salmonid interactions was able to demonstrate any substantial negative impact on salmonid populations solely as a result of predation by eels (Thomas 1962; Sinha and Jones 1967; Mann and Blackburn 1991; Butler and Marshall 1996). While there is a considerable niche overlap between eels and trout (e.g., *Salmo trutta*) both feeding on benthic invertebrates (Santamarina 1993), competition for invertebrate food has been found to be buffered by factors such as differences in feeding activity times (Sagar and Glova 1994) or habitat use (Cadwallader 1975; Sagar et al. 2005). Finally, as observed by Jellyman (1996), some of New Zealand's most productive trout fisheries occur in rivers where considerable numbers of large *A. dieffenbachii* coexist with large trout (Jellyman and Graynoth 1994).

Information limited species

As indicated in the introduction, the number of anguillid species for which a sufficient number of studies covering the entire post-elver feeding ecology has been accessible to us was limited to four (*A. anguilla*, *A. rostrata*, *A. australis*, and *A. dieffenbachii*). In this section we summarize the information obtained for other anguillid species (Table 2). Recently published studies on *A. japonica* dealt with feeding in rivers (Kaifu et al. 2013a; Itakura et al. 2015) and/or and coastal systems (Kaifu et al. 2013a,b). Eels in a size range of 40 cm to 60 cm total length were investigated. The diet composition of *A. japonica* was consistent with the general pattern described earlier in this chapter with crustaceans being the main food in coastal systems, while for freshwater habitats crustaceans, insects, and fish were most important. Pantulu (1957) intensively studied the diet of *A. bengalensis* and found it to be generally omnivorous. Major prey groups of *A. reinhardtii* in the Morass still water system were teleosts and insects with no apparent direct relationship between the size of a particular item taken and the size of the eel (Beumer 1979). Resh et al. (1999) investigated the diet of *A. marmorata* and *A. obscura* in Rivers in the tidal zone in Moorea Island, French Polynesia. The low number of individuals studied does not allow for any firm conclusions to be drawn on their diet. However, regular amphidromy was mentioned indicating that eels regularly migrate from freshwater to sea in order to feed. Further, a study of Butler and Marshall (1996) demonstrated that *A. labiata* in the upper Kairezi River, Zimbabwe predominantly fed on freshwater crustaceans.

Table 2. Diet of anguillid eels in different habitats. Prey organisms are listed in order of importance.

	A. japonica	
Lake	**Coastal/Estuary**	**River/Stream**
	Kojima Bay, Japan (Kaifu et al. 2013a)	Asahi River, Japan (Kaifu et al. 2013a)
	Mud shrimps (*Upogebia major*): 78% prey biomass, 56% prey occurrence; crabs (*Hemigrapsus penicillatus*): 3% prey biomass, 3% prey occurrence; unidentified crustaceans: 14% prey biomass, 31% prey occurrence; no fish. Eel body size: 58.9 ± 6.5 cm total length (n = 96)	Freshwater crayfish (*Procambarus clarkii*) 75% prey biomass, 55% prey occurrence; Insecta: 15% prey biomass, 28% prey occurrence; fish: 0.6% prey biomass, 3.4% prey occurrence. Eel body size 52.7 ± 8.6 cm total length (n = 50)
	Kojima Bay and estuary (Kaifu et al. 2013b)	River Tone, Japan (Itakura et al. 2015)
	Estuary: mud shrimp most important food with 70% of prey biomass and 4% fish, other crustaceans and sand worms. Eel body size: 43.1 ± 11.6 cm (n = 162). Bay: 96% prey biomass crustaceans and 0% fish. Eel body size: 55.0 ± 8.8 cm total length (n = 218)	Most important prey (biomass ratio): fish: 24 (26.3), shrimp: 22.9 (15.4), crayfish: 7.8 (17.6), chironomids 8.5 (2.1), clams: 7.8 (3.3), Oligochaeta: 12.4 (19.4). Eel body size: 46.4 ± 10.1 cm total length (n = 586)
	A. bengalensis	
Lake	**Coastal/Estuary**	**River/Stream**
		Channel of Calcutta Waterworks near Pulta, India (Pantulu 1957)
		Percentage of occurrence: fish 23.6%, prawns 22.8%, Megalopa (crabs) larvae 14.9%, crabs (adult) 14.5%, macro- and microphytes 6.9%, insects 6.9%, annelids 5.1%, misc. 5.4%. By volume: fish 40.3%, crabs 26.0%, prawns 20.7%. Eel body size: range 14 to 66 cm total length (n = 450)

A. reinhardtii		
Lake	**Coastal/Estuary**	**River/Stream**
Macleods Morass, Victoria, Australia (Beumer 1979) Most important prey groups (percentage of occurrence): teleosts, insects. No apparent relationship between the size of a particular item taken and the size of the eel. Body size: range 42.4 to 117 cm total length (n = 86)		
A. marmorata		
Lake	**Coastal/Estuary**	**River/Stream**
		Rivers in tidal zone in Moorea Island, French Polynesia (Resh et al. 1999) Food items (and number of eels containing those in brackets): Crustacea: lobster (1), shrimp (2), isopod type (1), Mollusca: octopus type (2), vertebrates: fish parts (1), Body size: range 28.5 to 85 cm total length (n = 6)
A. obscura		
Lake	**Coastal/Estuary**	**River/Stream**
		Rivers in tidal zone in Moorea Island, French Polynesia (Resh et al. 1999) Food items (and number of eels containing those in brackets): Mollusca: snails (2), insects: Diptera pupa and larvae (2). Body size: range 41.7 to 95 cm total length (n = 4)

Table 2. contd....

Table 2. contd.

Lake	Coastal/Estuary	River/Stream
		Upper Kairezi River, Zimbabwe (Butler and Marshall 1996)
		Most important prey (percentage of occurrence): 63% river crabs, 10% mountain catfish, 12% dragonfly larvae. Eel body size 99.2 ± 0.47 cm total length (range: 66.4 to 124 cm) (n = 13)

A. labiate

Summary

Concerted attempts to describe the diet of anguillid larvae (leptocephali) and glass eels (elvers) have only recently (within the last 25 years) been undertaken and still no definitive conclusions can be drawn. What is clear now is that the earlier theories on epithelial uptake of dissolved organic matter (DOM) can be rejected. Evidence points in the direction of particulate organic matter (POM) or "marine snow" as a major food component especially in the form of fecal pellets but other food sources such as larvacean houses or zooplankton may also play a role. During metamorphosis the leptocephali do not ingest food. The newly formed glass-eels will at first eat items similar to the leptocephali but upon reaching estuaries or coastal waters switch to benthic feeding. Anguillid yellow eels are poikilothermic and well adapted to warm water conditions. Generally, the most intense feeding occurs during the spring and summer periods. Seasonal differences in stomach fullness indices and feeding habits, however, can largely be attributed to availability of the different prey taxa. With the provision that not all studies can be directly compared, it seems that eels feed on a wide range of organisms and the relative importance of any food item varies from area to area. Comparing the main feeding habits of the four eel species *A. anguilla*, *A. rostrata*, *A. australis*, and *A. dieffenbachii* confirms a general pattern of eels being generalized nocturnal benthos feeders with diets reflecting the availability of prey in the benthic communities. For eels with a size of about ≥ 40 cm total length, fish and often crustaceans can reasonably be expected to form an increasingly important component of their diet. There is strong indication that eels have the potential to influence invertebrate macrozoobenthos and benthic fish communities in running waters and couple benthic and pelagic food webs in lakes. Eels have often been of major concern for recreational fishery managers as they were believed to have strong negative impacts on salmonid stocks through predation. However, none of the studies where eel—salmonid interactions were specifically investigated was able to demonstrate any substantial negative impact on salmonid populations solely as a result of predation by eels.

Acknowledgements

We wish to thank Dr. John Casey very much for linguistic revision and very helpful comments on previous versions of this book chapter.

Keywords: *Anguilla*, feeding, diet, generalist, opportunistic, food web, benthic, lepthocephali, elver, yellow eel, predator, fish, invertebrates, river, lake, coastal zone

References

Beumer, J.P. 1979. Feeding and movement of *Anguilla australis* and *A. reinhardtii* in Macleods Morass, Victoria, Australia. J. Fish Biol. 14: 573–592.

Bouchereau, J.-L., C. Marques, P. Pereira, O. Guélorget and Y. Vergnez. 2006. Trophic characterization of the Prévost lagoon (Mediterranean Sea) by the feeding habits of the European eel *Anguilla anguilla*. Cah. Biol. Mar. (2006) 47: 133–142.

Bouchereau, J.-L., C. Marques, P. Pereira, O. Guélorget and Y. Vergnez. 2009. Food of the European eel *Anguilla anguilla* in the Mauguio lagoon (Mediterranean, France). Acta Adriat. 50: 5–15.

Burnet, A.M.R. 1952. Studies on the ecology of the New Zealand longfinned eel *Anguilla dieffenbachii* Gray. Aust. J. Mar. Fresh. Res. 3: 33–63.

Burnet, A.M.R. 1968. A Study of the Relationships between Brown Trout and Eel in a New Zealand stream. Fisheries Tech. Report no. 26.

Butler, J.R.A. and B.E. Marshall. 1996. Resource use within the crab-eating guild of the upper Kairezi River, Zimbabwe. J. Trop. Ecol. 12: 475–490.

Cadwallader, P.L. 1975. Feeding Relationships of Galaxiids, Bullies, Eels and Trout in a New Zealand River. Aust. J. Mar. Fresh. Res. 26: 299–316.

Cairns, D. 1942. Life history of the two species of New Zealand freshwater eels. N.Z. J. Sci. Tech. B23: 132–148.

Cairns, D. 1950. New Zealand Fresh Water Eels. Tuatara 3(2): 43–52.

Chow, S., H. Kurogi, S. Katayama, D. Ambe, M. Okazaki, T. Watanabe, T. Ichikawa, M. Kodama, J. Aoyama, A. Shinoda, S. Watanabe, K. Tsukamoto, S. Miyazaki, S. Kimura, Y. Yamada, K. Nomura, H. Tanaka, Y. Kazeto, K. Hata, T. Handa, A. Tawa and N. Mochioka. 2010. Japanese eel *Anguilla japonica* do not assimilate nutrition during the oceanic spawning migration: evidence from stable isotope analysis. Mar. Ecol. Prog. Ser. 402: 233–238.

Costa, J.L., C.A. Assis, P.R. Almieda, F.M. Moreira and M.J. Costa. 1992. On the food of the European eel, *Anguilla anguilla* (L.), in the upper zone of the Tagus estuary, Portugal. J. Fish Biol. 41: 841–850.

Cullen, P. and T.K. McCarthy. 2007. Eels (*Anguilla anguilla* (L.)) of the lower River Shannon with particular reference to seasonality in their activity and feeding ecology. Biology and environment: P. Roy. Irish Acad. B107(2): 87–94.

Deelder, C.L. 1984. Synopsis of biological data on *A. anguilla* (Linnaeus, 1758). FAO Fisheries Synopsis no. 80.

Dekker, W. 2004. What caused the decline of the Lake IJsselmeer eel stock after 1960? ICES J. Mar. Sci. 61: 394–404. doi: 10.1016/j.icesjms.2004.01.003.

De Nie, H.W. 1987. Food, feeding periodicity and consumption of the eel *Anguilla anguilla* (L.) in the shallow eutrophic Tjeukemeer (the Netherlands). Arch. Hydrobiol. 109: 421–443.

Dörner, H. and J. Benndorf. 2003. Piscivory by large eels on young-of-the-year fishes: its potential as a biomanipulation tool. J. Fish Biol. 62: 491–494.

Dörner, H., C. Skov, S. Berg, T. Schulze, D.J. Beare and G. Van der Velde. 2009. Piscivory and trophic position of large eels (*Anguilla anguilla*) in two lakes: importance of macrozoobenthos density. J. Fish Biol. 74(9): 2115–2131.

Ezzat, A.E. and S.S. El-Seraffy. 1977. Food of *Anguilla anguilla* in Lake Manzalah, Egypt. Mar. Biol. 41: 287–291.

Fitzsimons, J.D., S.B. Brown, L.R. Brown, G. Verreault, R. Tardif, K.G. Drouillard, S.A. Rush and J.R. Lantry. 2013. Impacts of Diet on Thiamine Status of Lake Ontario American Eels. Trans. Am. Fish. Soc. 142(5): 1358–1369.

Folke, C.S., B. Carpenter, B. Walker, M. Scheffer, T. Elmqvist, L. Gunderson and C.S. Holling. 2004. Regime shifts, resilience, and biodiversity in ecosystem management. Annu. Rev. Ecol. Evol. S 35: 557–581.

Frost, W.E. 1946. Observations on the Food of Eels (*Anguilla anguilla*) from the Windermere Catchment Area. J. Anim. Ecol. 15(1): 43–53.

Gamito, S., A. Pipes, C. Pita and K. Erzini. 2003. Food Availability and the Feeding Ecology of Ichthyofauna of a Ria Formosa (South Portugal) Water Reservoir. Estuaries 26: 938–948.

Godfrey, H. 1957. Feeding of eels in four New Brunswick salmon streams. Progress Reports for the Atlantic Stations. Biological Station St. Andrews, N.B., Note No. 150. Ottawa.

Golani, D., D. Shefler and A. Gelman. 1988. Aspects of Growth and Feeding Habits of the Adult European Eel (*Anguilla anguilla*) in Lake Kinneret (Lake Tiberias), Israel. Aquaculture 74: 349–354.

Govoni, J.J. 2010. Feeding on protists and particulates by the leptocephali of the worm eels *Myrophis* spp. (Teleostei: Anguilliformes: Ophichthidae), and the potential energy contribution of large aloricate protozoa. Sci. Mar. 74: 339–344.

Helfman, G.S. and D.L. Winkelman. 1991. Energy trade-offs and foraging mode choice in American eels. Ecology 72(1): 310–318.

Hopkins, C.L. 1965. Feeding relationships in a mixed population of freshwater fish. New Zeal. J. Sci. 8: 149–157.

Hulet, W.H. 1978. Structure and functional development of the eel leptocephalus *Ariosoma balearicum* (DeLa Roche, 1809). Phil. Trans. Roy. Soc. London, Biol. Sci. 252: 107–138.

Itakura, H., T. Kaino, Y. Miyake, T. Kitagawa and S. Kimura. 2015. Feeding, condition, and abundance of Japanese eels from natural and revetment habitats in the Tone River, Japan. Environ. Biol. Fish. In press.

Jellyman, D.J. 1989. Diet of two species of freshwater eel (*Anguilla* spp.) in Lake Pounui, New Zealand. New Zealand journal of marine and freshwater research 23: 1–10.

Jellyman, D.J. 1996. Diet of longfinned eels, *Anguilla dieffenbachii*, in Lake Rotoiti, Nelson Lakes, New Zealand. New Zeal. J. Mar. Fresh. 30(3): 365–369.

Jellyman, D.J. and E. Graynoth. 1994. Headwater trout fisheries in New Zealand. New Zealand freshwater research report 12. NIWA Freshwater, Christchurch, New Zealand. 87pp.

Kaifu, K., S. Miyazaki, J. Aoyama, S. Kimura and K. Tsukamoto. 2013a. Diet of Japanese eels *Anguilla japonica* in the Kojima Bay-Asahi River system, Japan. Environ. Biol. Fish. 96: 439–446.

Kaifu, K., M.J. Miller, J. Aoyama, I. Washitani and K. Tsukamoto. 2013b. Evidence of niche segregation between freshwater eels and conger eels in Kojima Bay, Japan. Fish. Sci. 79: 593–603.

Kangur, K., A. Kangur and P. Kangur. 1999. A comparative study on the feeding of eel, *Anguilla anguilla* (L.), bream, *Abramis brama* (L.) and ruffe, *Gymnocephalus cernuus* (L.) in Lake Võrtsjärv, Estonia. Hydrobiologia 408/409: 65–72.

Kilner, A.R. and J.M. Akroyd. 1978. Fish and invertebrate macrofauna of Ahuriri Estuary, Napier. New Zealand Ministry of Agriculture and Fisheries. Fish. Tech. Rep. 153.

Kracht, R. and F.-W. Tesch. 1981. Progress report on the eel expedition of R.V. "Anton Dohrn" and R.V. "Friedrich Heincke" to the Sargasso Sea 1979. Env. Biol. Fish. 6: 371–375.

Liao, I.C. and S.L. Chang. 2001. Induced spawning and larval rearing of Japanese eel, *Anguilla japonica* in Taiwan. J. Taiwan Fish. Res. 9: 97–108.

Mann, R.H.K. and J.H. Blackburn. 1991. The biology of the eel *Anguilla anguilla* (L.) in an English chalk stream and interactions with juvenile trout *Salmo trutta* L. and salmon *Salmo salar* L. Hydrobiologia 218: 65–76.

Miller, M.J. 2009. Ecology of anguilliform leptocephali: remarkable transparent fish larvae of the ocean surface layer. Aqua-BioSci. Monogr. 2: 1–94.

Miller, M.J., Y. Chikaraishi, N.O. Ogawa, Y. Yamada, K. Tsukamoto and N. Ohkouchi. 2013. A low trophic position of Japanese eel larvae indicates feeding on marine snow. Biol. Lett. 9: 20120826. http://dx.doi.org/10.1098/rsbl.2012.0826.

Miyazaki, S., H.-Y. Kim, K. Zenimoto, T. Kitagawa, M.J. Miller and S. Kimura. 2011. Stable isotope analysis of two species of anguilliform leptocephali (*Anguilla japonica* and *Ariosoma major*) relative to their feeding depth in the North Equatorial Current region. Mar. Biol. 158: 2555–2564.

Mochioka, N. and M. Iwamizu. 1996. Diet of anguilloid larvae: leptocephali feed selectively on larvacean houses and fecal pellets. Mar. Biol. 125: 447–452.

Mochioka, N., M. Iwamizu and T. Kanda. 1993. Leptocephalus eel larvae will feed in aquaria. Environ. Biol. Fish. 36: 381–384.

Moore, J.W. and I.A. Moore. 1976. The basis of food selection in some estuarine fishes. Eels, *Anguilla anguilla* (L.), whiting, *Merlangius merlangus* (L.), sprat, *Sprattus sprattus* (L.) and stickleback, *Gastoreus aculeatus* L. J. Fish Biol. 9: 375–390.

Moriarty, C. 1972. Studies of the eel *Anguilla anguilla* in Ireland. 1. In the lakes of the Corrib System. Irish Fisheries Investigations, Series A 10: 3–39.

Moriarty, C. 1973. Studies of the eel *Anguilla anguilla* in Ireland. 2. In Lough Conn, Lough Gill and North Cavan Lakes. Irish Fisheries Investigations, Series A 13: 3–13.

Moriarty, C. and W. Dekker. 1997. Management of the European eel—enhancement of the European eel fishery and conservation of the species. Irish Fisheries Bulletin (Dublin) 15.

Moser, H.G. 1981. Morphological and functional aspects of marine fish larvae. pp. 90–131. *In*: R. Lasker (ed.). Marine Fish Larvae: Morphology, Ecology, and relation to Fisheries. Washington Sea Grant Program, University of Washington Press, Seattle, Washington, USA.

Ogden, J.C. 1970. Relative Abundance, Food Habits, and Age of the American Eel, *Anguilla rostrata* (LeSueur), in Certain New Jersey Streams. Trans. Am. Fish. Soc. 1: 54–59.

Otake, T., K. Nogami and K. Maruyaka. 1993. Dissolved and particulate organic matter as possible food sources for eel leptocephali. Mar. Ecol. Prog. Ser. 92: 27–34.

Pantulu, V.R. 1957. Studies on the biology of the Indian fresh-water eel, *Anguilla bengalensis* Gray. Ind. Nat. Sci. Acad. Proc. B22: 259–280.

Pfeiler, E. 1986. Toward an explanation of the developmental strategy in leptocephalus larvae of marine fish. Environ. Biol. Fish. 15: 3–13.

Radke, R.J. and R. Eckmann. 1996. Piscivorous eels in Lake Constance: can they influence year class strength of perch? Ann. Zool. Fennici 33: 489–494.

Rasmussen, G. and B. Therkildsen. 1979. Food, growth and production of *Anguilla anguilla* L. in a small Danish stream. Rapp. P.-v. Reún. Cons. Int. Explor. Mer 174: 32–40.

Resh, V.H., M. Moser and M. Poole. 1999. Feeding habits of some freshwater fishes in streams of Moorea, French Polynesia. Annls Limnol. 35(3): 205–210.

Riemann, L., H. Alfredsson, M.M. Hansen, T.D. Als, T.G. Nielsen, P. Munk, K. Aarestrup, G.E. Maes, H. Sparholt, M.I. Petersen, M. Bachler and M. Castonguay. 2010. Qualitative assessment of the diet of European eel larvae in the Sargasso Sea resolved by DNA barcoding. Biol. Lett. 6: 819–822.

Ryan, P.R. 1984. Diel and seasonal feeding activity of the short-finned eel, *Anguilla australis* schmidtii, in Lake Ellesmere, Canterbury, New Zealand. Env. Biol. Fish. 11(3): 229–234.

Ryan, P.R. 1986. Seasonal and size-related changes in the food of the short-finned eel, *Anguilla australis* in Lake Ellesmere, Canterbury, New Zealand. Env. Biol. Fish. 15(1): 47–58.

Sagar, P.M.E. and G.J. Glova. 1994. Food partitioning by small fish in a coastal New Zealand stream. New Zeal. J. Mar. Fresh. 28(4): 429–436.

Sagar, P.M.E. and G.J. Glova. 1998. Diel feeding and prey selection of three size classes of shortfinned eel (*Anguilla australis*) in New Zealand. Mar. Freshwater Res. 49: 421–428.

Sagar, P.M.E., E. Graynoth and G.J. Glova. 2005. Prey selection and dietary overlap of shortfinned (*Anguilla australis*) and longfinned (*A. dieffenbachii*) eels during summer in the Horokiwi Stream, New Zealand. New Zeal. J. Mar. Fresh. 39(4): 931–939.

Santamarina, J. 1993. Feeding ecology of a vertebrate assemblage inhabiting a stream of NW Spain (Riobo; Ulla basin). Hydobiologia 252: 175–191.

Schein, A., S.C. Courtenay, K.A. Kidd, K.A. Campbell and M.R. van den Heuvel. 2013. Food web structure within an estuary of the southern Gulf of St. Lawrence undergoing eutrophication. Can. J. Fish. Aquat. Sci. 70: 1805–1812.

Sinha, V.R.P. 1969. A Note on the Feeding of Larger Eels *Anguilla anguilla* (L.). J. Fish Biol. 1: 279–283.

Sinha, V.R.P. and J.W. Jones. 1967. On the food of the freshwater eels and their feeding relationship with the salmonids. J. Zool., Lond. 153: 119–137.

Smith, H.W. and J.W. Saunders. 1955. The American eel in certain freshwaters of the Maritime Provinces of Canada. J. Fish. Res. Bd. Can. 12: 238–269.

Stranko, S.A., M.J. Ashton, R.H. Hilderbrand, S.L. Weglein, D.C. Kazyak and J.V. Kilian. 2014. Fish and Benthic Macroinvertebrate Densities in Small Streams with and without American Eels. Trans. Am. Fish. Soc. 143(3): 700–708.

Tanaka, H., H. Kagawa, H. Ohta, K. Okuzawa and K. Hirose. 1995. The First Report of Eel Larvae Ingesting Rotifers. Fish. Sci. 61: 171–172.

Tanaka, H., H. Kagawa, H. Ohta, T. Unuma and K. Nomura. 2003. The first production of glass eel in captivity: fish reproductive physiology facilitates great progress in aquaculture. Fish Physiol. Biochem. 28: 493–497.

Terahara, T., S. Chow, H. Kurogi, S.-H. Lee, K. Tsukamoto, N. Mochioka, H. Tanaka and H. Takeyama. 2011. Efficiency of Peptide Nucleic Acid-directed PCR clamping and its application in the investigation of natural diets of the Japanese eel leptocephali. PlosONE 6: Article e25715, 7pp.

Tesch, F.-W. 2003. The Eel, 3rd edn. Blackwell Science Ltd., Oxford, United Kingdom.

Thomas, J.D. 1962. The Food and Growth of Brown Trout (*Salmo trutta* L.) and its Feeding Relationships with the Salmon Parr (*Salmo salar* L.) and the Eel (*Anguilla anguilla* (L.)) in the River Teify, West Wales. J. Anim. Ecol. 31(2): 175–205.

Waldt, E.M., R. Abbett, J.H. Johnson, D.E. Dittman and J.E. McKenna. 2013. Fall diel diet composition of American eel (*Anguilla rostrata*) in a tributary of the Hudson River, New York, USA. J. Freshw. Ecol. 28(1): 91–98.

Wasserman, R.J., L.L. Pereira-da-Conceicoa, N.A. Strydom and O.L.F. Weyl. 2012. Diet of *Anguilla mossambica* (Teleostei, Anguillidae) elvers in the Sundays River, Eastern Cape, South Africa. Afr. J. Aquat. Sci. 37: 347–349.

Woods, C. 1964. Fisheries aspects of the Tongariro Power development project. N.Z. Fish. Tech. Rep. 10, 214pp.

10

Adaptation to Varying Salinity

M. Giulia Lionetto, * M. Elena Giordano* and
Trifone Schettino

Introduction

Changes in the osmolarity of the environmental medium are a physiological stress faced by a variety of aquatic organisms. The physiological ability to cope with this environmental variability allowed a number of aquatic organisms to colonize osmotically changing habitats such as brackish waters or to span from fresh water (FW) to sea water (SW) environments. Many fishes undertake vast migrations between rivers and oceans in order to find food or breeding sites. Eels are one of the most well-known euryhaline migrating fish. They are born in sea water and pass through the leptocephalus stage, an extended larval phase, in the open ocean. Then, they undergo a juvenile stage that is a smaller version of the adult stage. The young eels swim upriver where they gradually develop into adults. They stay in the fresh water for several years before they become sexually mature, at which time they begin swimming down river towards the sea. Therefore, during their lifecycle the eels experience environmental osmolarity changes twice ranging from about 1100 mOsm in SW to nearly 0.5 mOsm in FW. They are able to maintain body fluid osmotic pressure and electrolyte concentration at levels independent of ionic concentration and osmolarity of the environmental medium. This is achieved by balancing the gain and loss of water and ions. In fresh water

Dipartimento di Scienze e Tecnolgie Biologiche e Ambientali (DiSTeBA), Università del Salento, Italy.
* Corresponding author: giulia.lionetto@unisalento.it

environment the animal continuously faces the osmotic gain of water and constantly needs to excrete the excess water. In sea water environment the eel needs to constantly acquire water to compensate for the osmotic water loss across the body surfaces.

The eels can survive in habitats with varying salinity thanks to the specific mechanisms of osmoregulation involving ion and water movements across specialized epithelia at the gill, kidney, oesophagus and intestinal level. Interestingly the function of these osmoregulatory epithelia must change according to changing environmental requirements during the transition from seawater to fresh water and *vice versa*.

How the eel copes with varying salinity

The role of the gastrointestinal tract

The gastrointestinal tract of the eel plays a pivotal role in the osmoregulation of the animal especially in SW environment. When the eels are in SW, they actively drink the surrounding medium as means of water intake. The eels always fill the oral cavity with seawater for breathing, and they can ingest water via the swallowing reflex (Nobata and Ando 2013). The animal can swallow if the upper oesophageal sphincter, which regulates the introduction of water into the gastrointestinal tract, relaxes. The drinking behaviour is finely regulated in relationship to the osmoregulatory requirements of the animal (Takei and Balment 2009), by integrated neural and hormonal signals, including angiotensin II, and natriuretic peptide. The ingested seawater is desalted and diluted to half-initial concentration in the oesophagus (Hirano and Mayer-Gostan 1976; Ando and Nagashima 1996). About 2/3 of the total desalinization is ouabain sensitive (Nagashima and Ando 1994), suggesting that active Na^+ intake significantly contributes towards the partial desalinization at the esophagous level. The oesophagus desalinization is made possible thanks to the very low water permeability of this epithelium corresponding to 2×10^{-4} cm/s (Nagashima and Ando 1994). Besides the oesophagus low water permeability, aquaporins AQP3 and AQP1 are expressed in this epithelium (Cutler and Cramb 2001; Cutler et al. 2007) although their function is not clear yet. Then, the ingested sea water is diluted to 1/3 in the stomach of the SW eel (Ando and Nagashima 1996). This dilution occurs mainly passively across the epithelium by water movement along the osmotic gradient (Hirano and Mayer-Gostan 1976).

At the intestinal level, water is absorbed across the intestinal epithelium under a nearly isosmotic condition according to the so called "solute-linked water flow" (Skadhauge 1974). *In vivo* flux experiments performed by Skadhauge (1969) in the intestine of the European eel and *in vitro* study of Ando et al. (1975) in the intestine of *Anguilla japonica* showed that Cl^- transport exceeds Na^+ transport in the eel intestine. The excess Cl^- absorption generates a transepithelial electrical potential which is negative on the serosal side and generates a measurable short circuit current (Ando et al. 1975). Simultaneous measurements of net sodium, potassium, chloride, and water fluxes have shown them to be tightly coupled (Ando 1985). The ion transport model proposed includes an apically located neutral Na^+-K^+-$2Cl^-$ cotransporter in series with a Cl^- conductance and presumably an electroneutral KCl cotransport, both localized on the basolateral side of the enterocytes. An apical barium sensitive K^+ channel permits recycling of K^+ into the lumen and contributes to the functional activity of the luminal Na^+-K^+-$2Cl^-$ cotransporter. The basolaterally located Na^+-K^+-ATPase, by generating an inwardly directed electrochemical gradient for Na^+, provides the driving force for the active intracellular accumulation of Cl^- by the electroneutral luminal symport (Schettino and Lionetto 2003). The transepithelial NaCl absorption is followed by water absorption across the intestine. Larsen et al. (2002, 2009) proposed the "sodium recirculation theory" by assuming the lateral intercellular space to be an osmotic coupling compartment which osmotically forces water through the epithelium energized by the lateral Na^+-K^+-ATPase. As indicated by Ando and Takei (2013) water may be drawn into the SW eel intestine though four pathways as demonstrated in the salmonid intestine such as (a) diffusion across the lipid bilayer of the intestinal cells, (b) paracellular diffusion through the tight junction, (c) symport with glucose by the Na^+-glucose cotransporter SGLT1, and (d) diffusion through aquaporins AQP1 and AQP3 which are both expressed in the eel intestine (Cutler and Cramb 2002b). In particular AQP1 expression increases during the transfer of the animal from FW to SW. On the other hand AQP3 is an aquaglyceroporin with low water permeability. Its mRNA expression does not differ between FW eels and SW eels.

The eel absorptive Na^+-K^+-$2Cl^-$ cotransporter was characterized in molecular terms by Cutler et al. (1996). cDNA for the homolog of the mammalian Na^+-K^+-$2Cl^-$ cotransporter (NKCC2) was cloned from the eel intestine and designated cot 2 (Cutler et al. 1996). Its expression predominates in the anterior and mid-regions of the eel gut, with reduced levels in the posterior intestine and rectum. Cutler and Cramb (2001, 2002a) also reported the expression of the NKCC1, the secretory form of the Na^+-K^+-$2Cl^-$ cotransporter, in the eel intestinal epithelium, suggesting also the presence of a NaCl secretion in eel intestine.

When eels are in FW they do not need to drink because of the osmotic influx of water across the gills. Therefore, it has been long accepted that FW eels rarely drink. However, it has been later recognized that body fluid balance seems to be maintained irrespective of drinking in FW eels (Nobata and Ando 2013). In fact, in FW excess water can be overcome by an increase of hypotonic urine production, while hyponatremia can be compensated by ion uptake at the gill levels.

In the eels in FW the ion transport mechanism described above in the intestine of SW eels is also expressed, although to a less extent (Trischitta et al. 1992; Cutler et al. 2007). This means that salt absorption, accomplished by the operation of the luminal Na^+-K^+-$2Cl^-$ cotransporter, is not silent in FW eels. The presence of this transport system in FW eels, even though reduced as compared to its rate in SW, suggests that the intestinal epithelium, in cooperation with the branchial epithelium, plays a role in absorbing salt that is passively lost to the FW environment (Schettino and Lionetto 2003). Osmotic water permeability across the FW eel intestine is six-times lower than that of the SW eel (Ando et al. 1975).

When the eel starts its migration towards a new environment, its osmoregulatory mechanisms start changing in order to prepare the organism to cope with the new osmotic requirement of the new environmental medium. In fact, the pre-migrating silver eel behaves as an atypical freshwater fish since it drinks much more than the freshwater yellow eel (Gaitskell and Chester 1971). Its catadromous migration is preceded by a change of its osmoregulatory capacity as demonstrated by the increased Cl⁻ serum concentration and by the increased Cl⁻ intestinal absorption (Lionetto et al. 1996) still persisting in FW. This suggests that during the pre-migrating phase the animal prepares its osmoregulatory mechanisms for the acquirement of the SW tolerance, and expresses an increased activity of the Na$^+$-K$^+$-2Cl as an intrinsic characteristic at this developmental stage.

When the migrating silver eel reaches the estuarine waters at the river delta it recognises the changes in Cl⁻ concentration of the water. The stimulation of chemoreceptors present in the oral cavity are able to evoke the so called "chloride response" consisting of a burst of drinking when the oral cavity is stimulated by chloride ions in SW (Hirano 1974). Therefore, migrating silver eels through the chloride response are able to forestall future dehydration in seawater and prevent it by increasing their drinking behaviour.

The role of gills

The gills are the first osmoregulatory surface coming into contact with the environmental medium and are thus the major site of ion and water flux (Ogasawara and Hirano 1984). Interestingly, the directions of ion transport in gills are reversed during the transfer of the eel to different salinity environment. In SW environment the gill epithelium is the main site of excess salt excretion while in FW it represents the main site of salt absorption to balance the passive loss of ions to the environment.

The gill epithelium of the eel, like other teleost fish, contains pavement cells and mitochondrion-rich (MR) cells, often referred to as chloride cells or ionocytes. Mitochondrion-rich cells in the gills are the major site of ion absorption and secretion, and are thus important in both FW and SW acclimatization. The MR cells are exposed to the external environments through their apical membrane, which is located at the boundary of pavement cells. The MR cells are characterized by the presence of a rich population of mitochondria. In addition an extensive tubular system in the cytoplasm is evident. As observed in the Japanese eel gills, two types of MR cells were previously identified via immunocytochemical analysis (Seo et al. 2009): MR cells in the gill filaments were most developed in SW, whereas MR cells in the lamellae were preferentially observed in FW. This observation suggests that filament and lamellar MR cells are responsible for ion secretion and absorption, respectively. In eels acclimated to FW, the apical membrane of MR cells appeared as a flat disk with a mesh-like structure when observed by scanning electron microscopy. Such a structure that enlarges the apical surface is thought to facilitate ion absorption (Katoh and Kaneko 2003; Inokuchi et al. 2008). On the other hand, in eels acclimated to SW the apical membrane of MR cells showed a slightly concave surface without a mesh-like structure; this appearance is characteristic of ion secreting cells in SW environment (Kaneko et al. 2008; Hwang et al. 2011). Accessory cells were identified to be adjacent to SW MR cells. Although the identity of accessory cells is still being debated (Evans et al. 2005) the multicellular complex of an accessory cell and a MR cell in SW environment is known to form apical membrane interdigitations with leaky junctions, thus providing a paracellular route for Na^+ extrusion down its electrical gradient in SW environment (Hwang and Hirano 1985).

In general in SW fish the current model of NaCl secretion across the gills proposes that NaCl extrusion is mediated by the basolateral cotransport Na^+-K^+-$2Cl^-$ down the Na^+ electrochemical gradient provided by the Na^+-K^+-ATPase. The operation of the Na^+-K^+-$2Cl^-$ is coupled with the apical exit of Cl^- via channels and the paracellular extrusion of Na^+ down its electrochemical gradients. Hence, the key transporters associated with the NaCl transport process are thought to be the Na^+-K^+-ATPase, the Na^+-K^+-$2Cl^-$ transporter and the cystic fibrosis trans-membrane conductance regulator (CFTR) Cl^- channel (Marshall et al. 2002; Hirose et al. 2003; Evans et al. 2005).

Eel, like other euryhaline fish such as salmon and tilapia, responds to salinity increase with higher Na^+-K^+-ATPase activity (Marsigliante et al. 1997) as well as MR cell densities (Hwang and Lee 2007) suggesting the primary role of Na^+-K^+-ATPase in NaCl secretion in SW environment.

An increase in the mRNA expression of gill NKCC, reflecting stimulation of Cl^- secretion, was found in the European eel after transfer to SW (Cutler and Cramb 2002a). More recently, Seo et al. (2013) obtained a partial cDNA encoding the secretory NKCC1a in Japanese eel gills and the relative protein expression was identified at the basolateral membrane of MR cells. The NKCC1 expression increases as the salinity increases, in accordance with previous results reported in the European eel (Cutler and Cramb 2002a), indicating that NKCC1 is one of the key ion-transport molecules responsible for ion secretion in the eel in SW environment. Although reduced with respect to the SW acclimated eel, the expression of NKCC1 detected by immunofluorescence at the basolateral membrane of MR cells is also appreciable in the FW acclimated eel. The presence of NKCC1 in MR cells in FW environments suggests that NKCC1 has some physiological ion transporting function in FW eel. A possible explanation is that NKCC1 can contribute to the transport of ammonia in FW eel. In fact NH_4^+ is able to replace K^+ on NKCC1 in fish acclimated to hypotonic environments (Nawata et al. 2010).

CFTR has been identified in the Japanese eel (*Anguilla japonicus*), where upregulation of CFTR mRNA was reported following SW exposure (Tse et al. 2006). These results demonstrate the prominent role of these channels in the salt excretion in SW environment.

In general in FW fish two current models have been proposed for the salt absorption at the gill level: the first model includes an apical V-type H^+-ATPase electrically linked with Na^+ absorption via the epithelial Na^+ channel (ENaC), while the second includes an electroneutral exchange of Na^+ and H^+ via an apical Na^+/H^+ exchanger (NHE) (Hwang and Lee 2007). However, in the eel the exact molecular mechanism accounting for salt excretion at the gill level in FW environment is still debated. A partial cDNA encoding NHE3 in the gill of Japanese eel was also identified. Its expression increased significantly with decreasing environmental salinity, suggesting that NHE3 is involved in ion uptake, presumably Na^+ uptake in FW. Therefore, these data indicate that NHE3 is important for acclimation to hypotonic environments in the eel. However, its role in SW cannot be neglected, since it could be related to proton excretion and pH regulation (Seo et al. 2013). In addition a cDNAs encoding vacuolar-type H^+-ATPase (V-ATPase) A-subunit was cloned and sequenced in the eel gills. However, contrasting data are reported about its role in FW salt absorption since Seo et al. (2013) failed to find any change in protein expression with varying external salinity as assessed by immunofluorescence analysis, in contrast to previous data of Tse and Wong (2011) who found H^+-ATPase mRNA and protein expressions to increase after seawater to freshwater transfer by Western blot analysis.

The role of kidney

Kidney plays an important role in the regulation of ion and water balance in both freshwater and seawater. Interestingly, as observed for the other osmoregulatory epithelia, its role changes in the animal during the transfer form freshwater to seawater and vice versa.

In general in FW teleost the kidney is known to produce a large amount of hypotonic urine by filtering the blood in glomeruli and reabsorbing ions such as Na^+ and Cl^- from the filtrate (Nishimura et al. 1983; Miyazaki et al. 2002). Conversely, in SW environment the kidney excretes excess divalent ions and produces a relatively small amount of isotonic urine in order to minimize water loss (Marshall and Grosell 2006).

Regarding the molecular mechanisms accounting for the monovalent-ion reabsorption in the kidney, it is known in case of mammalian kidneys that monovalent ions such as Na^+ and Cl^- are reabsorbed from primitive urine through the Na^+/H^+ exchanger 3 (NHE3) in the proximal tubule, Na^+-K^+-$2Cl^-$ cotransporter 2 (NKCC2) in the thick ascending limb of Henle's loop, Na^+-Cl^- cotransporter (NCC) in the distal convoluted tubule, and epithelial Na^+ channel (ENaC) in the collecting duct (Teranishi et al. 2013). In the European eel *Anguilla anguilla* two NKCC2 isoforms (NKCC2α and NKCC2β) and two NCC isoforms (NCCα and NCCβ) have been identified (Cutler and Cramb 2008). European eel NKCC2α and NCCα mRNAs are expressed only in the renal tissue. The NCCα expression exhibited a tendency to increase as the environmental salinity decreased, whereas the NKCC2α expression did not significantly differ with varying salinity. In addition a partial sequence of cDNA encoding NHE3 in the kidney of the Japanese eel *Anguilla japonica* was identified (Teranishi et al. 2013). Immunohistochemical studies showed that NHE3 was localized at the apical membrane of the epithelial cells composing the second segments of the proximal renal tubule in seawater-acclimated eel. Meanwhile, the apical membranes of epithelial cells in the distal renal tubule and collecting duct showed more intense immunoreactions of NKCC2α and NCCα, respectively, in freshwater eel than in seawater eel. These findings suggest that renal monovalent-ion reabsorption is mainly mediated by NKCC2α and NCCα in freshwater eel and by NHE3 in seawater eel (Teranishi et al. 2013).

In SW environment the eel kidney plays a key role in the excretion of excess divalent ions, such as SO_4^{2-}, Mg^{2+} and Ca^{2+} which enter the body *via* the gills and the digestive tracts across their concentration gradient. In the apical membrane of the proximal tubule, Cl^-/SO_4^{2-} exchanger (solute carrier family 26 member family 6, S1c26a6) (Kato et al. 2009; Watanabe and Takei 2011), Na^+/Mg^{2+} exchanger (Beyenbach 2004), and Na^+/Ca^{2+} exchanger (NCX2) (Islam et al. 2011) are involved in the secretion of SO_4^{2-}, Mg^{2+}, and Ca^{2+}. The expression of S1c26a6 is strongly reduced during FW acclimatization. In FW eels the proximal tubule is the site of SO_4^{2-} reabsorption (Nakada et al. 2005). FW eels have a system for renal SO_4^{2-} reabsorption which is mediated by S1c13a1, an electrogenic Na^+-SO_4^{2-} cotransporter (Nakada et al. 2005; Watanabe and Takei 2011). The eel S1c13a1 is specifically expressed in the kidney of the FW eel and is localized on the apical membrane of the proximal tubule cells. The expression is reduced to undetectable levels 1–3 days after transfer from FW to SW, and no expression was observed in the kidney of SW eel.

Endocrine control of the eel osmoregulatory responses

The extraordinary capability of the eel to cope with changes of environmental salinity and to overcome the osmotic challenges is sustained by a complex endocrine system.

The established osmoregulatory hormones such as cortisol, adrenocorticotropic hormone (ACTH), and prolactin play a pivotal role in eel osmoregulation by modulating ion and water fluxes at the osmoregulatory epithelia level in response to internal and/or external osmotic stimuli (Evans 2002). ACTH and cortisol are well known as SW adaptation hormones and they increase ion and water transport in the intestine and ion excretion at the gill level. Plasma cortisol levels are known to rise following the transfer of eels from FW to SW (WendelaarBonga 1997). The duration of the elevated plasma cortisol level is correlated with morphological and physiological changes in the gill and intestinal epithelium. It is known to increase Na⁺-K⁺-ATPase in branchial MR cells and angiogenesis in the intestinal tract in order to increase the active transport of water and ions. On the other hand prolactin antagonizes cortisol's effect by lowering water and ion permeabilities in the gastrointestinal tract, promoting FW acclimatization. In the Japanese eel, plasma prolactin levels increased after transfer from sea water to fresh water and decreased when the fish were returned to sea water (Suzuki et al. 1991). Cortisol, ACTH and prolactin remodel the osmoregulatory epithelia, promote the *de novo* synthesis of ion transporters and their action is slow and long term when the animal faces salinity changes.

More recently, the pivotal role of the renin-angiotensin system (RAS) and natriuretic peptides (NPs) in the osmoregulatory responses of the eel to varying salinity has emerged. These hormones respond to rapid changes of environmental salinity. They operate rapidly on various organs including the brain, gill, gastrointestinal tract, and kidney to modulate the activity of ion-channels and transporters, and to control the drinking behavior (Tsukada et al. 2013).

The renin-angiotensin-system (RAS) is involved mainly in SW acclimatization because FW to SW transfer rapidly increases plasma renin activity (Wong and Takei 2012). The RAS action on the osmoregulatory responses of the eel includes (1) decrease in the renal net sodium reabsorption, which is an important feature in SW adaptation facilitating ion excretion; (2) regulation of cortisol secretion, since renin injection increased the circulating levels of cortisol in control eel but not in hypophysectomised eel, suggesting that the stimulation of RAS on cortisol secretion could be via the pituitary-ACTH-inter-renal axis (Hirano 1969); (3) a direct stimulatory effect on Na⁺-K⁺-ATPase activity in both isolated branchial and renal tissue (Marsigliante et al. 2000); (4) stimulation of the drinking behavior in SW (Ando et al. 2000).

Natriuretic peptides (NP) are peptide hormones with 22–36 amonoacid residues known to play an important role in cardiovascular functions and body fluid homeostasis. In the eel, a NP was isolated from the atrium (Takei et al. 1989), the so called eel atrial natriuretic peptide (eANP). Later a distinct NP with long C-terminal tail was isolated and sequenced from the eel ventricle and called eel ventricular natriuretic peptide (eVNP) (Takei et al. 1991). NPs secretion responds primarily to an acute increase in the plasma osmolarity. The major mode of action of these peptides is represented by the reduction of salt loading by inhibition of oral ingestion and intestinal ionic uptake. In SW eels the intra-arterial injection of NP strongly inhibits drinking (Miyanishi et al. 2011). It is thought that the site of action of NP for this antidispogenic effect is represented by medulla oblongata (hindbrain) (Nobata and Takei 2011), which plays an important role in the regulation of drinking in eels. Moreover, NPs are thought to act directly on the intestinal epithelium inhibiting Na⁺-K⁺-2Cl⁻ cotransporter on the apical membrane through a cGMP-dependent intracellular pathway (Loretz and Takei 1997). The effects of NP are fast but short-lived and it is thought to be important in SW acclimatization in order to prevent the risk of hypernatremia. In fact the elevated drinking after SW transfer, the so called "chloride response", is decreased soon after the plasma increase of NP (Takei et al. 1994). Therefore, the role of NP seems to be aimed towards counteracting the abrupt increase in plasma Na$^+$ concentration by reducing oral and intestinal Na$^+$ uptake in the initial phase of acclimation. Moreover, NP induces cortisol secretion from the inter-renal tissue (Li and Takei 2003) contributing to promote long-term SW acclimation in the eel.

Cell volume regulation

When the eel moves from SW to FW and *vice versa*, the fish faces abrupt transitory changes of plasma osmolarity before the acclimation process in the new osmotic environment is completed in a period ranging from some days to few weeks. During the transfer from FW to SW, plasma osmolarity rapidly increases by about 50% (Lionetto et al. 2001), mostly due to the loss of water from all the permeable body surfaces. On the other hand during the transition from seawater to fresh water, plasma osmolarity decreases by about 20% because of the osmotic intake of water (Lionetto et al. 2005). Over a period of 4–5 days plasma osmolarity slowly tends to return to the initial level, due to the activation of osmoregulatory processes at the intestinal, gill and kidney level (see above). Therefore, when the fish experiences environmental salinity changes, its cells must face dramatic variations in the extracellular osmolarity, which in turn are responsible for abrupt cell volume perturbation. Under these conditions well developed cell volume regulation mechanisms are triggered, enabling the cells to counteract the cell volume changes that jeopardize the constancy of their intracellular milieu.

Most of the studies on cell volume regulation in the eel have focused on the intestinal epithelium (Lionetto et al. 2001, 2002, 2005, 2006, 2008, 2010), which is one of the main interfaces of fish with the environment. As previously demonstrated the eel intestinal cells are sensitive to changes in the osmolarity of the external medium, either hypertonic (Lionetto et al. 2001) or hypotonic stress (Lionetto et al. 2005) and are able to activate cell volume regulatory responses. When the isolated eel intestine is experimentally exposed to hypertonic stress by 25% increase of the osmolarity of the perfusion solution, it initially shrinks, as indicated by the decrease in epithelium height (about 32% decrease after 5 min exposure). After about 45 min, the epithelium height increases again in the continued presence of the osmotic stress, suggesting that the eel intestine is able to perform a regulatory volume increase (RVI) response following hypertonic shrinkage (Lionetto et al. 2001). When the isolated eel intestine is experimentally exposed to hypotonic stress the epithelium swells as indicated by the increase in epithelium height after 5 min exposure. After about 45 min, the epithelium height starts decreasing again towards the initial value. These results indicate that the eel intestinal epithelium is able to regulate its volume after hypotonic swelling (Lionetto et al. 2005) through a regulatory volume decrease (RVD) response.

The mechanisms for the regulation of cell volume after osmotic stress are highly conserved and are principally similar in cells from various tissues as well as between evolutionary distant species (Lang et al. 1998). Following a hypotonic cell swelling, they consist of the release of osmolytes from the cell followed by loss of osmotically obliged water (RVD) (Hoffmann et al. 2009). Following hypertonic cell shrinkage the cell increases the intracellular osmolarity via intake of osmolytes, and hence influx of water (RVI) takes place. The most rapid mechanism to change intracellular osmolarity and accomplish cell volume regulation is ion transport across cell membrane. In the RVD response, the cells release KCl by the activation of K^+ and Cl^- efflux through independent K^+ and anion channels (Hoffmann et al. 2009). Electroneutral K^+-Cl^- cotransporter is an alternative system contributing to RVD in some cell types. In the RVI response, the cells generally initiate a net gain of ions via the activation of Na^+-K^+-$2Cl^-$ cotransport, Na^+/H^+ exchange, and/or nonselective cation channels (Hoffmann et al. 2009).

In the slightly longer term, changes in non ionic osmolyte catabolism/metabolism occur contributing to osmoregulation. This is accounted by altered transcription of a number of osmoregulatory genes most of which are involved in the catabolism/metabolism of methylamines, polyalcohols, aminoacids and their derivatives (Hoffmann et al. 2009).

In the eel intestinal epithelium the stimulation of the luminal Na^+-K^+-$2Cl^-$ transporter has been demonstrated as the main ion transport mechanism responsible for RVI following hypertonic stress (Lionetto et al. 2001). The actin-based cytoskeleton has been proposed to play a key role in Na^+-K^+-$2Cl^-$ stimulation via cell shrinkage, since pre-treatment with 20 µM cytochalasin D or 0.5 µM latrunculin A, known to induce F-actin depolymerization by different molecular mechanisms, almost completely obliterates the Na^+-K^+-$2Cl^-$ cotransporter stimulation induced by hypertonic stress exposure, as assessed on the *ex vivo* using chamber mounted epithelium (Lionetto et al. 2002). In several cell types the functional involvement of the cytoskeleton in response to osmotic stress has been shown to be associated with changes in the polymerization state of F-actin or with F-actin rearrangement (Pedersen et al. 2001). However, in the eel intestine, no detectable alteration in F-actin organization after hypertonic stress exposure was observed by confocal imaging (Lionetto et al. 2002) and no significant change in F-actin content was measured by using a quantitative F-actin assay (Lionetto et al. 2002). Therefore, in the eel intestine the role of actin cytoskeleton in the Na^+-K^+-$2Cl^-$ transporter stimulation following hypertonic stress requires the presence of an intact and organized microfilament system, but does not involve detectable changes in the polymerization state of F-actin or F-actin remodelling (Lionetto et al. 2002).

The hypotonicity induced activation of BK and SK Ca^{2+} activated K^+ channels on the apical and the basolateral membranes of the eel intestine has been demonstrated as the main ion transport mechanism responsible for RVD following hypotonic stress (Lionetto et al. 2005, 2008). On each membrane the

K^+ efflux is accompanied by a parallel anion efflux through volume activated anion channel (Lionetto et al. 2005).

Conclusion

Due to its diadromous migration the eels experience drastic osmotic changes of the environmental medium during their life cycle, and as such they have evolved specialized physiological mechanisms to cope with this variability. These mechanisms, which represent key aspects of the eel adaptive physiology, include the activity of specialized ion and transport epithelia (gills, gastrointestinal tract and kidney), which are able to change their functions according to changing environmental osmotic demands, and a complex endocrine control of the osmoregulatory responses.

Keywords: Osmoregulation, salinity, gastrointestinal tract, gills, kidney, hormones, freshwater, seawater, ion transport mechanisms, cell volume

References

Ando, M. 1985. Relationship between coupled Na^+-K^+-Cl^- transport and water absorption across the seawater intestine. J. Comp. Physiol. 155: 311–317.

Ando, M. and K. Nagashima. 1996. Intestinal Na^+ and Cl^- levels control drinking behaviour in the sea-water adapted eel *Anguilla japonica*. J. Exp. Biol. 199: 711–716.

Ando, M. and Y. Takei. 2013. Intestinal absorption of salts and water. pp. 160–177. *In*: F. Trischitta, Y. Takei and P. Sébert (eds.). Eel Physiology. CRC Press, Boca Raton, FL, USA.

Ando, M., S. Utida and H. Nagahama. 1975. Active transport of chloride in eel intestine with special reference to seawater adaptation. Comp. Biochem. Physiol. 51A: 27–32.

Ando, M., Y. Fujii, T. Kadota, T. Kozaka, T. Mukuda, I. Takase and A. Kawahara. 2000. Some factors affecting drinking behavior and their interactions in seawater-acclimated eels, *Anguilla japonica*. Zoological Science 17(2): 171–178.

Beyenbach, K.W. 2004. Kidneys sans glomeruli. A. J. Physiol. 286: F811–F827.

Cliff, W.H. and K.W. Beyenbach. 1992. Secretory renal proximal tubules in seawater- and freshwater-adapted killifish. Am. J. Physiol. Renal Physiol. 262: F108–F116.

Cutler, C.P. and G. Cramb. 2001. Molecular physiology of osmoregulation in eels and other teleosts: the role of transporter isoforms and gene duplication. Comparative Biochemistry and Physiology Part A 130: 551–564.

Cutler, C.P. and G. Cramb. 2002a. Two isoforms of the Na^+/K^+/$2Cl^-$ cotransporter are expressed in the European eel (*Anguilla anguilla*). Biochimica et Biophysica Acta 1566: 92–103.

Cutler, C.P. and G. Cramb. 2002b. Branchial expression of an aquaporin 3 (AQP-3) homologue is downregulated in the European eel *Anguilla anguilla* following seawater acclimation. J. Exp. Biol. 205: 2643–2651.

Cutler, C.P. and G. Cramb. 2008. Differential expression of absorptive cation-chloride-cotransporters in the intestinal and renal tissues of the European eel (*Anguilla anguilla*). Comp. Biochem. Physiol., Part A 149: 63–73.

Cutler, C.P., I.L. Sanders, G. Luke, N. Hazon and G. Cramb. 1996. Ion transport in teleosts: identification and expression of ion transporting proteins in the branchial and intestinal epithelia of the European eel (*Anguilla anguilla*). pp. 43–74. *In*: S.K. Ennion and G. Goldspink (eds.). Society of Experimental Biologists, Seminar Series, 58. Cambridge University Press, Cambridge.

Cutler, C.P., A.S. Martinez and G. Cramb. 2007. The role of aquaporin 3 in teleost fish. Comp. Biochem. Physiol. Part A 205: 2643–2651.

Evans, D.H. 2002. Cell signaling and ion transport across the fish gill epithelium. J. Exp. Zool. 293: 336–347.

Evans, D.H., P.M. Piermarini and K.P. Choe. 2005. The multifunctional fish gill: dominant site of gas exchange, osmoregulation, acid-base regulation, and excretion of nitrogenous waste. Physiol. Rev. 85: 97–177.

Gaitskell, R.E. and J.L. Chester. 1971. Dinking and urine production in the European eel (*Anguilla anguilla*). Gen. Comp. Endocrinol. 16: 478–483.

Hirano, T. 1969. Effects of hypophysectomy and salinity change on plasma cortisol concentration in the Japanese eel, *Anguilla japonica*. Endocrinol. Jpn. 16: 557–560.

Hirano, T. 1974. Some factors regulating water intake by the eel, *Anguilla japonica* J. Exp. Biol. 61: 737–747.

Hirano, T. and N. Mayer-Gostan. 1976. Eel esophagous as an osmoregulatory organ. Proc. Natl. Acad. Sci. USA 73: 1348–1350.

Hirose, S., T. Kaneko, N. Naito and Y. Takei. 2003. Molecular biology of major components of chloride cells. Comp. Biochem. Physiol. B 136: 593–620.

Hoffmann, E.K., I.H. Lambert and S.F. Pedersen. 2009. Physiology of cell volume regulation in vertebrates. Physiol. Rev. 89: 193–277.

Hwang, P.P. and T. Hirano. 1985. Effect of environmental salinity on intracellular organization and junctional structure of chloride cells in early stages of teleost development. Journal of experimental Zoology 236: 115–126.

Hwang, P.P. and T.H. Lee. 2007. New insights into fish ion regulation and mitochondrion-rich cells. Comparative Biochemistry and Physiology, Part A 148: 479–497.

Hwang, P.P., T.H. Lee and L.Y. Lin. 2011. Ion regulation in fish gills: recent progress in the cellular and molecular mechanisms. Am. J. Physiol. 301: R28–R47.

Inokuchi, M., J. Hiroi, S. Watanabe, K.M. Lee and T. Kaneko. 2008. Gene expression and morphological localization of NHE3, NCC and NKCC1a in branchial mitochondriarich cells of Mozambique tilapia (*Oreochromis mossambicus*) acclimated to a wide range of salinities. Comp. Biochem. Physiol. A151: 151–158.

Islam, Z., A. Kato, M.F. Romero and S. Hirose. 2011. Identification and apical membrane localization of an electrogenic Na^+/Ca^{2+} exchanger NCX2a likely to be involved in renal Ca^{2+} excretion by seawater fish. Am. J. Physiol. 301: R1427–R1439.

Kaneko, T., S. Watanabe and K.M. Lee. 2008. Functional morphology of mitochondrionrich cells in euryhaline and stenohaline teleosts. Aqua Biosci. Monogr. 1: 1–62.

Kato, A., M.H. Chang, Y. Kurita, T. Nakada, M. Ogoshi, T. Nakazato, H. Doi, S. Hirose and M.F. Romero. 2009. Identification of renal transporters involved in sulfate excretion in marine teleost fish. Am. J. Physiol. Regul. Integr. Comp. Physiol. 297: R1647–R1659.

Katoh, F. and T. Kaneko. 2003. Short-term transformation and long-term replacement of branchial chloride cells in killifish transferred from seawater to freshwater, revealed by morphofunctional observations and a newly established 'time differential double fluorescent staining' technique. J. Exp. Biol. 206: 4113–4123.

Lang, F., G.L. Busch, H. Volkl. 1998. The Diversity of Volume Regulatory Mechanisms. Cell Physiol. Biochem. 8: 1–45.

Larsen, E.H., J.B. Sørensen and J.N. Sørensen. 2002. Analysis of the sodium recirculation theory of solute-coupled water transport in small intestine. J. Physiol. 542: 33–50.

Larsen, E.H., N.J. Willumsen, N. Møbjerg and J.N. Sørensen. 2009. The lateral intercellular space as osmotic coupling compartment in isotonic transport 195: 171–186.

Lionetto, M.G. and T. Schettino. 2006. The Na^+-K^+-$2Cl^-$ cotransporter and the osmotic stress response in a model salt transport epithelium. Acta Physiol. 187: 115–24.

Lionetto, M.G., M. Maffia, F. Vignes, C. Storelli and T. Schettino. 1996. Differences in intestinal electrophysiological parameters and nutrient transport rates between eels (*Anguilla anguilla*) at yellow and silver stages. J. Exp. Zool. 275: 399–405.

Lionetto, M.G., M.E. Giordano, G. Nicolardi and T. Schettino. 2001. Hypertonicity Stimulates Cl^- transport in the intestine of fresh water acclimated eel, *Anguilla anguilla*. Cell Physiol. Biochem. 11: 41–54.

Lionetto, M.G., S.F. Pedersen, E.K. Hoffmann, M.E. Giordano and T. Schettino. 2002. Roles of the cytoskeleton and of protein phosphorilation events in the osmotic stress response in eel intestinal epithelium. Cell Physiol. Biochem. 12: 163–178.

Lionetto, M.G., M.E. Giordano, F. De Nuccio, G. Nicolardi, E.K. Hoffman and T. Schettino. 2005. Hypotonicity induced K^+ and anion conductive pathways activation in eel intestinal epithelium. J. Exp. Biol. 208: 749–760.

Lionetto, M.G., A. Rizzello, M.E. Giordano, M. Maffia, F. De Nuccio, G. Nicolardi, E.K. Hoffman and T. Schettino. 2008. Molecular and functional expression of high conductance Ca^{2+} activated K^+ channels in the eel intestinal epithelium. Cell Physiol. Biochem. 21: 361–337.

Lionetto, M.G., M.E. Giordano, A. Calisi, R. Caricato, E.K. Hoffmann and T. Schettino. 2010. Role of BK channels in the Apoptotic Volume Decrease in native eel intestinal cells. Cell Physiol. Biochem. 25: 733–744.

Lorez, C.A. and Y. Takei. 1997. Natriuretic peptide inhibition of intestinal salt absorption in the Japanese eel: physiological significance. Fish Physiol. Biochem. 17: 319–324.

Marshall, W.S. and M. Grosell. 2006. Ion transport, osmoregulation, and acid–base balance. pp. 177–230. *In*: D.H. Evans and J.B. Claiborne (eds.). The Physiology of Fishes, 3rd ed. CRC Press, New York.

Marshall, W.S., E.M. Lynch and R.R. Cozzi. 2002. Redistribution of immunofluorescence of CFTR anion channel and NKCC cotransporter in chloride cells during adaptation of the killifish *Fundulus heteroclitus* to seawater. J. Exp. Biol. 205: 1265–1273.

Marsigliante, S., A. Muscella, G.P. Vinson and C. Storelli. 1997. Angiotensin II receptors in the gill of sea water- and freshwater-adapted eel. J. Mol. Endocrinol. 18: 67–76.

Marsigliante, S., A. Muscella, S. Greco, M.G. Elia, S. Vilella and C. Storelli. 2000. Na^+/K^+ATPase activity inhibition and isoform-specific translocation of protein kinase C following angiotensin II administration in isolated eel enterocytes. J. Endocinol. 168: 339–346.

Miyanishi, H., S. Nobata and Y. Takei. 2011. Relative antidipsogenic potencies of six homologous natriuretic peptides in eels. Zool. Sci. 28: 719–726.

Miyazaki, H., T. Kaneko, S. Uchida, S. Sasaki and Y. Takei. 2002. Kidney-specific chloride channel, OmClC-K, predominantly expressed in the diluting segment of freshwater adapted tilapia kidney. Proc. Natl. Acad. Sci. USA 99: 15782–15787.

Nagashima, K. and M. Ando. 1994. Characterization of esophageal desalination in the seawater eel, *Anguilla japonica*. J. Comp. Physiol. B 164: 47–54.

Nakada, T., K. Zandi-Nejad, Y. Kurita, H. Kudo, V. Broumand, C.Y. Kwon, A. Mercado, D.B. Mount and S. Horose. 2005. Roles of S1c13a1 and S1c26a1 sulphate transporters of eel kidney in sulphate homeostasis and osmoregulation in freshwater. Am. J. Physiol., Regul. Integr. Comp. Physiol. 289: R575–R585.

Nawata, C.M., S. Hirose, T. Nakada, C.M. Wood and A. Kato. 2010. Rh glycoprotein expression is modulated in pufferfish (Takifugurubripes) during high environmental ammonia exposure. J. Exp. Biol. 213: 3150–3160.

Nishimura, H., M. Imai and M. Ogawa. 1983. Sodium chloride and water transport in the renal distal tubule of the rainbow trout. Am. J. Physiol. 244: F247–F254.

Nobata, S. and Y. Takei. 2011. The area postrema in hindbrain is a central player for regulation of drinking behavior in Japanese eels. Am. J. Physiol. 300: R1569–R1577.

Nobata, S. and M. Ando. 2013. Regulation of drinking. pp. 225–248. *In*: F. Trischitta, Y. Takei and P. Sébert (eds.). Eel Physiology. CRC Press, Boca Raton, FL.

Ogasawara, T. and T. Hirano. 1984. Changing in osmotic water permeability of the eels during seawater and freshwater adaptation. J. Comp. Physiol. B 154: 3–11.

Pedersen, S.F., E.K. Hoffmann and J.W. Mills. 2001. The cytoskeleton and cell volume regulation. Comp. Biochem. Physiol. 130: 385–399.

Schettino, T. and M.G. Lionetto. 2003. Cl⁻ absorption in European eel intestine and its regulation. J. Exp. Zool. 300A: 63–68.

Seo, M.Y., K.M. Lee and T. Kaneko. 2009. Morphological changes in gill mitochondria-rich cells in cultured Japanese eel *Anguilla japonica* acclimated to a wide range of environmental salinity. Fish. Sci. 75: 1147–1156.

Seo, M.Y., M. Mekuchi, K. Teranishi and T. Kaneko. 2013. Expression of ion transporters in gill mitochondrion-rich cells in Japanese eel acclimated to a wide range of environmental salinity. Comp. Biochem. Physiol. Part A 166: 323–332.

Skadhauge, E. 1969. The mechanisms of salt and water absorption in the intestine of the eel (*Anguilla anguilla*) adapted to waters of various salinities. J. Physiol. London 204: 135–158.

Skadhauge, E. 1974. Coupling of Transmural Flows of NaCl and Water in the Intestine of the eel (*Anguilla Anguilla*). J. Exp. Biol. 60: 535–546.

Suzuki, R., A. Yasuda, J. Kondo, H. Kawauchi and T. Hirano. 1991. Isolation and characterization of Japanese eel prolactins. Gen. Comp. Endocrinol. 161: 147–153.

Takei, T., A. Takahashi, T.X. Watanabe, K. Nakajima and S. Sakakibara. 1991. A novel natriuretic peptide isolated from eel cardiac ventricles. FEBS Letters 282: 317–320.

Takei, Y. and R.J. Balment. 2009. The neuroendocrine regulation of fluid intake and fluid balance. Fish Physiology 28: 365–419.

Takei, Y., A. Takahashi, T.X. Watanabe, K. Nakajima and S. Sakakibara. 1989. Amino acid sequence and relative biological activity of eel atrial natriuretic peptide. Biochemical and Biophysical Research Communications 164: 537–543.

Takei, Y., M. Ueki and T. Nishizawa. 1994. Eel ventricular natriuretic peptide: cDNA cloning and mRNA expression. J. Mol. Endocrinol. 13: 339–345.

Teranishi, K., M. Mekuchi and T. Kaneko. 2013. Expression of sodium/hydrogen exchanger 3 and cation-chloride cotransporters in the kidney of Japanese eel acclimated to a wide range of salinities. Comparative Biochemistry and Physiology, Part A 164(2013): 333–343.

Trischitta, F., M.G. Denaro, C. Faggio and T. Schettino. 1992a. Comparison of Cl⁻ absorption in the intestine of the seawater and freshwater adapted eel, *Anguilla anguilla*: evidence for the presence of an Na-K-Cl cotransport system on the luminal membrane of the enterocyte. J. Exp. Zool. 263: 245–253.

Tse, W.K.F. and C.K.C. Wong. 2011. Nbce1 and H⁺-ATPase mRNA expression are stimulated in the mitochondria-rich cells of freshwater-acclimating Japanese eels (*Anguilla japonica*). Can. J. Zool. 89: 348–35.

Tse, W.K.F., D.W.T. Au and C.K.C. Wong. 2006. Characterization of ion channel and transporter mRNA expressions in isolated gill chloride and pavement cells of seawater acclimating eels. Biochem. Biophys. Res. Commun. 346: 1181–1190.

Tsukada, T., M.K.S. Wong, M. Ogoshi and S. Yuge. 2013. Endocrine control of osmoregulation. pp. 178–224. *In*: F. Trischitta, Y. Takei and P. Sébert (eds.). Eel Physiology. CRC Press, Boca Raton, FL, USA.

Watanabe, T. and Y. Takei. 2011. Molecular physiology and functional morphology of SO_4^{2-} excretion by the kidney of seawater-adapted eels. J. Exp. Biol. 214: 1783–1790.

WendelaarBonga, S.E. 1997. The stress response in fish Physiol. Rev. 77: 591–625.

Wong, M.K.S. and T. Takei. 2012. Changes in plasma angiotensin subtypes in Japanese eel acclimated to various salinities from deionized water to double-strength seawater. Gen. Comp. Endocrinol. 178: 250–258.

11

Migration, Gamete Biology and Spawning

P. Mark Lokman

Introduction

An extensive body of research has been dedicated towards understanding the breeding biology of freshwater eels (*Anguilla* spp.). For decades and perhaps even centuries, the enigmatic marine phase of these fish, with its long-distance migration and mysterious spawning biology, has been a source of fascination for biologists; hence, numerous studies have sought to reveal the specifics of anguillid migration, their rendezvous at remote locations and their spawning behaviour and physiology-culminating in the release of fertilizable gametes and the start of a new generation.

More recently, curiosity and fascination have been joined by an urgent need to complete the life cycle of the eel in captivity for conservation purposes; thus, alarms about the dramatic decreases in the abundance, recruitment and distribution of several eel species have been rung during the last 10–15 years, although declines for some species started well before that time, reductions in glass eel recruitment for *Anguilla anguilla* being evident before 1983 (ICES 2013). The reasons for these declines remain unclear (Hanel et al. 2014); possible causes include any or a combination of the following factors: habitat destruction, pollution, overfishing, pathogen- or parasite-mediated reductions in fitness of migrants on their return journey (e.g., *Anguillicola crassus* in *A. anguilla*; Pelster 2015) and climate-change related changes in oceanic conditions (see Hanel et al. (2014) for further discussions on this topic)—

Department of Zoology, University of Otago, PO Box 56, Dunedin 9054, New Zealand.
E-mail: mark.lokman@otago.ac.nz

recently, these findings were further reinforced by the low abundances of eel larvae at the spawning area of the Atlantic eels (Hanel et al. 2014). Accordingly, several species are now listed on the IUCN Red List as endangered (*A. rostrata, A. japonica*), critically endangered (*A. anguilla*), or vulnerable (*A. borneensis*). This, in turn, has obvious implications for ecosystem functioning and biodiversity, and also for the eel aquaculture industry, which even today depends on the capture of wild juveniles for stocking purposes.

The dire situation of Northern Hemisphere temperate eels (*A. anguilla, A. rostrata, A. japonica*)—much less is known about tropical eels, but their situation is quite possibly no better than that of their temperate congeners (see www.iucnredlist.org)—has resulted recently in a notable increase in the volume of research papers that focus on the final stages of the eel's life, whether *in situ* (sampling of eels at the spawning site) or by inference (tracking or experimental studies). Many of these studies have been reviewed in several excellent treatises, including (but not limited to) "Eel Biology" (Aida et al. 2003), "Spawning Migration of the European Eel" (Van den Thillart et al. 2009) and "Eel Physiology" (Trischitta et al. 2013). In this Chapter, I aim to focus on the most recent findings—findings that I have aimed to place in context of earlier reviews and also extended with information from recent papers in order to give an up-to-date outline of the migratory and reproductive (gamete production and spawning) biology of eels belonging to the genus *Anguilla*.

Eel migration

Historical perspective

Notwithstanding the gaps in our understanding of eel biology that persist even today, tremendous advances have been made towards unravelling the eel life cycle in the past 150 years. It was in 1856, when Kaup described leptocephali from the Mediterranean Sea and assigned them the scientific name *Leptocephalus brevirostris*, recognizing their similarity to leptocephali from other fish. However, the link between *L. brevirostris* and the European eel was not made until 1896, when Grassi and Calandruccio collected these fish from the Mediterranean and allowed them to metamorphose into freshwater eels, which eventually were to be named *Anguilla anguilla*. The origin of their existence remained unknown, however, and was initially pledged to be somewhere in the Mediterranean, an idea that can be easily defended in light of the presence of eels in advanced stages of development, as occasionally reported from the Mediterranean (c.f., Capoccioni et al. 2014). However, this belief was later challenged, when Petersen (1905) and Hjort (1910) collected leptocephali from along the Atlantic coast and the central North Atlantic, South of the Azores, respectively. Hjort (1910) proposed (page 106) that "…I should be inclined to regard the continental slope as the area where the transformation of the larvae takes place, and the southern central part of the North Atlantic ocean as the probable spawning area of the eel", speculation that would, notwithstanding a more westerly spawning location, turn out to be quite prophetic (see below).

Spawning ecology

To date, the spawning areas of only a few eel species have been conclusively identified. Early charting was conducted by the Danish researcher Johannes Schmidt, who, on the basis of the capture of increasingly smaller larvae, deduced that the approximate spawning area for the Atlantic eels would have to be in the Sargasso Sea (Schmidt 1912, 1923). Subsequent cruises in the West Atlantic, many years later, have confirmed Schmidt's ground-breaking revelations (e.g., Schoth and Tesch 1982; Castonguay and McCleave 1987; Hanel et al. 2014).

More recently, the application of Schmidt's approach of back-tracking increasingly smaller larvae and taking into account oceanic currents has pin-pointed the spawning area for *A. japonica* in the North-West Pacific (Tsukamoto 1992), along the West Mariana Ridge (Kurogi et al. 2011). Tsukamoto's discovery was reinforced by the collection of spawned eggs (Aoyama et al. 2014), newly-hatched larvae (Kurogi et al. 2011) and fully mature or spent adults (Chow et al. 2009; Kurogi et al. 2011) nearly 20 years later. Northern Hemisphere temperate eel species are all believed to spawn panmictically (e.g., Atlantic eels: Als et al. 2011) at a single spawning site with a defined seasonality (e.g., Aoyama et al. 2003; Righton et al. 2012; Arai 2014).

Panmictic spawning events at a single spawning location have also been proposed for *A. australis* and *A. dieffenbachii*, both of which display seasonal arrival (spring) of glass eels (Jellyman et al. 1999) and an autumn-focused adult out-migration (Todd 1981b). However, the spawning areas of these eel species, and that of *A. reinhardtii*, remain unknown, although estimated spawning locations northeast of New Caledonia have been proposed on the basis of hydrological, hydrographical and climatological conditions (Jellyman and Bowen 2007).

Tropical anguillid eel species typically do not display some or all of the temperate eel traits, like panmixia, seasonality of spawning and single spawning locations. Furthermore, their migrations may be short (also see the section on 'Tropical Eels' below) and hence, dispersal is likely to be limited, resulting in a more restricted distributional range (c.f., Aoyama et al. 2003). For example, Aoyama and colleagues (2003) deduced that *A. borneensis* spawn in the Celebes Sea (predicted migration distance 480–650 km), whilst *A. celebesensis* is likely to reproduce in the even closer Tomini Bay (80–300 km). Moreover, a second spawning area for *A. celebesensis* is likely to exist in the Celebes Sea (Aoyama et al. 2003). Whilst these two spawning locations for *A. celebesensis* are geographically close, the different spawning locations of *A. bicolor* are distant, one being in the Pacific (c.f., Tanaka et al. 2014), the other off the coast of east Africa (e.g., Pous et al. 2010). Multiple spawning locations are also evident for *A. marmorata*, for which four genetically different populations have been identified (Minegishi et al. 2008); spawning is known to occur in the Northwest Pacific (capture of mature adults in the same area as where mature *A. japonica* have been captured: Chow et al. 2009) and is further thought to take place elsewhere in the North Pacific, somewhere in the South Pacific, and in the Indian Ocean (Minegishi et al. 2008), quite possibly off the African coast (c.f., Pous et al. 2010), for the other three populations.

In keeping with a generic spawning area in the south-west Indian Ocean for *A. marmorata* and *A. bicolor*, analysis of early life history traits of *A. mossambica* similarly suggested an area somewhere north-east of Madagascar (Reveillac et al. 2009). A prolonged spawning period for this species (Reveillac et al. 2009) at a single location (Pous et al. 2010) is likely.

Interestingly, there is increasing evidence of overlapping spawning areas for several species; thus, *A. anguilla* and *A. rostrata* spawn in partially overlapping parts of the North Atlantic (Schmidt 1922), the four African species may all spawn in a generically similar region between 13° S and 19° S westwards of 60.5° E (Pous et al. 2010), at least two species are likely to spawn near the West Mariana Ridge in the North West Pacific (e.g., Chow et al. 2009; Han et al. 2012) and several species are sympatric in the Indonesian Archipelago and, given their limited distributions, may have overlapping spawning grounds. Similarly, the species from Oceania may well share a spawning locale northeast of New Caledonia (c.f., Jellyman and Bowen 2007). These overlaps prompt interesting questions about larval dispersal and differences in distributions between species, as recently posed for *A. japonica* and *A. marmorata* (Han et al. 2012); differential temperature tolerance and preference between these two species, together with interfering coastal currents, were identified as possible mechanisms responsible for differences in distribution (Han et al. 2012).

Final destination

With elucidation of the spawning grounds, and indeed the capture of some mature adults, parts of the mystery of the eel life cycle have been unveiled, although natural spawning has not, to date, been observed for any species. Other notable questions remain (also see Aoyama et al. 2014)—how do the eels get to their distant destinations (also see next section) and what are the cues to relay that the final destination has been reached? The collection of larvae of the Atlantic eels is closely associated with temperature gradients, and hence, the temperature cline associated with the frontal zone within the North Atlantic Subtropical Convergence has been considered important as a cue for spawning locale (McCleave 2003). For the Japanese eel, a series of sea mounts near the West Mariana Ridge have been proposed as landmarks that may facilitate the formation of spawner aggregations (Tsukamoto et al. 2003), a notion that is supported by the capture of mature individuals from near this location (Chow et al. 2009). More recently, Aoyama and co-workers proposed that the sea mounts may act as a longitudinal landmark, whereas salinity fronts appear to be important as latitudinal cues (Aoyama et al. 2014).

Physiological preadaptations for migration

In order for temperate anguillid eels to successfully migrate to far-away spawning areas, numerous and comprehensive changes to the physiology of the fish are required—these include adjustments to the visual system, to metabolism and lipid physiology, to intestinal function, to swimbladder physiology and to skin structure, to name but a few. Furthermore, the animal transforms into an "endurance athlete", as is reflected by the increases in heart size and red muscle mass. Together, these changes are known as the yellow-to-silver transformation, or 'silvering' for short. A series of studies and reviews have highlighted the specifics of many of the silvering-associated changes that occur (Lokman et al. 2003; Durif et al. 2009; Rousseau et al. 2009; Righton et al. 2012)—changes that are not limited to physiology or morphology, but that extend to behaviour as well (Bruijs and Durif 2009). For example, eels cease feeding and move downstream during the silvering process.

Co-ordination of the silvering transformation

The multitude of changes that occur as eels undergo silvering have been likened to metamorphosis, although Aroua and colleagues (2005) concluded that it should not qualify as such. Regardless of whether the transformation is considered a "true" metamorphosis or not, the many dramatic changes lead to tremendous differences in the function and ecology of the eel (also see above), and appropriate timing and co-ordination of tissue re-modelling is therefore essential. The most likely candidate co-ordinating all the changes is the steroid hormone 11-ketotestosterone, a conclusion that is supported by a combination of observations, namely (i) circulating levels of this androgen are greatly increased in wild silver eels (Lokman and Young 1998; Lokman et al. 1998); (ii) levels increase in both male and female fish, presenting a rare occurrence of significant amounts of 11-ketotestosterone in a female fish (Lokman et al. 2002) and reinforcing its importance in "doing something unique"; and (iii) 11-ketotestosterone treatment of non-migratory yellow eels *in vivo* induces many of the morphological (increased eye size, increased countershading, changes to gut, muscle, heart, swimbladder, etc.; Rohr et al. 2001; Setiawan et al. 2012a,b; Sudo et al. 2012; Lokman et al. 2015) and physiological changes (liver and intestinal function; Forbes 2013) associated with silvering (see Fig. 1). Likewise, there is some evidence that has implicated 11-ketotestosterone in modulating eel behaviour (Setiawan et al. 2012a).

Fig. 1. Effects of treatment with sustained-release implants of 11-ketotestosterone on the external morphology of female short-finned eel, *Anguilla australis* in the yellow stage. Fish were treated with implants that contained no (top) or 1 mg 11-ketotestosterone (bottom) and they were sampled two weeks later. Note the darkened pectoral fins and the changes in head shape and eye placement. Photo courtesy of Dr. Alvin Setiawan and Mr. Ken Miller.

Timing of the silvering transformation

It has long been known that temperate eels tend to move downstream during autumn (e.g., Todd 1981b), particularly during dark nights (new moon, rough weather)—clearly, these environmental conditions act as proximate signals. But what are the ultimate signals that turn a feeding yellow eel into a fasting, migratory silver eel?

It has been suggested that the growth, reproductive and/or corticotropic axes (Rousseau et al. 2009) could be involved in initiating silvering; however, irrespective of which of these endocrine cascades is the primary driver, it is likely that nutritional status is the ultimate key. Indeed, without sufficient energetic resources, the eels would not be able to successfully travel to the faraway spawning areas (Larsson et al. 1990). In other words, nutritional status would act as an adaptive, selective pressure that would shape the timing and activation of the appropriate endocrine cascade—this, in turn, most likely leads to 11-ketotestosterone production and subsequent induction of the suite of silvering-related changes, as outlined above.

Energetics

As eels leave their growth habitats, they cease feeding and dramatically change their lipid physiology (e.g., Damsteegt et al. 2015b); dietary energy is no longer available, but instead, the fish depend on stored energy which they need to mobilise in order to allow migration, gametogenesis and spawning to take place. The supply of energy to muscle is quite possibly straight-forward, since major storage depots (intramuscular fat) are located in the immediate vicinity of the engines that need to be supplied with fuel (skeletal red muscle) for the long-distance journey. Studies on the European eel have demonstrated that the amount of energy that is stored is normally sufficient to fuel the trip to the Sargasso Sea especially as eels have proven to be very efficient swimmers (Van Ginneken et al. 2005a). Furthermore, it seems very plausible that eels might ride on subsurface currents that flow in the opposite direction to surface currents, using them as a "conveyor belt" (e.g., Aarestrup et al. 2009; Righton et al. 2012); indeed, the positive rheotaxis ('movement with the current') seen in silver American eels (Hain 1975) is suggestive of such tail current-facilitated movement and would indicate that the energy needs estimated from the use of swim tunnels that have employed a head current (described in Van den Thillart et al. 2004) could be notably overestimated. However, such experimental systems cannot account for the daily vertical movements that eels display during their journey (e.g., Jellyman and Tsukamoto 2010; Aarestrup et al. 2009; Schabetsberger et al. 2013); indeed substantial energy is required for migrating eels to move up and down between the upper layers of the water column during the night (typically several hundred meters) and the deeper layers during the day time (often 600–1000 m), as suggested by Aarestrup et al. (2009) and Trancart et al. (2015).

Whilst lipids used for oxidation in muscle are intuitively obtained from adjacent stores, the lipids that are needed for reproduction may need to be moved some considerable distance from the depots to the gonad; this travel of lipids is likely to be facilitated by their packaging into lipoprotein particles in the liver, thereby enabling these nutrients to be transported through the aqueous body fluids towards the more distant targets, notably the ovary (Damsteegt et al. 2015a). Here, the uptake of lipids is likely to be dependent on binding of the lipoprotein particle to the low-density lipoprotein receptor, enabling the eel oocyte to sequester its lipids via any number of possible mechanisms, as reported recently by Damsteegt et al. (2015b)—part of this lipid accumulation in the oocyte occurs prior to migration, but a substantial amount of lipid uptake is realized after the onset of migration (c.f., Divers et al. 2010).

Navigation

Upon commencing their oceanic journey, eels inevitably must swim back to the spawning grounds—in order for this to be possible, the geomagnetic field has been identified as the most likely guiding navigational cue (reviewed in Hunt et al. 2013).

Tropical eels

On the basis of molecular phylogenies, the ancestry of anguillid eels is proposed to be in the tropics, possibly in the Indonesian archipelago (Aoyama et al. 2001; also see Minegishi et al. 2005), where several eel species occur sympatrically. These revelations have generated notable interest, prompting research to be conducted on these tropical eel species in more detail. Findings from Prof Katsumi Tsukamoto's group accordingly have pointed out the basic life history of tropical eels as being largely comparable to that of the much better-studied temperate eels—notwithstanding the notable differences that seem to relate primarily to (i) the closer distance of tropical eels to their marine spawning grounds (see Aoyama et al. 2003; Arai 2014) which is associated with larval-to-juvenile metamorphosis at a smaller size and younger age (Kuroki et al. 2006), and to (ii) the less-defined seasonality of recruitment—although evidence for some degree of seasonality of spawning was reported for *A. celebesensis* by Wouthuyzen et al. (2009). Despite the relative proximity of the spawning grounds to the freshwater growth habitats seen in some tropical eel species, many of the morphological changes associated with silvering still occur (see Hagihara et al. 2010), suggesting that this is a universal trait among anguillid eels.

Gamete biology

The growth of gametes is under the control of the endocrine system. In the context of reproduction, the complex interactions between environment, brain, pituitary glands and gonads are pivotal, providing a fine-tuned system that is largely conserved across the vertebrate clade (e.g., Norris 2007). Regulation of the early stages of germ cell growth in fish is not well understood, but is likely to depend to great extent on growth factors, many of which may be locally produced. Once puberty is initiated, the reproductive axis is activated, resulting in the release of gonadotropins from the pituitary gland into the systemic circulation. These gonadotropins (follicle-stimulating hormone, FSH, and luteinizing hormone, LH), chemical signals that can be detected by specific receptors on target cells in ovary and testis, affect somatic support cells around the developing germ cells—in turn, the somatic support cells can produce and release steroid hormones and/or growth factors that stimulate germ cell development.

Gamete production in anguillid eels has received ample attention from biologists, but due to the effective disappearance of silver eels soon after the onset of migration, our understanding of gametogenesis in eels is, by necessity, limited—limited to what has been puzzled together from observations of wild-caught migrants in the early stages of reproductive development and from the findings from eels in captivity.

Only very recently have samples from wild eels in peri-spawning condition been successfully collected (Chow et al. 2009, 2010; Kurogi et al. 2011; Izumi et al. 2015), opening new avenues for gaining baseline data, both for fundamental studies and for application in (fine-tuning of) induced maturation experiments. A brief outline of the current understanding of the reproductive physiology of anguillids (wild eels, artificially matured eels) is given in the following sections.

Gametogenesis in wild eels

Irrespective of species, yellow eels collected from the wild are always prepubertal—this is reflected in the presence of oocytes that contain lipid droplets, but in which yolk protein deposits (visible as granules) are not encountered. In contrast, silver eels have already entered puberty, yet there are vast differences in the developmental stages of the gonads between species at the time that the eels start to embark on their oceanic journey; indeed, gonadosomatic indices may reach 10% or more in female *A. celebesensis* (Hagihara et al. 2010) and *A. dieffenbachii* (Todd 1981a), whereas values of around 2% are typical for *A. anguilla* (Durif et al. 2005).

In the European eel, arguably the least sexually developed species, oocytes typically do not contain visible yolk granules, whereas small peripheral granules are evident in *A. australis* (Todd 1981a; Lokman et al. 1998). As development progresses, yolk is increasingly accumulated and can be seen in the form of 'platelets' in the ooplasm, towards the nucleus, indicative of midvitellogenesis, as seen for example, in *A. dieffenbachii* (Todd 1981a; Lokman et al. 1998) and *A. celebesensis* (Arai 2014). These changes are associated with changes in the endocrine milieu (e.g., increased plasma levels of estradiol-17β, testosterone and 11-ketotestosterone), as documented by a number of authors (Lokman et al. 1998; Sbaihi et al. 2001; Han et al. 2003).

Eels with ovaries in more advanced stages of development are not accessible from the wild, notwithstanding the recent catches of mature *A. japonica* (Kurogi et al. 2011) and *A. marmorata* (Chow et al. 2009) off the West Mariana Ridge.

It has been suggested that relative gonadal size (i.e., gonadosomatic index, an index that reflects gonadal stage) is inversely proportional to the migratory distance to the spawning ground (e.g., Hagihara et al. 2010), an idea that has gained traction with the relatively advanced stage of the development of some tropical eel species, and their nearby spawning grounds, for example with regard to *A. celebesensis* in Sulawesi.

Differences in gonadal development between males of different anguillid species reflect the patterns found in females; thus, in yellow males, the testes are poorly developed, the most advanced germ cell stage typically being limited to Type A and the occasional Type B spermatogonium. With entry into puberty, however, gonadotropin-mediated actions result in initiation of spermatogonial proliferation, increasing the abundance of Type B spermatogonia, coinciding with increasing blood levels of the androgens testosterone and 11-ketotestosterone (e.g., Lokman and Young 1998); this early stage of spermatogenesis is seen in silver males at the onset of migration in a number of eel species (e.g., *A. australis*, *A. japonica*). In several other species, amongst which *A. dieffenbachii* is a notable example, spermatogenesis is much further advanced (Todd 1981a; Lokman and Young 1998), and all stages of spermatogenesis can be encountered in the testes, again in association with increased levels of androgens.

Tremendous progress has been made in our understanding of the molecular mechanisms controlling spermatogenesis in teleost fish in the last 20 years—and much of this has come from studies on anguillids, especially the research conducted on the Japanese eel by Prof. Takeshi Miura and his co-workers (e.g., Miura et al. 2007; Schulz et al. 2010).

Hormone-induced re-initiation of gonadal growth

The first experiments aimed at artificial induction of maturation (or better: *re-initiation of gonadal growth*, because full maturity—the acquisition of fertilizable gametes—is not attained in many such studies) in eel date back to the 1930s. Since that time, numerous studies have been conducted, mostly, but not exclusively, on temperate eels. Spawning females were first obtained in the 1960s (Fontaine et al. 1964), whereas the first hatching of eel larvae was reported in 1974 (Yamamoto and Yamauchi 1974) and acquisition of the first fingerlings (glass eels) was first documented in 2001 (Tanaka et al. 2003).

So far, only Japanese researchers have developed the technology to complete the life cycle of the eel in captivity (Tanaka et al. 2003; Ijiri et al. 2011; Okamura et al. 2014; Tanaka 2015), and even though this milestone was achieved some 15 years ago, the mass production of glass eels has remained an elusive goal. However, recent sentiments (e.g., Tanaka 2015) are upbeat and suggestive that this 'problem' may get solved in the near future as the needs of larvae (nutrition, husbandry, etc.) are increasingly being understood. Indeed, it seems that larval feeding is currently the last remaining bottleneck preventing the large-scale production of juvenile eels (see Okamura et al. 2014; Tanaka 2015).

In captive eels, gametogenesis does not proceed beyond the peripubertal stage (e.g., Sudo et al. 2011). Moreover, gametogenesis in eels in relatively advanced stages of gametogenesis (such as *A. dieffenbachii*) is arrested following their transfer to captivity, and yolky oocytes (developing 'eggs') become atretic (e.g., Lokman et al. 2001). Dufour and colleagues (1989) attributed this generic failure to develop beyond the peripubertal stage to dopaminergic inhibition of gonadotropin synthesis and release—therefore, hormonal manipulations aimed at overriding the developmental arrest have typically involved the use of gonadotropin preparations. The current state-of-the-art (*A. japonica*) is excellently summarized by Tanaka (2015), and therefore, only a brief account of hormonal overriding of gametogenic arrest is given in this chapter (below).

Induction of spermatogenesis

The most commonly adopted hormone for induction of gonadal development in male eels is human chorionic gonadotropin (hCG), a hormone found in the urine of pregnant women. Even a single injection with hCG can induce widespread changes in the eel testes, but fertilizing ability is often very limited. For that reason, weekly injections with hCG tend to be preferred, often in combination with a priming dose on the day prior to spawning, as published for the European (Pérez et al. 2000) and Japanese eel (Ohta et al. 1997b). Alternatively, sperm quality may be improved via incubation of milt in artificial seminal fluid containing potassium and bicarbonate ions (Ohta et al. 1997a).

Induction of oogenesis

Most researchers aiming to induce full ovarian maturation have employed carp or salmon pituitary homogenates (or extracts), either singly, or in combination with hCG. This approach is now routinely used for acquisition of eel larvae in Japan by several groups—indeed, the protocol from Ohta et al. (1997b) appears to be the common template for induction of oogenesis in other eel species (e.g., Lokman and Young 2000; Palstra et al. 2005). A main challenge has been to keep hormone levels elevated for prolonged periods of time, which is achieved mostly by repeated (weekly) injections of hormonal preparations. However, sustained-release approaches, including water-in-oil-in-water emulsions (Sato et al. 1997) and osmotic minipumps (Kagawa et al. 2013b) have also been trialled, yielding promising results overall.

The use of pituitary homogenates, also known as hypophysation, is a well-known method of fertility treatment in the aquaculture industry, dating back to the 1930s (see review by Yaron et al. 2015). The application of this method in female eels in present-day experiments is not fundamentally different from the methodology applied by Fontaine and his co-workers in the 1960s—however, a suite of studies have aimed to optimize the induced spawning protocol, taking into account temperature, current, water pressure and the presence/absence of conspecifics. The inclusion of other fertility drugs in the artificial maturation protocol, has also been trialled—in this regard, it is of interest to note that the treatment of silver eels with 11-ketotestosterone stimulated lipid uptake (Fig. 2; Lokman et al. 2015), although such treatment resulted in decreased levels of estradiol-17β compared to vehicle-implanted controls (unpublished observations).

New developments have seen the production of recombinant eel Fsh and Lh for use in artificial maturation experiments, for example, in *A. japonica*. This approach is likely to yield valuable insights, especially with regard to the functions of Fsh (it only activates the Fsh receptor, Fshr; Kazeto et al. 2012) and Lh (it can activate both Fshr and Lhr; Kazeto et al. 2012) during gametogenesis. Of further interest is the notion that hCG only activates Lhr (Kazeto et al. 2012; Minegishi et al. 2012), enabling future studies to specifically address the respective roles of these receptors in mediating reproduction in anguillid eels.

Induction of ovulation

Partly due to the joining of forces by many eel researchers in Japan (Tanaka 2015), Japanese scientists have perfected the artificial propagation of eels to

Fig. 2. Effects of treatment with sustained-release implants of 11-ketotestosterone in female shortfinned eel, *Anguilla australis* in the silver stage. Fish were treated with implants that contained no (top) or 1 mg 11-ketotestosterone (bottom) and they were sampled 7 weeks later. Note the increase in the size of the ovaries (cream-coloured tissue running the length of the body cavity) by over 50%.

a fine art, but in order to do so, they needed to address a key problem: how to obtain haploid eggs of good quality? Whereas pituitary homogenates are effective tools for the induction of ovarian growth, numerous issues have been encountered with regard to the induction of final oocyte maturation (resumption of meiosis and cytoplasmic maturation, including water uptake to ensure buoyancy) and ovulation. This event is typically associated with the so-called Lh surge (well-described especially in mammals)—and indeed, a priming injection with pituitary homogenates has been known to yield fertilizable eggs, but this approach by itself has been unreliable (see review by Tanaka 2015). After the realization that Lh induces the production of a steroidal mediator, namely 17,20β-dihydroxy-4-pregnen-3-one (DHP), in salmon (Nagahama and Adachi 1985), subsequent basic science explorations succeeded in identifying DHP as the maturation-inducing hormone in eel also (Kagawa et al. 1995)—this, in turn, led to the optimization of artificial maturation protocols with inclusion of a DHP administration step (or its precursor, 17-hydroxyprogesterone), an approach that now seems to be universally employed (see Kagawa et al. (2013a) for a review on the hormonal induction of final stages of oogenesis in *A. japonica*).

Despite the progress in the artificial propagation of *A. japonica*, issues with other eel species remain, especially with regard to the timing of DHP injection and egg retrieval (also see next section on 'Spawning'). Further improvements may be needed in order to reliably obtain good quality eggs from these species of *Anguilla* to complete their life cycle in the future.

Spawning

Spawning of anguillid eels has not been observed in the natural environment, but ecological and experimental studies have pieced together a story that is likely to represent a reasonably accurate picture—thus, the finding of eggs and mature fish just before or during the new moon, and the capture of eel larvae just after the new moon provide compelling evidence that the timing of spawning is likely to be associated with the new moon in late spring/early summer, at least in the Japanese eel (Aoyama et al. 2014). In this species, spawning probably takes place at depths of 160–250 m (Aoyama et al. 2014) and at temperatures that are likely to be in the 20–23 C range (c.f., Unuma et al. 2012). It is plausible that spawning episodes may be repeated during several consecutive new moons—support for this idea comes from induced spawning studies in which *A. australis* have been repeatedly spawned on several occasions (Lokman and Young 2000). Successive monthly spawnings provide opportunities for the fish to recruit additional batches of eggs/sperm and to maximize parentage and genetic variation, and thus, ensure the likelihood of survival of at least some of the offspring.

Observations on artificially matured eels have yielded a more or less complete behavioural spawning repertoire for both sexes, although differences were evident between the European and Japanese eels; in the former species, hCG-injected males were found to approach the head region, touch the operculum, or approach the urogenital area of matured females (n = 2) subjected to treatment with DHP (Van Ginneken et al. 2005b). In contrast, in *A. japonica*, no such extensive interactions were observed between males and females, although some degree of "schooling" just prior to egg release was observed at times, which could be suggestive of group mating (Dou et al. 2007). Proof that this spawning behaviour in captivity reflects what happens in the wild is currently missing; regardless, induced spawning experiments have increasingly shifted towards spontaneous spawning in tanks, as egg quality is typically better than that from strip-spawned, artificially fertilized eggs (Horie et al. 2008; Di Biase et al. 2015)—it seems reasonable that the observations of captive fish are therefore indicative of the behaviours that occur during spawning in the wild.

Summary

Some of the mysteries surrounding the reproductive biology of freshwater eels have been solved in recent years, most notably for the Japanese eel. Voids in knowledge remain for other temperate eel species, and even more so for tropical eel species. The increasing understanding of the life history of the eel will hopefully contribute towards the conservation of these exceptional animals—this may be further helped by the promising progress being made towards mass production of eels in captivity, thus reducing fishing pressures on natural stocks. Artificial propagation may be essential for the future survival of some of the anguillids on this planet—that the survival of the eel could ever be in doubt would have been unthinkable in the days of Schmidt, whose conclusion still rings true today:

> *"Altogether, the whole story of the eel and its spawning has come to read almost like a romance, wherein reality has far exceeded the dreams of phantasy"*. (Johs Schmidt, Nature 1912)

Acknowledgements

I am indebted to Prof. Takaomi Arai for the invitation to contribute a chapter to this book. I further would like to thank Dr. Erin Damsteegt and Ms. Jolyn Chia, both members of my research group, for providing critical comments on an earlier draft of this chapter, and to Dr. Alvin Setiawan (NIWA, Bream Bay) and Mr. Kenneth Miller (Department of Zoology, University of Otago) for contributing to one of the figures in this work.

Keywords: eel, spawning, migration, artificial maturation, landmarks

References

Aarestrup, K., F. Okland, M.M. Hansen, D. Righton, P. Gargan, M. Castonguay, L. Bernatchez, P. Howey, H. Sparholt, M.I. Pedersen and R.S. McKinley. 2009. Oceanic spawning migration of the European eel (*Anguilla anguilla*). Science 325: 1660.

Aida, K., K. Tsukamoto and K. Yamauchi. 2003. Eel Biology. Springer, Tokyo.

Als, T.D., M.M. Hansen, G.E. Maes, M. Castonguay, L. Riemann, K. Aarestrup, P. Munk, H. Sparholt, R. Hanel and L. Bernatchez. 2011. All roads lead to home: panmixia of European eel in the Sargasso Sea. Mol. Ecol. 20: 1333–1346.

Aoyama, J., M. Nishida and K. Tsukamoto. 2001. Molecular phylogeny and evolution of the freshwater eel, genus *Anguilla*. Mol. Phyl. Evol. 20: 450–459.

Aoyama, J., S. Wouthuyzen, M.J. Miller, T. Inagaki and K. Tsukamoto. 2003. Short-distance spawning migration of tropical freshwater eels. Biol. Bull. 204: 104–108.

Aoyama, J., S. Watanabe, M.J. Miller, N. Mochioka, T. Otake, T. Yoshinaga and K. Tsukamoto. 2014. Spawning sites of the Japanese eel in relation to oceanographic structure and the West Mariana Ridge. PLoS ONE 9: e88759.

Arai, T. 2014. Evidence of local short-distance spawning migration of tropical freshwater eels, and implications for the evolution of freshwater eel migration. Ecol. Evol. 4: 3812–3819.

Aroua, S., M. Schmitz, S. Baloche, B. Vidal, K. Rousseau and S. Dufour. 2005. Endocrine evidence that silvering, a secondary metamorphosis in the eel, is a pubertal rather than a metamorphic event. Neuroendocrinology 82: 221–232.

Bruijs, M.C.M. and C.M.F. Durif. 2009. Silver eel migration and behaviour. pp. 65–95. *In:* G. Van den Thillart, S. Dufour and J.C. Rankin (eds.). Spawning Migration of the European Eel. Springer.

Capoccioni, F., C. Costa, E. Canali, J. Aguzzi, F. Antonucci, S. Ragonese and M.L. Bianchini. 2014. The potential reproductive contribution of Mediterranean migrating eels to the *Anguilla anguilla* stock. Sci. Rep. 4: 7188.

Castonguay, M. and J.D. McCleave. 1987. Vertical distributions, diel and ontogenetic vertical migrations and net avoidance of leptocephali of *Anguilla* and other common species in the Sargasso Sea. J. Planton Res. 9: 195–214.

Chow, S., H. Kurogi, N. Mochioka, S. Kaji, M. Okazaki and K. Tsukamoto. 2009. Discovery of mature freshwater eels in the open ocean. Fish. Sci. 75: 257–259.

Chow, S., H. Kurogi, S. Katayama, D. Ambe, M. Okazaki, T. Watanabe, T. Ichikawa, M. Kodama, J. Aoyama, A. Shinoda and K. Tsukamoto. 2010. Japanese eel *Anguilla japonica* do not assimilate nutrition during the oceanic spawning migration: evidence from stable isotope analysis. Mar. Ecol. Prog. Ser. 402: 233–238.

Damsteegt, E.L., H. Mizuta, N. Hiramatsu and P.M. Lokman. 2015a. How do eggs get fat? Insights into ovarian fatty acid accumulation in the shortfinned eel, *Anguilla australis*. Gen. Comp. Endocrinol. In press.

Damsteegt, E.L., A. Falahatimarvast, S.P.A. McCormick and P.M. Lokman. 2015b. Triacylglyceride physiology in the short-finned eel, *Anguilla australis*—changes throughout early oogenesis. Am. J. Physiol. Reg. Integr. Comp. Physiol. In press.

Di Biase, A., A. Casalini, P. Emmanuele, M. Mandelli, P.M. Lokman and O. Mordenti. 2015. Controlled reproduction in *Anguilla anguilla* (L.): comparison between spontaneous spawning and stripping-insemination approaches. Aquacult. Res. In press.

Divers, S.L., H.J. McQuillan, H. Matsubara, T. Todo and P.M. Lokman. 2010. Effects of reproductive stage and 11-ketotestosterone on LPL mRNA levels in the ovary of the shortfinned eel. J. Lipid Res. 51: 3250–3258.

Dou, S.Z., Y. Yamada, A. Okamura, S. Tanaka, A. Shinoda and K. Tsukamoto. 2007. Observations on the spawning behavior of artificially matured Japanese eels *Anguilla japonica* in captivity. Aquaculture 266: 117–129.

Dufour, S., N. Le Belle, S. Baloche and Y.-A. Fontaine. 1989. Positive feedback control by the gonads on gonadotropin (GTH) and gonadoliberin (GnRH) levels in experimentally matured female silver eels, *Anguilla anguilla*. Fish Physiol. Biochem. 7: 157–162.

Durif, C., S. Dufour and P. Elie. 2005. The silvering process of *Anguilla anguilla*: a new classification from the yellow resident to the silver migrating stage. J. Fish Biol. 66: 1025–1043.

Durif, C.M.F., V. Van Ginneken, S. Dufour, T. Muller and P. Elie. 2009. Seasonal evolution and individual differences in silvering eels from different locations. pp. 13–38. *In*: G. Van den Thillart, S. Dufour and J.C. Rankin (eds.). Spawning Migration of the European Eel. Springer.

Fontaine, M., E. Lopez, O. Callamand and E. Bertrand. 1964. Sur la maturation des organs genitaux de l'anguille femelle (*Anguilla anguilla* L.) et l' emission spontanee des oeufs en aquarium. C.R. Heb. Seances Acad. Sci. 259: 2907–2910.

Forbes, E.L. 2013. Apolipoprotein B and triacylglyceride physiology in the shortfinned eel, *Anguilla australis*. Ph.D. thesis, University of Otago, Dunedin, New Zealand.

Grassi, B.G. and S. Calandruccio. 1896. The reproduction and metamorphosis of the common eel (*Anguilla vulgaris*). Proc. R. Soc. Lond. 60: 260–271.

Hagihara, S., J. Aoyama, D. Limbong and K. Tsukamoto. 2010. Morphological and physiological changes of female tropical eels, *Anguilla celebesensis* and *Anguilla marmorata*, in relation to downstream migration. J. Fish Biol. 81: 408–426.

Hain, J.H.W. 1975. The behavior of migratory eels, *Anguilla rostrata*, in response to current, salinity and lunar period. Helgol. Wiss. Meeresunt. 27: 211–233.

Han, Y.-S., I.-C. Liao, W.-N. Tzeng, Y.-S. Huang and J.Y.-L. Yu. 2003. Serum estradiol-17β and testosterone levels during silvering in wild Japanese eel *Anguilla japonica*. Comp. Biochem. Physiol. 136B: 913–920.

Han, Y.-S., A.V. Yambot, H. Zhang and C.-L. Hung. 2012. Sympatric spawning but allopatric distribution of *Anguilla japonica* and *Anguilla marmorata*: temperature- and oceanic current-dependent sieving. PLoS ONE 7: e37484.

Hanel, R., D. Stepputis, S. Bonhommeau, M. Castonguay, S. Schaber, K. Wysujack, M. Voback and M.J. Miller. 2014. Low larval abundance in the Sargasso Sea: new evidence about reduced recruitment of the Atlantic eels. Naturwissenschaften 101: 1041–1054.

Hjort, J. 1910. Eel-larvae (*Leptocephalus brevirostris*) from the Central North Atlantic. Nature 85: 104–106.

Horie, N., T. Utoh, N. Mikawa, Y. Yamada, A. Okamura, S. Tanaka and K. Tsukamoto. 2008. Influence of artificial fertilization methods of the hormone-treated Japanese eel *Anguilla japonica* upon the quality of eggs and larvae (comparison between stripping-insemination and spontaneous spawning methods). Nippon Suisan Gakkaishi 74: 26–35 (in Japanese with English abstract).

Hunt, D.M., N.S. Hart and S.P. Collin. 2013. Sensory systems. pp. 118–159. *In*: F. Trischitta, Y. Takei and P. Sébert (eds.). Eel Physiology. CRC Press, Taylor & Francis Group, Boca Raton, USA.

ICES. 2013. Report of the Joint EIFAAC/ICES Working Group on Eels (WGEEL), 18–22 March 2013 in Sukarietta, Spain, 4–10 September 2013 in Copenhagen, Denmark. ICES CM 2013/ACOM: 18: 851pp.

Ijiri, S., K. Tsukamoto, S. Chow, H. Kurogi, S. Adachi and H. Tanaka. 2011. Controlled reproduction in the Japanese eel (*Anguilla japonica*), past and present. Aquacult. Eur. 36: 13–17.

Izumi, H., S. Hagihara, H. Kurogi, S. Chow, K. Tsukamoto, H. Kagawa, H. Kudo, S. Ijiri and S. Adachi. 2015. Histological characteristics of the oocyte chorion in wild post-spawning and artificially matured Japanese eels *Anguilla japonica* Fish. Sci. 81: 321–329.

Jellyman, D.J. and M.M. Bowen. 2007. Modelling larval migration routes and spawning areas of anguillid eels of New Zealand and Australia. *In*: A. Haro, K.L. Smith, R.A. Rulifson, C.M. Moffitt, R.J. Klauda, M.J. Dadswell, R.A. Cunjak, J.E. Cooper, K.L. Beal and T.S. Avery (eds.). Challenges for Diadromous Fishes in a Dynamic Global Environment. American Fisheries Society Symposium 69: 255–274.

Jellyman, D.J. and K. Tsukamoto. 2010. Vertical migrations may control maturation in migrating female *Anguilla dieffenbachii*. Mar. Ecol. Prog. Ser. 404: 241–247.

Jellyman, D.J., B.L. Chisnall, M.L. Bonnett and J.R.E. Sykes. 1999. Seasonal arrival patterns of juvenile freshwater eels (*Anguilla* spp.) in New Zealand. N.Z. J. Mar. Freshwater Res. 33: 249–261.

Kagawa, H., H. Tanaka, H. Ohta, K. Okuzawa and K. Hirose. 1995. *In vitro* effects of 17α-hydroxy progesterone and 17, 20β-dihydroxy-4-pregnen-3-one on final maturation of oocytes at various developmental stages in artificially matured Japanese eel (*Anguilla japonica*). Fish. Sci. 61: 1012–1015.

Kagawa, H., Y. Sakurai, R. Horiuchi, Y. Kazeto, K. Gen, H. Imaizumi and Y. Masuda. 2013a. Mechanism of oocyte maturation and ovulation and its application to seed production in the Japanese eel. Fish Physiol. Biochem. 39: 13–17.

Kagawa, H., N. Fujie, H. Imaizumi, Y. Masuda, K. Oda, J. Adachi, A. Nishi, H. Hashimoto, K.J. Teruya and S. Kaji. 2013b. Using osmotic pumps to deliver hormones to induce sexual maturation of female Japanese eels, *Anguilla japonica*. Aquaculture 388-391: 30–34.

Kaup, J.J. 1856. Catalogue of Apodal Fish in the Collection of the British Museum. Published by order of the Trustees, London, UK.

Kazeto, Y., M. Kohara, R. Tosaka, K. Gen, M. Yokoyama, C. Miura, T. Miura, S. Adachi and K. Yamauchi. 2012. Molecular characterization and gene expression of Japanese eel (*Anguilla japonica*) gonadotropin receptors. Zool. Sci. 29: 204–211.

Kazeto, Y., R. Ito, Y. Ozaki, H. Suzuki, T. Tanaka, H. Imaizumi and K. Gen. 2014. Mass production of Japanese eel recombinant follicle-stimulating hormone and luteinizing hormone by a stable expression system: they fully induced ovarian development at a differential mode. *In*: Abstract Book, 10th International Symposium on the Reproductive Physiology of Fish, Olhão, Portugal, 25–30 May 2014. p. 67.

Kuroki, M., J. Aoyama, M.J. Miller, S. Wouthuyzen, T. Arai and K. Tsukamoto. 2006. Contrasting patterns of growth and migration of tropical anguillid leptocephali in the western Pacific and Indonesian Seas. Mar. Ecol. Prog. Ser. 309: 233–246.

Kurogi, H., M. Okazaki, N. Mochioka, N. Jinbo, H. Hashimoto, M. Takahashi, A. Tawa, J. Aoyama, A. Shinoda and K. Tsukamoto. 2011. First capture of post-spawning female of the Japanese eel *Anguilla japonica* at the southern West Mariana Ridge. Fish. Sci. 77: 199–205.

Larsson, P., S. Hamrin and L. Okla. 1990. Fat content as a factor inducing migratory behavior in the eel (*Anguilla anguilla* L.) to the Sargasso Sea. Naturwissenschaften 77: 488–490.

Lokman, P.M. and G. Young. 1998. Gonad histology and plasma steroid profiles in wild New Zealand freshwater eels (*Anguilla dieffenbachii* and *A. australis*) before and at the onset of the natural spawning migration. II. Males. Fish Physiol. Biochem. 19: 339–347.

Lokman, P.M. and G. Young. 2000. Induced spawning and early ontogeny of New Zealand freshwater eels (*Anguilla dieffenbachii* and *A. australis*). N.Z. J. Mar. Freshwater Res. 34: 135–145.

Lokman, P.M., G.J. Vermeulen, J.G.D. Lambert and G. Young. 1998. Gonad histology and plasma steroid profiles in wild New Zealand freshwater eels (*Anguilla dieffenbachii* and *A. australis*) before and at the onset of the natural spawning migration. I. Females. Fish Physiol. Biochem. 19: 325–338.

Lokman, P.M., R.T. Wass, H.C. Suter, S.G. Scott, K.F. Judge and G. Young. 2001. Changes in steroid hormone profiles and ovarian histology during salmon pituitary-induced vitellogenesis and ovulation in female New Zealand longfinned eels, *Anguilla dieffenbachii* Gray. J. Exp. Zool. 289: 119–129.

Lokman, P.M., B. Harris, M. Kusakabe, D.E. Kime, R.W. Schulz, S. Adachi and G. Young. 2002. 11-Oxygenated androgens in female teleosts: prevalence, abundance, and life history implications. Gen. Comp. Endocrinol. 129: 1–12.

Lokman, P.M., D.H. Rohr, P.S. Davie and G. Young. 2003. The physiology of silvering in anguillid eels—androgens and control of metamorphosis from the yellow to the silver stage. pp. 331–349. *In*: K. Aida, K. Tsukamoto and K. Yamauchi (eds.). Eel Biology. Springer, Tokyo.

Lokman, P.M., M.J. Wylie, M. Downes, A. Di Biase and E.L. Damsteegt. 2015. Artificial induction of maturation in female silver eels, *Anguilla australis*: The benefits of androgen pre-treatment. Aquaculture 437: 111–119.

McCleave, J.D. 2003. Spawning areas of the Atlantic eels. pp. 141–155. *In*: K. Aida, K. Tsukamoto and K. Yamauchi (eds.). Eel Biology. Springer, Tokyo.

Minegishi, Y., J. Aoyama, J.G. Inoue, M. Miya, M. Nishida and K. Tsukamoto. 2005. Molecular phylogeny and evolution of the freshwater eels genus *Anguilla* based on the whole mitochondrial genome sequences. Mol. Phyl. Evol. 34: 134–146.

Minegishi, Y., J. Aoyama and K. Tsukamoto. 2008. Multiple population structure of the giant mottled eel, *Anguilla marmorata* Mol. Ecol. 17: 3109–3122.

Minegishi, Y., R.P. Dirks, D.L. de Wijze, S.A. Brittijn, E. Burgerhout, H.P. Spaink and G.E.E.J.M. van den Thillart. 2012. Quantitative bioassays for measuring biologically functional gonadotropins based on eel gonadotropic receptors. Gen. Comp. Endocrinol. 178: 145–152.

Miura, C., T. Higashino and T. Miura. 2007. A progestin and an estrogen regulate early stages of oogenesis in fish. Biol. Reprod. 77: 822–828.

Nagahama, Y. and S. Adachi. 1985. Identification of maturation-inducing steroid in a teleost, the amago salmon (*Oncorhynchus rhodurus*). Dev. Biol. 109: 428–435.

Norris, D.O. 2007. Vertebrate Endocrinology, 4th ed., Elsevier Academic Press, Amsterdam, Boston.

Ohta, H., K. Ikeda and T. Izawa. 1997a. Increases in concentrations of potassium and bicarbonate ions promote acquisition of motility *in vitro* by Japanese eel spermatozoa. J. Exp. Zool. 277: 171–180.

Ohta, H., H. Kagawa, H. Tanaka, K. Okuzawa, N. Iinuma and K. Hirose. 1997b. Artificial induction of maturation and fertilization in the Japanese eel, *Anguilla japonica*. Fish Physiol. Biochem. 17: 163–169.

Okamura, A., N. Horie, N. Mikawa, Y. Yamada and K. Tsukamoto. 2014. Recent advances in artificial production of glass eels for conservation of anguillid eel populations. Ecol. Freshw. Fish 23: 95–110.

Palstra, A.P., E.G.H. Cohen, P.R.W. Niemantsverdriet, V.J.T. Van Ginneken and G.E.E.J.M. Van den Thillart. 2005. Artificial maturation and reproduction of European silver eel: Development of oocytes during final maturation. Aquaculture 249: 533– 547.

Pelster, B. 2015. Swimbladder function and the spawning migration of the European eel *Anguilla anguilla*. Front. Physiol. 5: 486.

Pérez, L., J.F. Asturiano, A. Tomás, S. Zegrari, R. Barrera, F.J. Espinós, J.C. Navarro and M. Jover. 2000. Induction of maturation and spermiation in the male European eel: assessment of sperm quality throughout treatment. J. Fish Biol. 57: 1488–1504.

Petersen, C.G.J. 1905. Larval Eels (*Leptocephalus brevirostris*) of the Atlantic coasts of Europe. Medd. Kom. Fisk. Havunders. 1: 1–5.

Pous, S., E. Feunteun and C. Ellien. 2010. Investigation of tropical eel spawning area in the South-Western Indian Ocean: Influence of the oceanic circulation. Prog. Oceanogr. 86: 396–413.

Reveillac, É., T. Robinet, M.-W. Rabenevanana, P. Valade and É. Feunteun. 2009. Clues to the location of the spawning area and larval migration characteristics of *Anguilla mossambica* as inferred from otolith microstructural analyses. J. Fish Biol. 74: 1866–1877.

Righton, D., K. Aarestrup, D. Jellyman, P. Sébert, G. Van den Tillart and K. Tsukamoto. 2012. The *Anguilla* spp. migration problem: 40 million years of evolution and two millennia of speculation. J. Fish Biol. 81: 365–386.

Rohr, D.H., P.M. Lokman, P.S. Davie and G. Young. 2001. 11-Ketotestosterone induces silvering-related changes in immature female short-finned eels, *Anguilla australis*. Comp. Biochem. Physiol. 130A: 701–714.

Rousseau, K., S. Aroua, M. Schmitz, P. Elie and S. Dufour. 2009. Silvering: metamorphosis or puberty? pp. 39–63. *In*: G. Van den Thillart, S. Dufour and J.C. Rankin (eds.). Spawning Migration of the European Eel. Springer.

Sato, N., I. Kawazoe, Y. Suzuki and K. Aida. 1997. Development of an emulsion prepared with lipophilized gelatin and its application for hormone administration in the Japanese eel *Anguilla japonica*. Fish Physiol. Biochem. 17: 171–178.

Sbaihi, M., M. Fouchereau-Peron, F. Meunier, P. Elie, I. Mayer, E. Burzawa-Gerard, B. Vidal and S. Dufour. 2001. Reproductive biology of the conger eel from the south coast of Brittany, France and comparison with the European eel. J. Fish Biol. 59: 302–318.

Schmidt, J. 1912. The reproduction and spawning-places of the fresh-water eel (*Anguilla vulgaris*). Nature 89: 633–636.

Schmidt, J. 1922. The breeding places of the eel. Phil. Trans. Royal Soc. London 211B: 179–208.

Schmidt, J. 1923. Breeding places and migrations of the eel. Nature 111: 51–54.

Schoth, M. and F.-W. Tesch. 1982. Spatial distribution of 0-group eel larvae (*Anguiilla* sp.) caught in the Sargasso Sea in 1979. Helgolander Meeresunters 35: 309–320.

Schulz, R.W., L.R. De Franca, J.J. Lareyre, F. Le Gac, H. Chiarini-Garcia, R.H. Nobrega and T. Miura. 2010. Spermatogenesis in fish. Gen. Comp. Endocrinol. 165: 390–411.

Setiawan, A.N., M.J. Wylie, E.L. Forbes and P.M. Lokman. 2012. The effects of 11-ketotestosterone on occupation of downstream location and seawater in the New Zealand shortfinned eel, *Anguilla australis*. Zool. Sci. 29: 1–5.

Setiawan, A.N., Y. Ozaki, A. Shoae, Y. Kazeto and P.M. Lokman. 2012. Androgen-specific regulation of FSH signalling in the previtellogenic ovary and pituitary of the New Zealand shortfinned eel, *Anguilla australis*. Gen. Comp. Endocrinol. 176: 132–143.

Schabetsberger, R., F. Økland, K. Aarestrup, D. Kalfatak, U. Sichrowsky, M. Tambets, G. Dall'Olmo, R. Kaiser and P.I. Miller. 2013. Oceanic migration behaviour of tropical Pacific eels from Vanuatu. Mar. Ecol. Prog. Ser. 475: 177–190.

Sudo, R., R. Tosaka, S. Ijiri, S. Adachi, H. Suetake, Y. Suzuki, S. Horie, J. Aoyama and K. Tsukamoto. 2011. Effect of temperature decrease on oocyte development, sex steroids, and gonadotropin β-subunit mRNA expression levels in female Japanese eel *Anguilla japonica*. Fish. Sci. 77: 575–582.

Sudo, R., R. Tosaka, S. Ijiri, S. Adachi, J. Aoyama and K. Tsukamoto. 2012. 11-Ketotestosterone synchronously induces oocyte development and silvering-related changes in the Japanese eel, *Anguilla japonica*. Zool. Sci. 29: 254–259.

Tanaka, C., F. Shirotori, M. Sato, M. Ishikawa, A. Shinoda, J. Aoyama and T. Yoshinaga. 2014. Genetic identification method for two subspecies of the Indonesian short-finned eel, *Anguilla bicolor*, using an allelic discrimination technique. Zool. Stud. 53: 57.

Tanaka, H. 2015. Progression in artificial seedling production of Japanese eel *Anguilla japonica*. Fish. Sci. 81: 11–19.

Tanaka, H., H. Kagawa, H. Ohta, T. Unuma and K. Nomura. 2003. The first production of glass eel in captivity: fish reproductive physiology facilitates great progress in aquaculture. Fish Physiol. Biochem. 28: 493–497.

Todd, P.R. 1981a. Morphometric changes, gonad histology, and fecundity estimates in migrating New Zealand freshwater eels (*Anguilla* spp.). N.Z. J. Mar. Freshwater Res. 15: 155–170.

Todd, P.R. 1981b. Timing and periodicity of migrating New Zealand freshwater eels (*Anguilla* spp.). N.Z. J. Mar. Freshwater Res. 15: 225–235.

Trancart, T., C. Tudorache, G.E.E.J.M. van den Thillart, A. Acou, A. Carpentier, C. Boinet, G. Gouchet and E. Feunteun. 2015. The effect of thermal shock during diel vertical migration on the energy required for oceanic migration of the European silver eel. J. Exp. Mar. Biol. Ecol. 463: 168–172.

Trischitta, F., Y. Takei and P. Sébert. 2013. Eel Physiology. CRC Press, Taylor & Francis Group, Boca Raton.

Tsukamoto, K. 1992. Discovery of the spawning area for the Japanese eel. Nature 356: 789–791.

Tsukamoto, K., T. Otake, N. Mochioka, T.-W. Lee, H. Fricke, T. Inagaki, J. Aoyama, S. Ishikawa, S. Kimura, M.J. Miller, H. Hasumoto, M. Oya and Y. Suzuki. 2003. Seamounts, new moon and eel spawning: the search for the spawning site of the Japanese eel. Env. Biol. Fish. 66: 221–229.

Unuma, T., S. Sawaguchi, N. Hasegawa, N. Tsuda, T. Tanaka, K. Nomura and H. Tanaka. 2012. Optimum temperature of rearing water during artificial induction of ovulation in Japanese eel. Aquaculture 358-359: 216–223.

Van den Thillart, G., V. Van Ginneken, F. Körner, R. Heijmans, R. Van der Linden and A. Gluvers. 2004. Endurance swimming of European eel. J. Fish Biol. 65: 312–318.

Van den Thillart, G., S. Dufour and J.C. Rankin. 2009. Spawning Migration of the European Eel. Springer.

Van Ginneken, V., E. Antonissen, U.K. Müller, R. Booms, E. Eding, J. Verreth and G. van den Thillart. 2005a. Eel migration to the Sargasso: remarkably high swimming efficiency and low energy costs. J. Exp. Biol. 208: 1329–1335.

Van Ginneken, V., G. Vianen, B. Muusze, A. Palstra, L. Verschoor, O. Lugten, M. Onderwater, S. Van Schie, P. Niemantsverdriet, R. Van Heeswijk, E. Eding and G. van den Thillart. 2005b. Gonad development and spawning behaviour of artificially matured European eel (*Anguilla anguilla* L.). Animal Biology 55: 203–218.

Wouthuyzen, S., J. Aoyama, Y. Sugeha, M.J. Miller, M. Kuroki, Y. Minegishi, S.R. Suharti and K. Tsukamoto. 2009. Seasonality of spawning by tropical anguillid eels around Sulawesi Island, Indonesia. Naturwissenschaften 96: 153–158.

Wysujack, K., H. Westerberg, K. Aarestrup, J. Trautner, T. Kurwie, F. Nagel and R. Hanel. 2015. The migration behaviour of European silver eels (*Anguilla anguilla*) released in open ocean conditions. Mar. Freshwater Res. 66: 145–157.

Yamamoto, K. and K. Yamauchi. 1974. Sexual maturation of Japanese eel and production of eel larvae in the aquarium. Nature 251: 220–222.

Yaron, Z., A. Bogolomonaya, S. Drori, I. Biton, J. Aizen, Z. Kulikovsky and B. Levavi-Sivan. 2009. Spawning induction in the carp: past experience and future prospects—a review. Isr. J. Aquacult. Bamidgeh 61: 5–26.

12

Contaminants in Eels and their Role in the Collapse of the Eel Stocks

Claude Belpaire,[1,*] *José Martin Pujolar,*[2]
Caroline Geeraerts[3] and *Gregory E. Maes*[4,5]

Introduction

Anguillid species of temperate regions have been under considerable pressure resulting in an overall decline in recruitment levels since the 1980s. Similar population collapses have also been reported in other vertebrate species, and in some cases experimental research has clearly shown that the population crashes could be attributed to the negative impact of pollutants, the most famous case probably being the collapse of birds of prey following the increasing presence of DDTs in the environment. In the case of eels, despite

[1] Research Institute for Nature and Forest (INBO), Duboislaan 14, B-1560 Hoeilaart, Belgium.
[2] Department of Bioscience, Aarhus University, Ny-Munkegade 114, Bldg. 1540, DK-8000 Aarhus C, Denmark.
  E-mail: jmartin@biology.au.dk
[3] Research Institute for Nature and Forest (INBO), Gaverstraat 4, B-9500 Geraardsbergen, Belgium.
  E-mail: caroline.geeraerts@inbo.be
[4] Centre for Sustainable Tropical Fisheries and Aquaculture, Comparative Genomics Centre, College of Marine and Environmental Sciences, James Cook University, Townsville, 4811 QLD, Australia.
  E-mail: gregory.maes@jcu.edu.au
[5] Laboratory of Biodiversity and Evolutionary Genomics, University of Leuven (KU Leuven), B-3000 Leuven, Belgium.
* Corresponding author: claude.belpaire@inbo.be

the fact that eels are commonly found in highly polluted waters and that the species is prone to bioaccumulate toxic chemicals, the eel decline was initially not believed to be related to pollution. The eel stock collapse was initially attributed to other anthropogenic disturbances (such as overfishing, habitat loss through degradation or fragmentation) or even natural selective agents (e.g., increase of the populations of cormorants).

Compared to other fish species, performing classical ecotoxicological research on eels is a challenging task. First, eels show a very long and complex life cycle that includes two metamorphoses and separate marine and continental stages, making realistic experimental conditions challenging. Second, until quite recently, the artificial reproduction of eels was not possible; resulting in a lack of inter-generational and larval toxicity studies (but see Palstra et al. 2006). Overall, these challenges made it difficult to thoroughly and reliably test the effects of pollutants on eels and also explain why it took so many years before research gathered enough convincing evidence to suggest that pollution could be implicated in the decline of eels.

In the case of the European eel (*Anguilla anguilla*), the first reports were restricted to the description of high levels of selected pollutants in local stocks and its implications for human health and environmental monitoring (e.g., De Boer and Hagel 1994). One of the first reviews of the possible effects of contaminants on the reproductive biology and physiology of eels was elaborated by Bruslé (1991), who concluded that eels might be more vulnerable than other common fish species, in contrast to what was originally assumed. Larsson et al. (1991) discussed the importance of persistent pollutants in European eel and suggested that during migration to the spawning grounds, lipid deposits are depleted and all lipophilic contaminants are released in the blood and may reach the vital organs and gonads. This is similar to what has been described in birds (Krom 1986); as birds starve, POPs are released from their fat reserves and poison the animals, leading to their death. Nevertheless, in the case of eels, this hypothesis has not been fully demonstrated experimentally. The catadromous life-history of eels does make them very vulnerable to many pollutants at early and adult life-stages.

While scientific debate is still open on the causes of the eel decline, an increasing number of recent reports support the idea that pollutants might be one of the possible synergistic causes contributing to the collapse of eel stocks. Recent advances in reproductive and molecular techniques have already shed new light on the role that pollution has played in the eel decline.

In this sense, significant advances have been made recently in the artificial reproduction of anguillids. In the case of the Japanese eel, Masuda et al. (2012) succeeded in completing the life cycle in captivity, although due to several technical difficulties, the authors have not yet established techniques for the mass production of glass eels. In the European eel, the full life cycle has not been completed yet in captivity but early stage larvae (up to 20 days) can be produced in high numbers (Butts et al. 2014). The production of eel larvae in captivity opens new possibilities for experimental work in many areas, including toxicology, as now researchers can test the effect of pollutants in reproduction, something that was untestable a few years ago, since eels only reproduce naturally in the remote Sargasso Sea.

Moreover, recent advances in genetics and genomic tools have the potential to advance our knowledge on the effect of pollutants on eels. Along with gene expression profiling tools (e.g., micro-arrays), the application of next generation sequencing methods allows for the parallel sequencing of thousands of genes at relatively affordable costs. This is a major step forward in comparison with previous studies based on a few candidate genes and allows researchers to interrogate the whole genome, with the potential to identify new biomarkers and new pathways related to toxicology.

However, the effects of pollutants on the eel's reproductive biology have mainly been hypothesized on the basis of the experimental knowledge of single pollutant exposure gathered from other fish species. Therefore it is important to review the recent developments and knowledge on the effects of pollutants specifically on the reproductive potential of eels under natural conditions of a cocktail of pollutants, considering that the species' spawner quality is crucial for its long-term resilience.

This chapter aims to briefly summarize the current status of contaminants in the eel and discuss the potential impact of pollution on the resilience of stocks and its implications for the international management of the eel.

Bioaccumulation in eels

All aquatic organisms are susceptible to the organic and inorganic contaminants present in the surrounding water. Due to their aquatic life, fish are constantly exposed to chemicals, taken up via gills and skin, or via their diet. They may accumulate these compounds in their bodies where those chemicals start to exert their toxic effects. The accumulation of chemicals in the tissue of an organism is termed bioaccumulation and may happen through respiration, ingestion, or through direct contact with the contaminated water or sediment. Bioconcentration is the process of uptaking chemicals through non-dietary exposure, while biomagnification results in an increase of contaminant levels as they pass up the food chain through two or more trophic levels.

The toxic effect of chemical substances is widely variable and depends on the intrinsic toxic properties of the toxicant, but also on processes within the body of the fish such as metabolism, storing, biotransformation, excretion, or other defense mechanisms. Apart from acute fish toxicity, i.e., the effects of short-term exposure to a chemical, often associated with sudden fish casualties, bioaccumulation of chemicals inside the fish may result in chronic fish toxicity with sublethal effects being apparent only at specific periods of the fish life (e.g., during maturation of the gonads (endocrine disruption), starvation, reproduction and offspring development).

Under the conditions of worldwide contamination of aquatic ecosystems, various fish species have been regarded as sentinels of environmental pollution, showing relatively clear and predictable biological responses at different pollutant levels. The European eel has been considered as a candidate sentinel species (Belpaire and Goemans 2007b). Due to their specific ecological and physiological traits, anguillids are particularly prone to bioaccumulation. As benthic species, eels live in or nearby the sediment (where many contaminants are bound). They are quite resistant to degradation of water quality, resulting in many eels living in highly polluted habitats. Being benthic carnivorous predators, eels are positioned at high trophic levels, and biomagnification processes govern the uptake of chemicals. Eels are long-lived and compared with many other fish species are very old before being able to reproduce, hence they may accumulate chemicals during the many years (5 to 20) spent in continental waters. Their large size enables the accumulation of substantial net amounts of chemicals. Muscle lipid content in eels is much higher compared to other fish species, facilitating bioaccumulation especially for the lipophilic contaminants. Moreover, being semelparous, female eels are not able to decrease their body burden of chemicals through multiple reproductions (Belpaire and Goemans 2007b).

Current pollution levels in the North Atlantic eels

In a review, Robinet and Feunteun (2002) reported concentrations of some PCBs (polychlorinated biphenyls) and pesticides in yellow European and American eel tissues. In 2007, an overview of reports presented bioaccumulation data of various chemicals in *A. anguilla* across 11 EC countries and the EU (Belpaire and Goemans 2007a). This list included data reports on PCBs, OCPs (organochlorine pesticides), heavy metals, dioxines, BFRs (brominated flame retardants), PAHs (polyaromatic hydrocarbons), VOCs (volatile organic compounds), and PFCs (perfluorinated compounds). Moreover, other compounds have also been reported bioaccumulating in wild or exposed eels, for example dyes (Schuetze et al. 2008), drugs such as cocaine (Capaldo et al. 2012), the Roundup[(R)]-aglyphosate-based herbicide (Guilherme et al. 2010) and musk compounds (Leonards and De Boer 2004).

Since then, many new papers have been published describing and discussing the presence of specific toxic compounds in eels. Comprehensive overviews of the new information available in ICES countries have been made available in the annual ICES WGEEL reports (between 2008 and 2013). Considering the increasing amount of data becoming available, many countries have become aware of the importance of monitoring contamination in the eel. However, as pointed out recently by ICES (2015), this monitoring is seldom tailored towards the evaluation of the quality of yellow eel or silver eel (spawner) stocks, but in most cases is targeted towards the assessment of human health risks related to the human consumption of polluted eel, or the assessment of habitat quality and the environmental status of contaminants, e.g., as requested by the European Water Framework Directive (for example Jürgens et al. 2013; Kammann et al. 2014).

Some of this information has been provided for international assessment in the Eel Quality Database (Belpaire et al. 2011), and is used for international advice on the European eel through ICES WGEEL. However, sampling, analysis and reporting procedures for monitoring contamination in the eel are not harmonized and this lack of standardization hinders stock wide assessment. Therefore, ICES (2015) recently developed a framework to increase harmonization and advised on the best practices to (sub)sample, analyze, report and visualize contaminants in the eel, including possible ways to integrate data. Recognizing the need for simple indicators for eel quality estimates, quality indices have been developed, using contaminant classes for a number of chemicals of concern (ICES 2013, 2015).

Reports presenting data over multiple locations within a wide geographical range clearly showed an extreme variability in body burden, which holds true for most of the contaminants studied. For example, Sum 7 PCB levels in eel muscle tissue vary between 3.5 and 12,455 ng/g wet weight in Belgian eels from 365 sample sites (Maes et al. 2008), but also within one river system large variations may occur between upstream and downstream sites (Belpaire et al. 2008; De Boer et al. 2010). More examples illustrating the high spatial variation of other chemicals is presented in Table 1. Note that similar variability has been reported in the European eel in other countries and also in the American eel (e.g., Byer et al. 2013; Ashley et al. 2007).

It is clear from this data that eels are exposed to local pollution pressures which may be very different between areas or parts of water basins. Belpaire et al. (2008) demonstrated the suitability of the eel as a model for fingerprinting the local pollution. For example, eels from rural areas presented higher loads of crop protectors (pesticides/herbicides), while individuals from industrialized areas were characterized by higher levels of PCBs and heavy metals. Eels also

Table 1. Examples of ranges of selected chemicals in the European eel as reported from Belgian eel studies. Concentrations are given in ng/g wet weight, except for the chemicals indicated by * which are given as ng/g lipid weight and ** as pg g^{-1} fresh weight.

Chemical	Range	Period	N sites	Reference
PCB 153	1–5,099	1994–2005	365	Maes et al. 2008
Lindane	0.01–22,225	1994–2005	365	Maes et al. 2008
p,p'-DDE	0.10–3,423	1994–2005	365	Maes et al. 2008
Cadmium	1.0–2,474	1994–2005	365	Maes et al. 2008
ΣHBCDs*	16–4,397	2000–2006	50	Roosens et al. 2010
ΣPBDEs*	10–5,811	2000–2006	50	Roosens et al. 2010
ΣPCDD/Fs**	1–110	2000–2007	38	Geeraerts et al. 2011

may show deviating pollution profiles for particular contaminants allowing the detection of specific pollution sources. This was shown for pentachlorophenol in Ireland (White et al. 2014) and 1,2-dichlorobenzene in Belgium (Belpaire et al. 2007b), and many other examples exist.

The extent of spatial variation in pollution profiles and intensity advocates for the need to take account of those differences in the management of the eel. Individuals from a local population within one catchment or within one Eel Management Unit (as defined by the EU Eel Regulation), may be quite different in quality, and for example, 40% of the individuals might be heavily loaded with PCBs while 20% of the individuals might show rather low PCB levels. Therefore, international stock assessment requires addressing such heterogeneity in the eel. However, considering the number of chemicals of concern, integrating eel quality indicators in quantitative stock assessment models remains challenging.

Long term monitoring studies of contamination in eel have demonstrated that historical contaminants such as PCBs and DDTs are decreasing, after their ban 40 years ago. In the Netherlands, a slow decrease in PCBs since 1977 was shown through a 30 year data collection (de Boer et al. 2010). In Belgium, PCB concentration in eels decreased with a modeled rate of 15% per year, and also some pesticides and metals tended to decrease following the environmental management of these chemicals (Maes et al. 2008). The concentration of lead in eel muscle tissue consistently decreased between 1994 and 2005 in Belgium, which was attributed to the gradual changeover from leaded to unleaded fuels and a reduction of industrial emissions. Significant decreases of PCBs and DDTs were also reported in *A. rostrata* (Byer et al. 2013). On the other hand, many other emerging chemicals have been reported in eels, but the data series are too restricted to allow trend analysis.

Despite these decreases, the status of pollution in eels from many areas is still a matter of high concern. Pollution levels in eels are much higher than in other fish species from the same habitat. While most data have been measured in immature yellow eels, contaminant levels generally increase with age, reaching a maximum just prior to silvering and spawning migration. Considering the high levels of pollution reported, contaminants in eel are expected to exhibit adverse sublethal effects in most of its distribution area. Moreover, as many of these chemicals are correlated, eels are often polluted by a combination of contaminants that might impact animals in a synergetic way.

Effects of pollutants on eels

Toxic effects of contaminants have been reported for many fish species, both under laboratory conditions and from field assessments (for reviews see, e.g., Kime 1995; Lawrence and Hemingway 2003; Bernanke and Köhler 2009). Effects can vary widely and will depend on the chemical, but also on its interaction with other environmental factors and on the developmental stage and the condition of the organism. Contaminants may impact the various levels of biological organisation from the molecular, individual to population and community levels; hence the nature of the effects varies to a very wide extent.

Lethal effects

Most obvious cases of contaminants impacting eel populations are direct fish mortalities occurring after spills or accidents. These casualties may result from inadvertent management of land users (e.g., spill) on a local scale, but sometimes severe and extended fish mortalities have been reported as a consequence of discharge or leakage on industrial sites producing or processing pesticides or other chemical compounds. A well-known example is the Basel accident (November 1986): after a fire in the chemical company Sandoz, a variety of pesticides were spilled in the River Rhine, and half a million eels (*ca*. 200 tonnes) were killed. The Rhine population was affected for years up to 650 km downstream (Christou 2000). Another example concerns the insecticide driven eel devastation in Lake Balaton in 1991 and 1995, which killed 300 and 30 tons of eels, respectively (Balint et al. 1997). Lethal concentrations of specific chemicals in the European eel have been identified through toxicity studies under controlled conditions (see for an overview Geeraerts and Belpaire 2010).

Sublethal effects

In contrast to these evident cases, demonstrating the clear effects of pollution on eels is not obvious when the impact is of sublethal nature. Eels store lipophilic xenobiotics in their fat, and the effects of these compounds will only emerge when they are released from the adipose tissue, i.e., when fat is metabolized for powering migration or reproduction. For eels, this means principally during the adult oceanic phase, hence this impact is very difficult to assess. Nevertheless, during the last 30 years the number of studies assessing the effects of toxic chemicals on eels increased exponentially, and we can only refer to some of the many examples here.

Bruslé (1990, 1991) described and discussed the effects of heavy metals, pesticides and PCBs on eels, listing experimental concentrations of various contaminants over different life stages of the European eel. Further reviews were made by Robinet and Feunteun (2002) and Geeraerts and Belpaire (2010). These made clear that a variety of contaminants affect eels and the effects were reported at several levels of biological organization, from subcellular, organ, individual up to even population level. Reports documented disturbances of the immune system, the reproduction system, the nervous system and the endocrine system. For a list of documented work on the effects of various compounds we refer to Geeraerts and Belpaire (2010).

Lawrence and Elliott (2003) presented a conceptual model to illustrate recognized and potential links between pollution pressure and effects on fish. The model showed possible mechanistic relations between the various hierarchical levels of biological response to pollution, from the molecular to the population level. We simplified this model and adapted it towards eels (Fig. 1), identifying the potential mechanistic links between pollution pressure and the responses, providing examples of evidence of pollution effects for different kinds of chemical compounds as reported by scientific literature. These studies may thus describe processes at various levels of the conceptual model presented in Fig. 1. Such a model has the potential to become a framework for the development of an advanced mathematical model with predictive capability.

Several chemicals were demonstrated as causing organ injuries and tissue damage. Pesticides may disrupt the integrity of eel gills and may impair respiration (Sancho et al. 1997). Roche et al. (2002) reported tumours in the liver and spleen of eels caused by long-lasting exposure to a combination of potentially carcinogenic pollutants (such as PAH, lindane and dieldrin). Also exposure to metals, such as cadmium can cause structural or functional perturbations in the gills and damage in liver (Gony 1990). Mercury exposure has been shown to induce immunotoxic effects including depressed haematopoiesis and enzyme activity to enhanced immune cell death (Carlson and Zelikoff 2008). Perfluorooctane sulfonic acids (PFOS) induced liver damage in eels under field conditions (Hoff et al. 2005).

Some chemicals may alter osmoregulation. Lionetto et al. (1998) reported that cadmium could alter both acid–base balance and osmoregulation in eel, by inhibiting the activity of carbonic anhydrase and Na^+-K^+-ATPase enzymes in the intestine and gills. Fabbri et al. (2003) showed that Cd and Hg may impair a crucial intracellular transduction pathway involved in the adrenergic control of glucose metabolism, but also in several other routes of hormonal regulation of liver functions.

Eel exposure to the insecticide fenitrothion resulted in a decrease in protein content, indicating the physiological adaptability to compensate for chemical stress (Sancho et al. 1997). To overcome the stress the animals require high energy. This energy demand might have led to the stimulation of protein catabolism and may disturb fat metabolism.

PCBs are one of the most toxic groups of compounds suggested to cause detrimental impact in eels. PCB mixtures and individual congeners have various endocrine-disrupting effects. These chemicals have a very slow elimination time in fish and may exert their toxicity in the organism even many years after being banned. Silver eels exposed to PCBs and subjected to simulated migration through experiments in swimming tunnels, showed several physiological and metabolic impacts, suggesting that the current levels of PCBs and other dioxin-like compounds in eels may impair the survival and reproductive potential of the European eel (van Ginneken et al. 2009b). PCBs also lead to impaired larval survival and development (Palstra et al. (2006) see below).

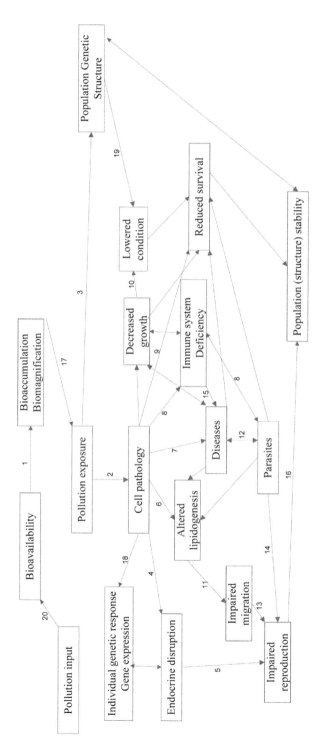

Fig. 1. A simplified conceptual model of the effects of pollution exposure on the population structure of the European eel, *A. anguilla*. Adapted from Lawrence and Elliott (2003). Numbers indicate references: (1) Bruslé 1991; Belpaire et al. 2007a; (2) Sancho et al. 1997; Pacheco and Santos 2002; (3) Nigro et al. 2002; Maes et al. 2005; Nogueira et al. 2006; (4) Versonnen et al. 2004; (5) Jobling et al. 1998; (6) Ceron et al. 1996; Sancho et al. 1998; Fernandez-Vega et al. 1999; Robinet and Feunteun 2002; Pierron et al. 2007; (7) Roche et al. 2002; (8) Sures and Knopf 2004; Sures 2006; (9) Sancho et al. 1997; (10) Gony 1990; (11) Sancho et al. 1998; van Ginneken et al. 2009b; Belpaire et al. 2009; (12) van Ginneken et al. 2005; (13) Palstra et al. 2006; (14) Sures 2006; (15) van Ginneken et al. 2005; (16) Corsi et al. 2003; (17) Van Campenhout et al. 2008; (18) Ahmad et al. 2006; Maes et al. 2013; Pujolar et al. 2012, 2013; (19) Maes et al. 2005; (20) Belpaire et al. 2007b.

Numerous biomarkers have been developed to monitor the exposure and effects of pollution on eels, at a variety of biological levels (ranging from behavioral, reproductive, tissue and organ, biochemical, enzymatic, proteomic and transcriptomic levels), see, e.g., Geeraerts and Belpaire (2010) and Maes et al. (2013). Having reviewed the current studies on biomarkers in eels, ICES (2015) concluded that current information on biomarkers is not able to reliably evaluate the potential of eels to successfully migrate and reproduce in their marine spawning habitats. However, given the demonstrated potential of these techniques, the further development of available (and novel) biomarkers applied to wild populations at field concentrations and in response to multiple stressors (combined abiotic and biotic) should be further promoted, in order to shed light on the effects of contaminants on the fitness and reproductive potential of eels.

Evolutionary and genotoxic effects - evidence from ecotoxicogenomic and transcriptomic studies

Assessing the long-term fitness effects of exposure to environmental pollutants in diadromous fish populations is of crucial importance to guarantee their evolutionary resilience under combined selective pressures in both freshwater and marine environments. Potential impact of pollution in fish also includes genetic damage and several authors have described the genotoxic effects of specific chemicals at the cellular level. Maria et al. (2002) reported a decrease in blood and liver DNA integrity in European eel exposed to benzo[a]pyrene. The combination of several contaminants may act synergistically and reinforce deleterious effects (see, e.g., the study of sequential exposure to PAHs and heavy metals of eel by Teles et al. (2005)). While the evolutionary consequences of pollutants have been well explored in natural populations of freshwater fishes using genomic and transcriptomic approaches (reviewed in Bozinovic and Oleksiak 2010), less attention has been paid to diadromous fish species that complete their life-cycle through ontogenic shifts between freshwater and marine habitats (e.g., salmonids, anguillids). The study of diadromous species is particularly relevant as they represent a natural model to understand the combined impact of continental anthropogenic and oceanic climate stressors. Unlike other model fish species, very little is known about the ecological and evolutionary consequences in species with a catadromous life-strategy, such as the European eel, which illustrates an example of a heavily threatened fish species, strongly affected by human stressors throughout its life cycle.

The use of molecular approaches and more recent "omics" (e.g., genomic, transcriptomic and proteomic) tools facilitates ecological studies exploring how natural populations respond to contemporary anthropogenic pressures such as pollution. For example Maes et al. (2005) studied the effects of heavy metals on the genome of wild eels with variable pollution load. A significant negative correlation between heavy metal pollution load and condition was observed (genetic erosion), suggesting an impact of pollution on the health of sub-adult eels. In general, a reduced genetic variability was observed in strongly polluted eels, as well as a negative correlation between level of bioaccumulation and allozymatic multi-locus heterozygosity. Using a targeted approach (real-time PCR analysis of two key detoxification genes, MT and CYP1A), Maes et al. (2013) showed that eels with high bioaccumulation levels of pesticides and heavy metals showed strongly down-regulated hepatic and gill gene transcription. High bioaccumulation levels were also correlated with poor health status in terms of condition factor and lipid reserves, suggesting an important negative impact on the health of eels. Also using a targeted approach, the recent real-time PCR study of Giuliani and Regoli (2014) showed catalase, glutathione peroxidase 1 and glutathione S-transferase pi to be up-regulated in response to oxidative stress.

In contrast to targeted approaches in which only a few genes are screened using real-time PCR (Nogueira et al. 2009; Maes et al. 2013; Giuliani and Regoli 2014), gene expression profiling, using either microarrays or RNA-seq, offers the opportunity to investigate the effects of pollutants at the genome-wide level. Hence, the analysis of transcriptome-wide responses could play an important role in identifying new toxicity genes and pathways as well as new biomarkers.

Two recent microarray studies have focused on understanding the effect of chronic exposure to environmental pollutants in the European eel. First, Pujolar et al. (2012) developed an eel-specific microarray of 14,913 annotated genes that was applied to detect differentially expressed genes in response to pollution. The study investigated the transcriptomic dynamics between individuals from highly polluted (River Tiber, Rome) vs. lowly polluted environments (Lake Bolsena, central Italy). Eels were measured for 36 PCBs, several pesticides and nine metals, which confirmed the large differences in pollution load between sites. Enrichment analysis showed a total of 168 transcripts differentially expressed between sites, 30 up-regulated and 138 down-regulated. Up-regulated genes were mostly involved in detoxification (drug metabolism). Those included genes encoding several members of the cytochrome P450 superfamily of enzymes (CYP2J23, CYP3A65, CYP46A1) that catalyze the oxidation of xenobiotic substances such as pollutants, drugs and toxins. On the other hand, down-regulated genes were mostly involved in energetic metabolism, including 32 genes from the oxidative phosphorylation pathway.

Using the same approach, Pujolar et al. (2013) measured multi-pollutant levels of bioaccumulation in eels from lowly, highly and extremely polluted environments in Belgium, which were compared to the lowly polluted habitat in Italy from the previous study. A parallel response to pollutant exposure was observed when comparing the highly polluted site in Belgium with the relatively clean sites in Belgium and Italy, with many shared up- and down-regulated genes. The observation of a similar response using two geographically distant reference populations suggests that the differences found are not random but due to pollution and points to shared mechanisms to cope with chronically polluted habitats. Shared up-regulated genes included enzymes involved in the metabolism of xenobiotic substances. This included a novel European eel CYP3A gene orthologous to zebrafish CYP3A65 that is involved in phase I of the xenobiotic metabolism and glutathione-5-transferase pi (GSTP1) that takes part in phase II. Aldehyde dehydrogenase and alcohol dehydrogenase were also up-regulated in the highly polluted site. On the other hand, several metabolic related genes were down-regulated. Many genes encoding enzymes in the mitochondrial respiratory chain and oxidative phosphorylation were down-regulated in samples from the highly polluted site in Belgium relative to the low polluted sites in Belgium and Italy. Those included NADH dehydrogenase, succinate dehydrogenase, ubiquinol-cytochrome c reductase, cytochrome c oxidase and ATP synthase. Although metabolism was not directly measured, results suggested a negative impact of pollutants at the metabolic level, seeing that many genes encoding key enzymes in the mitochondrial respiratory chain and oxidative phosphorylation were down-regulated, possibly resulting in a low energetic status of the individuals from polluted sites. This has important implications, since the suggested low energetic status of pre-migrating eels points to a poor quality of future spawners that could jeopardize spawning migration and reproduction in the Sargasso Sea. Poor quality of spawners due to the detrimental effect of pollutants on fitness and fecundity could lead to an impoverished stock in terms of recruitment and effective population size. Since the panmictic European eel lacks the capacity to adapt towards pollution tolerance, the only long-term evolutionary response to such highly variable and heavy habitat changes and alterations could be the development of an extreme and costly phenotypic plasticity in gene regulation.

The transcriptome-wide response to contaminants was also recently studied in both North Atlantic eels (European and American eel) using RNA-seq (Baillon et al. 2015). While microarrays are hybridization-based approaches that typically involve incubating fluorescently labelled cDNA, RNA-seq uses the capabilities of next-generation sequencing methodologies to generate millions of DNA sequence reads derived from the entire RNA molecule. The resulting reads can be aligned to a reference genome or assembled *de novo* so that full-length transcript sequences are reconstructed from the reads. Quantification is carried out by counting the number of reads aligning to each transcript. In the RNA-seq study of Baillon et al. (2015), several genes over-expressed in the livers of eels associated with individual contaminants were identified. For instance, genes of the p53 signaling pathway and lipid and protein metabolism were associated with cadmium contamination. In response to arsenic, genes involved in protein metabolism and protein folding were up-regulated, while genes involved in vasculotoxicity were down-regulated.

In addition to monitoring gene expression at the transcriptome level and given the various post-transcriptional regulation mechanisms, it is important to perform proteomic studies that quantify protein abundance. Proteomic techniques are being increasingly used in aquatic toxicology as a powerful tool to understand global regulatory mechanisms and complex metabolic pathways triggered by pollution exposure. Roland et al. (2013, 2014) have used the 2D-DIGE approach to study the effect of perfluorooctane sulfonate (PFOS) on eels. A total of 48 proteins were identified, with functions related to cytoskeleton, protein folding, cell signaling, proteolytic pathway and carbonate and energy metabolism.

Is it likely that the decline of the eel population is caused by pollution?

The potential causes of the stock declines observed in the European, American and Japanese eel are still not fully understood and many potential causes have been put forward, basically divided into environmental (i.e., changes in oceanic processes affecting larval migration and survival) and anthropogenic causes. The latter include overfishing, constructions of dams, pollution by hazardous substances, and introduction of non-native parasites and viruses.

The important role of contaminants as causative factor in the decline of eels has been debated a.o. by Robinet and Feunteun (2002), Palstra et al. (2006), Belpaire (2008), van Ginneken et al. (2009b), Byer (2013). Eels represent one of the most intriguing examples of how toxic stress may impact a species at the population level. Benefitting from new scientific evidence, pollution has received increasing attention as a possible cause for the decline of the eel. While processes may be complex and interrelated (see Fig. 1), two major mechanisms may be involved. Figure 2 presents a simplified concept describing how reprotoxic chemicals may influence the status of the stock of the European eel.

Hypothesis 1

During their growth in continental waters yellow eels accumulate toxic compounds. Many of these substances are localized in the fat stores, due to their high lipophilicity. Lipophilic pollutants can impact reproduction through direct damage to target organs after their remobilization from the fat. Larsson et al. (1991) suggested that while the lipid reserves are consumed during migration, contaminants are released into the blood and may damage reproductive organs and affect embryogenesis. They hypothesized that the eel stock decline could be the result of the negative impact of these highly contaminated lipid reserves in the eel. Palstra et al. (2006) studied the effects of toxic compounds on embryo survival time. They described teratogenic effects even at levels below the EU maximum consumption limit for dioxin in food, thus at the environmentally relevant concentrations. Pollutants can be passed on from the mother to the eggs and impair larval survival and development.

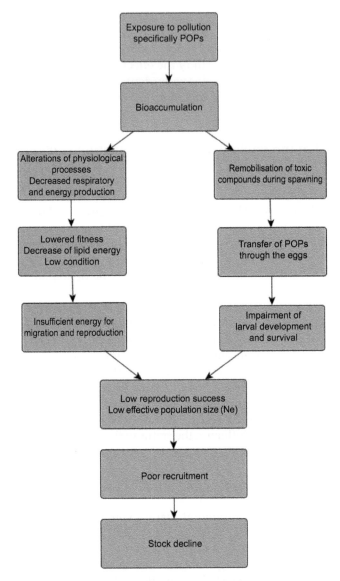

Fig. 2. Mechanistic model of the impact of reprotoxic and persistent organic pollutants (POPs) on the stock of the European eel. Environmental pressure by lipophilic contaminants leads to internal exposure through uptake from water, food and sediment. Eels are particularly sensitive to bioaccumulation of POPs and the concentrations of these compounds in eels attain much higher levels than in other fish species. The effects of the high body burden of contaminants may act through two possible pathways. One way, is through the disturbance of the lipid metabolism, resulting in lowered lipid reserves and a decrease in condition (fitness). As a result, energy stores are insufficient to fuel their reproductive migration or eels arrive in inadequate condition which does not allow normal reproduction. Alternatively, lipophilic contaminants are released from the fat during migration as the fat stores are gradually depleted. Their reprotoxic properties impede the quality and the survival of the developing eggs/larvae, resulting in diminished recruitment.

Elevated PCB-levels may also interfere with steroid hormone function and therefore with reproduction and hatching success of the larvae (van Ginneken et al. 2009b). PCB exposure also led to swelling of the yolk-sac, especially the pericardium (elevated wet weights of embryos/larvae), thus showing disturbed hydromineral balance (oedema) (van Ginneken et al. 2009a). The inverse relationship between the TEQ level in gonads and the survival period of the fertilized eggs suggested that the current levels of dioxin-like compounds impair the reproduction of the European eel. Palstra et al. (2006) suggested that these contaminants contributed significantly to the current collapse in the eel population. Note however, that the reported conclusions were deduced from a very limited set of data. Comparison with toxicological thresholds for normal reproduction gained from other species may have an indicative value. von Westernhagen et al. (2006) compared tissue burden and hatching success in whiting and proposed threshold values for PCBs. ICES 2010 collected PCB body burden in eels (from individual sites and country means) from eight European countries and compared those values with the threshold values of whiting ovary contamination above which impairment of reproductive success was likely to occur. In 63% of the cases, the whiting benchmark was exceeded. In case of the American eel, recent studies on Lake Ontario eels suggested that pre-2000 toxic equivalent (TEQ) concentrations in eels were above the threshold value for chronic toxicity for lake trout. The results indicated that embryotoxicity of maternally-derived dioxin-like compounds could have contributed to a possible lack of reproductive success in Lake Ontario eels (Byer 2013). Remobilization of pollutants to gonads might also occur with heavy metals, as demonstrated for cadmium. Pierron et al. (2008) reported that after 30 days of Cd exposure a significant metal accumulation was observed in the kidney, liver, gills and digestive tract of Cd exposed eels. Thereafter, during the maturation phase a significant increase in the Cd content in the gonads and kidney of Cd pre-contaminated eels was measured.

Hypothesis 2

Alternatively, pollutants may affect the eel stock through interaction with lipid physiology. The impact of pesticides on the lipid metabolism of eels has been studied under laboratory conditions by Ceron et al. (1996), Sancho et al. (1998) and Fernandez-Vega et al. (1999). Disturbance of the lipid accumulation was reported and was attributed to the inhibition of the acetylcholinesterase (AChE) activity after pesticide exposure. Similar effects on lipid storage efficiency have been described in eels for cadmium (Pierron et al. 2007). Geeraerts et al. (2007) analyzed an extensive dataset of contaminants in eels from Flanders by statistical modeling and concluded that PCBs, especially the higher chlorinated ones, and DDTs, have a negative impact on the lipid content of the eel. A significant and considerable decrease of the muscle lipid contents of yellow eels in Belgium and the Netherlands has been described

(Belpaire et al. 2009), namely from 20% to 12% over a 13-year period since 1994 in Belgium and from 21% to 13% in the Netherlands since around 1985. Also in the American eels sampled from Lake Ontario (known to have high concentrations of POPs), lipid levels and condition were significantly lower in 2008 as compared to 1988 or 1998 (Byer 2013). By contrast, a similar decrease in lipids did not occur in Scotland, where the eels experience an overall low level of pollution (Oliver et al. 2015). Establishing sufficient lipid energy is essential in order to fulfil the life cycle of eels. The silvering process, the subsequent downstream and transoceanic reproductive migration, as well as gonad maturation, can only take place if a sufficient quantity of energy is stored as lipids. The minimal lipid stores required to fuel migration and reproduction is dependent on the eel size, gender, and distance to the Sargasso Sea. ICES (2013) presented model estimates for reproductive potential of the variation in size (body mass), muscle lipids content and distance from the river catchment to the Sargasso Sea, incorporating body mass dependent cost of transport values and estimates of the number of eggs produced per unit of energy as available in literature (illustrated in Fig. 3, for details see ICES 2013).

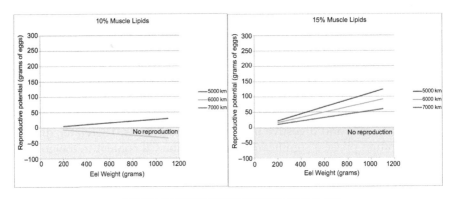

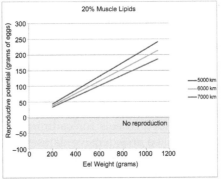

Fig. 3. Modeled impact of body mass and distance to the Sargasso Sea on the reproductive potential of female silver eels leaving a catchment as a function of increasing muscle lipid levels (using mean estimate of energetic costs of swimming) (from ICES 2013).

Empirical values showed a very high variability in reproductive potential in female silver eels from several catchments spread over the distribution range of the European eel (ICES 2013). Observations of lowered fat levels in local eel stocks question the ability of these eels to start silvering and to achieve their spawning migration. In a species such as the European eel, disturbed lipidogenesis may reduce the migration and reproduction success (Corsi et al. 2003; Belpaire et al. 2009). In eels, one gram of fat is required to produce 1.72 g eggs (as presented in van Ginneken and van den Thillart 2000). A deficiency of lipid reserves available for gonad maturation leads to a decrease in egg production (Henderson and Tocher 1987). Eels with low fat reserves might still silver and start their spawning migration but, due to insufficient energy stores migration and/or reproduction might not be successful, which would ultimately result in decreased recruitment and impaired stock.

Other arguments

Overall, we believe there is substantial evidence that pollution is one of the key factors contributing to the collapse of eels. As detailed above, eels are extremely prone to bioaccumulate high levels of many chemicals, with many detrimental effects at the cellular, tissue and organ level. Also genetic diversity seems to be lowered by pollution pressure, as shown in the study of Maes et al. (2005), where individuals with a high metal load showed a significantly lower genetic variability in comparison with low metal load individuals. Although metabolism was not directly measured, the transcriptomic studies of Pujolar et al. (2012, 2013) showed that genes involved in phosphorylation (energy production) were down-regulated in individuals from highly polluted areas relative to individuals from less polluted areas, suggesting a negative impact of pollution on metabolism.

Many other cases where species or stocks collapsed due to pollution have been reported. Contamination has been demonstrated as the cause of population collapse of many other biota from 1960s onwards (e.g., the collapse of several birds of prey due to DDT). To cite only some of the numerous cases documented in fish, for example Fairchild et al. (1999) explained the reduction in the population of the Atlantic salmon after large scale treatment of pine forests with the insecticide aminocarb and 4-nonylphenol. Salmon smolts exposed to the herbicide atrazine and to pentabromodiphenyl ether showed impaired osmoregulation, increased cortisol levels, a lowered survival rate in the sea and a decrease in migration activity (Moore and Lower 2001). Fertility is considered a sensitive indicator for exposure to xenobiotics, and reproductive disturbance as a biomarker is widely used because of its high ecological value and demonstrated association with anthropogenic stress (Greeley 2002). For example, the Canadian bullhead populations *Cottus cognatus* from agricultural environments showed reproductive disturbances and a reduction in the number of young fish, related to smaller eggs, lower fecundity and showed a reduction in size of the gonads (Gray and Munkittrick 2005).

This specially holds true for the incidence of endocrine disrupting effects reported in fish. Since the first description of intersex in *Aphyosemion punctatum* in 1931 (Goldschmidt 1931), during the last 15 years, reports of endocrine disruption in fish exponentially increased, especially after 1998 when Jobling et al. (1998) reported intersexuality in roach (*Rutilus rutilus*) associated with discharges of sewage treatment plants. Meanwhile, yearly, hundreds of reports document examples of endocrine disruption (see, e.g., Mills and Chichester (2005) and Bahamonde et al. (2013) for reviews). It is important to note that many of these cases occur in wild fish populations with body burdens of endocrine disrupting chemicals much lower than the levels reported in some eel stocks. Considering the high levels of contamination in eels from many areas, endocrine disruption in mature silver eels might be expected, jeopardizing normal reproduction. It is very plausible to assume that in eel endocrine disruption will also take place if the eels are highly loaded with endocrine disrupting compounds (EDCs) as was documented in many large rivers of industrialized areas. However, these disturbances are not easily discernible in eels, since they only emerge in the maturing silver eel stage during the oceanic migration towards the spawning grounds (Versonnen et al. 2004).

Many chemicals have been developed and brought into the market, simultaneously with the intensification of agricultural and industrial activities in the last century. While the timing of crashes in several populations attributed to pollution was most evident during the seventies, the population crash in eel occurred in the early eighties. This time lag may well be explained by the long generation time in eels. The timing of the increase in the production and release of some of the most toxic chemicals may fit with the timing of the decrease in recruitment from 1980 onwards. As an example the peaking of PCB levels in the environment occurred around 1970s, leading to high bioaccumulation during the growth of elvers and yellow eels at that time. Due to the high levels accumulated during this growth period of peaking PCBs, eels from this cohort would end up with extreme levels of PCBs accumulated in their fat stores, when starting their reproductive journey (on average late 70s). Due to their high body burden these eels would have an impaired migration or reproduction leading to recruitment collapse in the early 1980s. Also Byer (2013) concluded from analyzing time trends of POPs in *A. rostrata* that the decline in the recruitment of American eels to Lake Ontario corresponds to the time period when eels were highly exposed to dioxin-like compounds.

The more or less simultaneous decreases in recruitment in the Northern-hemisphere eel species (*A. anguilla*, *A. rostrata* and *A. japonica*), suggest that a common source or multiple causes are involved, reinforcing the argument that specific broadly distributed contaminants over the industrialized world are key elements in the decline (Geeraerts and Belpaire 2010).

Management measures

Assessing the quality of maturing silver eels leaving continental waters towards their spawning grounds is of vital importance not only for the assessment of the stocks, but also in order to understand how pollution affects eels and what consequences it has on the life cycle of the species. Minimal requirements for the estimation of the quality of silver eels have been sketched by ICES. Policy makers responsible for the international eel management have been encouraged to create the framework for the development of a harmonized monitoring of silver eel quality over the distribution area of the species.

Restocking of glass eel is considered to be a valuable management measure to enhance local eel stocks. However, as discussed above, it may be anticipated that eels growing up in contaminated environments will only have a modest contribution in terms of larval production. Therefore, considering the rather low quantity of glass eels available on the market, target waters for restocking should be prioritized taking into account the contaminant load of the receiving waters.

In many countries, eels are targeted for human consumption by professional and recreational fisheries. In most cases, these fisheries are scattered all over the country. Given the high degree of spatial variation, surveillance systems should be put in place to protect consumer health. In several habitats, especially in large western European rivers, the levels of specific chemicals of concern measured in eels are very high, making them improper for consumption (Bilau et al. 2007). As a result, eel fisheries have been closed in a substantial part of the riverine habitats in Western Europe. In Germany, fishing eels in some contaminated river stretches was already prohibited for management reasons before exceeded consumption levels (PCDD/Fs, dl-PCB) became known (i.e., Upper Rhine). In other affected areas (Lower Rhine, Elbe, Weser, Maas, Ems), the trade of eels was banned and advice against the consumption of eels was given. In the Netherlands, the eel fisheries on the main rivers have been closed since 2011 due to high levels of pollutants found in eels, affecting 50 fishing companies, catching 170 tonnes of eel in 2010, which represents roughly one-third of the annual eel landings. In Belgium, in Wallonia since 2006 eel fishing is prohibited in all waters due to high levels of PCBs, while in some other areas (i.e., Flanders) it is highly recommended not to consume eels. In France, a national plan against PCBs including eel sampling had been set up until 2008. Following those analyses the eel fisheries ban was put in place at an increasing number of fisheries. In Italy, some lake eel fisheries (Lago di Garda) had been closed in 2011 due to fish contamination by dioxin (ICES 2011, 2012, 2013). While these measures contribute towards safeguarding a significant number of eels and contribute towards the increase of the number of silver eels leaving continental waters for their journey to the spawning ground, it is doubtful whether these (highly polluted) eels make a substantial contribution for increasing offspring quantity in the Sargasso Sea. It may be reasonably anticipated that more areas will be closed for eel fisheries in the context of human health considerations, following the increasing sanitary control of local eel stocks targeted for fisheries.

Conclusion and perspectives

Thirty-five years after the dramatic crash of eel stocks, the causes for the decline have still not been elucidated. It is possible that several agents both environmental (i.e., changes in food availability for larvae in the Sargasso Sea) and anthropogenic (i.e., overfishing, pollution, construction of dams) acted together in a synergetic way, resulting in the collapse of the stocks. Dekker (2004) suggested that the most likely proximate cause of the collapse in recruitment observed in the European eel was an insufficient quantity of spawners. From the evidence available, *"it may be concluded that not only the quantity, but also the quality of the potential spawners leaving continental waters, is insufficient, and has contributed to the decline of the stock"* (Belpaire 2008). Contaminant pressure in continental waters seems to represent a major threat for the European eel stock that could jeopardize the possibilities of restoration of the stock. While substantial evidence has highlighted the negative impact of a variety of chemical compounds on the eel, international measures to restore the eel stock currently insufficiently take into account these pollution issues. Management measures and monitoring efforts, as prescribed by the European eel regulation, are tailored towards ensuring a sufficient quantity of silver eels escaping from continental waters. However, within the (inter) national eel restoration plans, measures to decrease contaminant pressure should be an essential issue.

Recognizing pollution as a thriving factor entails several possible eel management measures. First, international management should protect through priority, high quality spawners from healthy catchments. Second, habitat quality parameters (including contamination by POPs) should be used for prioritizing catchments for restocking. Third, international management should create the framework for a comprehensive and standardized monitoring of eel quality over the distribution area. Finally, eel quality indicators need to be included in the models of the international assessment of eel stocks.

Research directed towards the health and capacity of pre-migrating silver eels to reach the spawning grounds and produce offspring (of 'quality'), was considered essential to alleviate the ongoing decline (Geeraerts and Belpaire 2010) and from 2003, the urgent need for the development of novel biomarkers to monitor the reproductive success in fish was emphasized (van der Oost et al. 2003). Advances in these fields have evidenced strong reproductive effects in different fish species such as alteration of the endocrine system resulting in abnormal gonadal development, abnormalities in embryo development, and the subsequent hatching and growth of larvae (Geeraerts and Belpaire 2010). However, the effects of pollutants on eel's reproductive biology have mainly been hypothesized based on experimental knowledge of single pollutant exposure gathered from other fish species. In order to elucidate the impact of contaminants on the reproduction success of the species there is an urgent need for an internationally coordinated research project (as was proposed by ICES 2013). Significant gaps in scientific knowledge have been recognized, such as to what extent and at what level these contaminants affect the eel reproductive success. Such a research project could now take advantage of the recent progress in artificial reproduction of the eel and the advancements in the use of new biomarkers.

Keywords: PCBs, OCPs (organochlorine pesticides), heavy metals, bioaccumulation, biomagnification, genotoxicity, effects of contaminants, reproductive migration, restocking, human health, consumption level, ecotoxicogenomics, genetic erosion, effective population size, detoxification, transcriptomics

References

Ahmad, I., M.V.L. Oliveira, M. Pacheco and M.A. Santos. 2006. Oxidative stress and genotoxic effects in gill and kidney of *Anguilla anguilla* L. exposed to chromium with or without pre-exposure to beta-naphthoflavone. Mutat. Res. 608(1): 16–28.

Ashley, J.T.F., D. Libero, E. Halscheid, L. Zaoudeh and H.M. Stapleton. 2007. Polybrominated diphenyl ethers in American eels (*Anguilla rostrata*) from the Delaware river, USA. Bull. Environ. Contam. Toxicol. 79: 99–103.

Bahamonde, P.A., K.R. Munkittrick and C.J. Martyniuk. 2013. Intersex in teleost fish: are we distinguishing endocrine disruption from natural phenomena? Gen. Comp. Endocr. 192: 25–35. http://dx.doi.org/10.1016/j.ygcen.2013.04.005.

Baillon, L., F. Pierron, R. Coudret, E. Normendeau, A. Caron, L. Peluhet, P. Labadie, H. Budzinski, G. Durrieu, J. Sarraco, P. Elie, P. Couture, M. Baudrimont and L. Bernatchez. 2015. Transcriptome profile analysis reveals specific signatures of pollutants in Atlantic eels. Ecotoxicology 24: 71–84.

Balint, T., J. Ferenczy, F. Katai, I. Kiss, L. Kraczer, O. Kufcsak, G. Lang, C. Polyhos, I. Szabo, T. Szegletes and J. Nemcsok. 1997. Similarities and differences between the massive eel (*Anguilla anguilla* L.) devastations that occurred in Lake Balaton in 1991 and 1995. Ecotox. Environ. Safe. 37(1): 17–23.

Belpaire, C. 2008. Pollution in eel. A cause of their decline? Ph.D. Thesis, Catholic University Leuven, Belgium. Research Institute for Nature and Forest, Brussels, pp. 485, III annexes.

Belpaire, C. and G. Goemans. 2007a. The European eel (*Anguilla anguilla*) a rapporteur of the chemical status for the Water Framework Directive? Vie Milieu 57(4): 235–252.

Belpaire, C. and G. Goemans. 2007b. Eels: contaminant cocktails pinpointing environmental contamination. ICES J. Mar. Sci. 64: 1423–1436.

Belpaire, C., G. Goemans, C. Geeraerts, P. Quataert and K. Parmentier. 2008. Pollution fingerprints in eels as models for the chemical status of rivers. ICES J. Mar. Sci. 65: 1483–1491.

Belpaire, C., C. Geeraerts, D. Evans, E. Ciccotti and R. Poole. 2011. The European eel quality database: towards a pan-European monitoring of eel quality. Environ. Monit. Assess. 183(1-4): 273–284.

Belpaire, C.G.J., G. Goemans, C. Geeraerts, P. Quataert, K. Parmentier, P. Hagel and J. De Boer. 2009. Decreasing eel stocks: Survival of the fattest? Ecol. Freshw. Fish 18(2): 197–214.

Bernanke, J. and H.-R. Köhler. 2009. The impact of environmental chemicals on wildlife vertebrates. *In*: D.M. Whitacre (ed.). Reviews of Environmental Contamination and Toxicology, Vol. 198 1. doi: 10.1007/978-0-387-09646-9, Springer Science + Business Media, LLC 2008.

Bilau, M., I. Sioen, C. Matthys, A. De Vocht, G. Goemans, C. Belpaire, J.H. Willems and S. De Henauw. 2007. Polychlorinated biphenyl (PCB) exposure through eel consumption in recreation fishermen as compared to the general population, using a probabilistic approach. Food Addit. Contam. 24(12): 1386–1393.

Bozinovic, G. and M.F. Oleksiak. 2010. Genomic approaches with natural fish populations from polluted environments. Environ. Toxicol. Chem. 30(2): 283–289.

Bruslé, J. 1990. Effects of heavy metals on eels, *Anguilla* sp. J. Aquat. Living Resour. 3: 131–141.

Bruslé, J. 1991. The eel (*Anguilla* sp.) and organic chemical pollutants. Sci. Total Environ. 102: 1–19.

Butts, I.A., S.R. Sørensen, S.N. Politis, T.E. Pitcher and J. Tomkiewicz. 2014. Standardization of fertilization protocols for the European eel, *Anguilla anguilla*. Aquaculture 426: 9–13.

Byer, J.D. 2013. Organohalogenated persistent organic pollutants in American eel (*Anguilla rostrata*) captured in eastern Canada, Ph.D. Thesis, Queen's University Kingston, Ontario, Canada.

Byer, J.D., M. Lebeuf, M. Alaee, B.R. Stephen, S. Trottier, S. Backus, M. Keir, C.M. Couillard, J. Casselman and P.V. Hodson. 2013. Spatial trends of organochlorinated pesticides, polychlorinated biphenyls, and polybrominated diphenyl ethers in Atlantic Anguillid eels. Chemosphere 90(5): 1719–28, http://dx.doi.org/10.1016/j.chemosphere.2012.10.018.

Capaldo, A., F. Gay, M. Maddaloni, S. Valiante, M. De Falco, M. Lenzi and V. Laforgia. 2012. Presence of cocaine in the tissues of the European eel, *Anguilla anguilla*, exposed to environmental cocaine concentrations. Water Air Soil Poll. 223(5): 2137–2143.

Carlson, E. and J.T. Zelikoff. 2008. The immune system of fish: a target organ of toxicity. pp. 489–529. *In*: R.T. Di Giulio and D.E. Hinton (eds.). The Toxicology of Fishes. CRC Press, USA.

Ceron, J.J., M.D. Ferrando, E. Sancho, C. Gutierrez-Panizo and E. Andreu-Moliner. 1996. Effects of diazinon exposure on cholinesterase activity in different tissues of European eel (*Anguilla anguilla*). Ecotox. Environ. Safe. 35: 222–225.

Christou, M.D. 2000. Substances dangerous to the environment in the context of Council Directive 96/82/EC. Report by Technical Working Group 7, EUR 19651 EN. pp. 43.

Corsi, I., M. Mariottini, C. Sensini, L. Lancini and S. Focardi. 2003. Cytochrome P450. Acetylcholinesterase and gonadal histology for evaluating contaminant exposure levels in fishes from a highly eutrophic brackish ecosystem: the Orbetello Lagoon, Italy. Mar. Pollut. Bull. 46: 203–212.

de Boer, J. and P. Hagel. 1994. Spatial differences and temporal trends of chlorobiphenyls in yellow eel (*Anguilla anguilla*) from inland water of The Netherlands. Sci. Total Environ. 141: 155–174.

de Boer, J., Q.T. Dao, S.P. van Leeuwen, M.J. Kotterman and J.H. Schobben. 2010. Thirty year monitoring of PCBs, organochlorine pesticides and tetrabromodiphenylether in eel from The Netherlands. Environ. Pollut. 158: 1228–1236.

Dekker, W. 2004. Synthesis and discussion: population dynamics of the European eel. pp. 127–145. *In*: Slipping Through Our Hands—Population Dynamics of the European Eel. Ph.D. Thesis, University of Amsterdam, Amsterdam, The Netherlands.

Fabbri, E., F. Caselli, A. Piano, G. Sartor and A. Capuzzo. 2003. Cd^{2+} and Hg^{2+} affect glucose release and cAMP-dependent transduction pathway in isolated eel hepatocytes. Aquat. Toxicol. 62(1): 55–65.

Fairchild, W.L., E.O. Swansburg, J.T. Arenault and S.B. Brown. 1999. Does an association between pesticide use and subsequent declines in catch of Atlantic salmon (*Salmo salar*) represent a case of endocrine disruption? Environ. Health Perspect. 107: 349–358.

Fernandez-Vega, C., E. Sancho, M.D. Ferrando and E. Andreu-Moliner. 1999. Thiobencarb toxicity and plasma AchE inhibition in the European eel. J. Environ. Sci. Health B 34(1): 61–73.

Geeraerts, C. and C. Belpaire. 2010. The effects of contaminants in European eel: a review. Ecotoxicology 19: 239–266.

Geeraerts, C., G. Goemans and C. Belpaire. 2007. (In Dutch) Ecologische en ecotoxicologische betekenis van verontreinigende stoffen gemeten in paling. MIRA/2007/05; INBO/R/2007/40. INBO, Groenendaal-Hoeilaart, pp. 241.

Geeraerts, C., J.-F. Focant, G. Eppe, E. De Pauw and C. Belpaire. 2011. Reproduction of European eel jeopardised by high levels of dioxins and dioxin-like PCBs? Sci. Tot. Environ. 409: 4039–4047. Doi: 10.1016/j.scitotenv.2011.05.046.

Giuliani, M.E. and F. Regoli. 2014. Identification of the Nrf2-Keap1 pathway in the European eel *Anguilla anguilla*: role for a transcriptional regulation of antioxidant genes in aquatic organisms. Aquat. Toxicol. 150: 117–123.

Goldschmidt, R. 1931. Die Sexuellen Zwischenstufen. Monographien aus dem Gesamtgebiet der Physiologie der Pflanzen und der Tiere. Band 23: 230–265.

Gony, S. 1990. Short note on the effects of cadmium on the gill of the glass eel (*Anguilla anguilla*). Int. Rev. Gen. Hydrobiol. 75(6): 835–836.

Gray, M.A. and K.R. Munkittrick. 2005. An effects-based assessment of Slimy Sculpin (*Cottus cognatus*) populations in agricultural regions of Northwestern New Brunswick. Water Qual. Res. J. Canada 40(1): 16–27.

Greeley, M.S. 2002. Reproductive indicators of environmental stress. pp. 312–377. *In*: S.M. Adams (ed.). Biological Indicators of Aquatic Ecosystem Stress. American Fisheries Society, Bethesda, Md, USA.

Guilherme, S., I. Gaivão, M.A. Santos and M. Pacheco. 2010. European eel (*Anguilla anguilla*) genotoxic and pro-oxidant responses following short-term exposure to Roundup(R)-aglyphosate-based herbicide. Mutagenesis 25: 523–530.

Henderson, R.J. and D.R. Tocher. 1987. The lipid-composition and biochemistry of freshwater fish. Prog. Lipid Res. 26(4): 281–347.

Hoff, P.T., K. Van Campenhout, K. Van de Vijver, A. Covaci, L. Bervoets, L. Moens, G. Huyskens, G. Goemans, C. Belpaire, R. Blust and W. De Coen. 2005. Perfluorooctane sulfonic acid and organohalogen pollutants in liver of three freshwater fish species in Flanders (Belgium): relationships with biochemical and organismal effects. Environ. Pollut. 137(2): 324–333.

ICES. 2008. The report of the 2008 Session of the Joint EIFAC/ICES Working Group on Eels, September 2008; ICES CM 2008/ACOM:15. 192pp. and Country Reports.

ICES. 2009. Report of the 2009 Session of the Joint EIFAC/ICES Working Group on Eels, FAO European Inland Fisheries Advisory Commission; International Council for the Exploration of the Sea, Göteborg, 7–12 September 2009, EIFAC Occasional Paper No. 45, ICES CM 2009/ACOM: 15. Rome, FAO/Copenhagen, ICES. 2010. p. 540 (Online).

ICES. 2010. The report of the 2010 Session of the Joint EIFAC/ICES Working Group on Eels, September 2010; ICES CM 2009/ACOM: 18: 198pp. and Country Reports.

ICES. 2011. Report of the 2011 Session of the Joint EIFAAC/ICES Working Group on Eels Lisbon, Portugal, 5–9 September 2011; ICES CM 2011/ACOM: 18: 244p.

ICES. 2012. Report of the 2012 Session of the Joint EIFAAC/ICES Working Group on Eels, Copenhagen, Denmark, 3–9 September 2012; ICES CM 2012/ACOM:18, EIFAAC Occasional Paper 49: 828pp.

ICES. 2013. Report of the Joint EIFAAC/ICES Working Group on Eels (WGEEL), 18–22 March 2013 in Sukarrieta, Spain, 4–10 September 2013 in Copenhagen, Denmark. ICES CM 2013/ACOM: 18: 851pp.

ICES. 2014. Report of the Joint EIFAAC/ICES/GFCM Working Group on Eel, 3–7 November 2014, Rome, Italy. ICES CM 2014/ACOM: 18: 203pp.

ICES. 2015. Report of the Workshop of a Planning Group on the Monitoring of Eel Quality under the subject "Development of standardized and harmonized protocols for the estimation of eel quality" (WKPGMEQ), 20–22 January 2015, Brussels, Belgium. ICES CM 2014/SSGEF: 14. 274 pp.

Jobling, S., M. Nolan, C.R. Tyler, G. Brighty and J.P. Sumpter. 1998. Widespread sexual disruption in wild fish. Environ. Sci. Technol. 32(17): 2498–2506.

Jürgens, M.D., A.C. Johnson, K.C. Jones, D. Hughes and A.J. Lawlor. 2013. The presence of EU priority substances mercury, hexachlorobenzene, hexachlorobutadiene and PBDEs in wild fish from four English rivers. Sci. Total Environ. 461: 441–452.

Kammann, U., M. Brinkmann, M. Freese, J.-D. Pohlmann, S. Stoffels, H. Hollert and R. Hanel. 2014. PAH metabolites, GST and EROD in European eel (*Anguilla anguilla*) as possible indicators for eel habitat quality in German rivers. Environ. Sci. Pollut. Res. 21(4): 2519–2530.

Kime, D.E. 1995. The effects of pollution on reproduction in fish. Rev. Fish Biol. Fisher. 5: 52–96.

Krom, M.D. 1986. An Evaluation of the Concept of Assimilative Capacity as Applied to Marine Waters. Ambio 15(4): 208–214.

Larsson, P., S. Hamrin and L. Okla. 1991. Factors determining the uptake of persistent pollutants in an eel population (*Anguilla anguilla* L.). Environ. Pollut. 69(1): 39–50.

Lawrence, A.J. and K. Hemingway. 2003. Effects of Pollution on Fish: Molecular Effects and Population Responses, Blackwell Science, Oxford, U.K.

Lawrence, A.J. and M. Elliot. 2003. Introduction and conceptual model. pp. 1–13. *In*: A. Lawrence and K. Hemingway (eds.). Effects of Pollution on Fish: Molecular Effects and Population Responses. Blackwell Science Ltd., Oxford, U.K.

Leonards, P.E.G. and J. de Boer. 2004. Synthetic musks in fish and other aquatic organisms. pp. 49–84. *In*: G. Rimkus (ed.). The Handbook of Environmental Chemistry; Synthetic Musk Fragrances in the Environment. Vol. 3, Part X. Springer, Berlin/Heidelberg, New York.

Lionetto, M.G., M. Maffia, M.S. Cappello, M.E. Giordano, C. Storelli and T. Schettino. 1998. Effect of cadmium on carbonic anhydrase and Na$^+$-K$^+$-ATPase in eel, *Anguilla anguilla*, intestine and gills. Comp. Biochem. Phys. A 120(1): 89–91(3).

Maes, G.E., J.A.M. Raeymaekers, C. Pampoulie, A. Seynaeve, G. Goemans, C. Belpaire and F.A.M. Volckaert. 2005. The catadromous European eel *Anguilla anguilla* (L.) as a model for freshwater evolutionary ecotoxicology: relationship between heavy metal bioaccumulation, condition and genetic variability. Aquat. Toxicol. 73: 99–114.

Maes, G.E., J.A.M. Raeymaekers, B. Hellemans, C. Geeraerts, K. Parmentier, L. De Temmerman, F.A.M. Volckaert and C. Belpaire. 2013. Gene transcription reflects poor health status of resident European eel chronically exposed to environmental pollutants. Aquat. Toxicol. 126: 242–255.

Maes, J., G. Goemans and C. Belpaire. 2008. Spatial variation and temporal pollution profiles of polychlorinated biphenyls, organochlorine pesticides and heavy metals in European yellow eel (*Anguilla anguilla* L.) (Flanders, Belgium). Environ. Pollut. 153: 223–237.

Maria, V.L., A.C. Correia and M.A. Santos. 2002. *Anguilla anguilla* L. biochemical and genotoxic responses to benzo[a]pyrene. Ecotox. Environ. Safe. 53: 86–92.

Masuda, Y., H. Imaizumi, K. Oda, H. Hashimoto, H. Usuki and K. Teruya. 2012. Artificial completion of the Japanese eel, *Anguilla japonica*, life cycle: challenge to mass production. Bull. Fish. Res. Agen. 35: 111–117.

Mills, L.J. and C. Chichester. 2005. Review of evidence: Are endocrine-disrupting chemicals in the aquatic environment impacting fish populations? Sci. Total Environ. 343(1-3): 1–34.

Moore, A. and N. Lower. 2001. The effects of aquatic contaminants on Atlantic salmon (*Salmo salar* L.) smolts. 6th International Workshop on Salmonid Smoltification Westport Ireland 3–7 September 2001 30p.

Nigro, M., G. Frenzilli, V. Scarcelli, S. Gorbi and F. Regoli. 2002. Induction of DNA strand breaks and apoptosis in the eel *Anguilla anguilla*. Mar. Environ. Res. 54: 517–520.

Nogueira, P., J. Lourenço, E. Rodriguez, M. Pacheco, C. Santos, J.M. Rotchell and S. Mendo. 2009. Transcript profiling and DNA damage in the European eel (*Anguilla anguilla* L.) exposed to 7,12-dimethylbenz[a]anthracene. Aquat. Toxicol. 94: 123–130.

Nogueira, P.R., J. Lourenço, S. Mendo and J.M. Rotchell. 2006. Mutation analysis of Ras gene in the liver of European eel (*Anguilla anguilla* L.) exposed to benzo[a]pyrene. Mar. Pollut. Bull. 52: 1611–1616.

Oliver, I.W., K. Macgregor, J.D. Godfrey, L. Harris and A. Duguid. 2015. Lipid increases in European eel (*Anguilla anguilla*) in Scotland 1986–2008: an assessment of physical parameters and the influence of organic pollutants. Environ. Sci. Pollut. Res. DOI 10.1007/s11356-015-4116-4.

Pacheco, M. and M.A. Santos. 2002. Biotransformation, genotoxic, and histopathological effects of environmental contaminants in European eel (*Anguilla anguilla* L.). Ecotox. Environ. Safe. 53: 331–347.

Palstra, A.P., V.J.T. Van Ginneken, A.J. Murk and G.E.E.J.M. Van Den Thillart. 2006. Are dioxin-like contaminants responsible for the eel (*Anguilla anguilla*) drama? Naturwissenschaften 93(3): 145–148.

Pierron, F., M. Baudrimont, A. Bossy, J.-P. Bourdineaud, D. Brèthes, P. Elie and J.-C. Massabuau. 2007. Impairment of lipid storage by cadmium in the European eel (*Anguilla anguilla*). Aquat. Toxicol. 81(3): 304–311.

Pierron, F., M. Baudrimont, S. Dufour, P. Elie, A. Bossy, S. Baloche, N. Mesmer-Dudons, P. Gonzalez, J.P. Bourdineaud and J.C. Massabuau. 2008. How cadmium could compromise the completion of the European eel's reproductive migration. Environ. Sci. Technol. 42(12): 4607–4612.

Pujolar, J., I. Marino, M. Milan, A. Coppe, G.E. Maes, F. Capoccioni, E. Ciccotti, L. Bervoets, A. Covaci, C. Belpaire, G. Cramb, T. Patarnello, L. Bargelloni, S. Bortoluzzi and L. Zane. 2012.

Surviving in a toxic world: transcriptomics and gene expression profiling in response to environmental pollution in the critically endangered European eel. BMC Genomics 13: 507.

Pujolar, J.M., M. Milan, I.A.M. Marino, F. Capoccioni, E. Ciccotti, C. Belpaire, A. Covaci, G. Malarvannan, T. Patarnello, L. Bargelloni, L. Zane and G.E. Maes. 2013. Detecting genome-wide gene transcription profiles associated with high pollution burden in the critically endangered European eel. Aquat. Toxicol. 132-133: 157–64.

Robinet, T. and E. Feunteun. 2002. Sublethal effects of exposure to chemical compounds: a cause for the decline in Atlantic eels? Ecotoxicology 11: 265–277.

Roche, H., A. Buet and F. Ramade. 2002. Accumulation of lipophilic microcontaminants and biochemical responses in eels from the Camargue Biosphere Reserve. Ecotoxicology 11: 155–164.

Roland, K., P. Kestemont, L. Hénuset, M.-A. Pierrard, M. Raes, M. Dieu and F. Silvestre. 2013. Proteomic responses of peripheral blood mononuclear cells in the European eel (*Anguilla anguilla*) after perfluorooctane sulfonate exposure. Aquat. Toxicol. 128-129: 43–52.

Roland, K., P. Kestemont, R. Loos, S. Tavazzi, B. Paracchini, C. Belpaire, M. Dieu, M. Raes and F. Silvestre. 2014. Looking for protein expression signatures in European eel peripheral blood mononuclear cells after *in vivo* exposure to perfluorooctane sulfonate and a real world field study. Sci. Total Environ. 468-469: 958–967.

Roosens, L., C. Geeraerts, C. Belpaire, I. Van Pelt, H. Neels and A. Covaci. 2010. Spatial variations in the levels and isomeric patterns of PBDEs and HBCDs in the European eel in Flanders. Environ. Int. 36: 415–423. Doi: 10.1016/j.envint.2010.03.001.

Sancho, E., M.D. Ferrando and E. Andreu. 1997. Sublethal effects of an organophosphate insecticide on the European eel, *Anguilla anguilla*. Ecotox. Environ. Safe. 36: 57–65.

Sancho, E., M.D. Ferrando and E. Andreu. 1998. Effects on sublethal exposure to a pesticide on levels of energetic compounds in *Anguilla anguilla*. J. Environ. Sci. Part B33 (4): 411–424.

Schuetze, A., T. Heberer and S. Juergensen. 2008. Occurrence of residues of the veterinary drug malachite green in eels caught downstream from municipal sewage treatment plants. Chemosphere 72: 1664–1670. Doi: 10.1016/j.chemosphere.2008.05.036.

Sures, B. 2006. How parasitism and pollution affect the physiological homeostasis of aquatic hosts. J. Helminthol. 80(2): 151–157.

Sures, B. and K. Knopf. 2004. Individual and combined effects of cadmium and 3,3′,4,4′,5-pentachlorobiphenyl (PCB126) on the humoral immune response in European eel (*Anguilla anguilla*) experimentally infected with larvae of *Anguillicola crassus* (Nematoda). Parasitology 128: 445–454.

Teles, M., M. Pacheco and M.A. Santos. 2005. Physiological and genetic responses of European eel (*Anguilla anguilla* L.) to short-term chromium or copper exposure—Influence of preexposure to a PAH-like compound. Environ. Toxicol. 20(1): 92– 99.

Van Campenhout, K., H. Goenaga Infante, G. Goemans, C. Belpaire, F. Adams, R. Blust and L. Bervoets. 2008. A field survey of metal binding to metallothionein and other cytosolic ligands in liver of eels using an on-line isotope dilution method in combination with size exclusion (SE) high pressure liquid chromatography (HPLC) coupled to Inductively Coupled Plasma time-of-flight Mass Spectrometry (ICP-TOFMS). Sci. Total Environ. 394(2-3): 379–389.

van der Oost, R., J. Beyer and N.P. Vermeulen. 2003. Fish bioaccumulation and biomarkers in environmental risk assessment: a review. Environ. Toxicol. Pharmacol. 13(2): 57–149.

van Ginneken, V.J.T. and G.E.E.J.M. van den Thillart. 2000. Eel fat stores enough to reach the Sargasso. Nature 403: 156–157.

van Ginneken, V.J.T., B. Ballieux, R. Willemze, K. Coldenhoff, E. Lentjes, E. Antonissen, O. Haenen and G.E.E.J.M. van den Thillart. 2005. Hematology patterns of migrating European eels and the role of EVEX virus. Comp. Biochem. Phys. C 140(1): 97–102.

van Ginneken, V.J.T., M. Bruijs, T. Murk, A.P. Palstra and G.E.E.J.M. van den Thillart. 2009a. The effect of PCBs on the spawning migration of European silver eel (*Anguilla anguilla* L.). pp. 363–384. *In*: G.E.E.J.M. van den Thillart, S. Dufour and J.C. Rankin (eds.). Spawning Migration of the European Eel. Reproduction Index, a Useful Tool for Conservation Management. Springer, New York.

van Ginneken, V., A. Palstra, P. Leonards, M. Nieveen, H. van den Berg, G. Flik, T. Spanings, P. Niemantsverdriet, G. van den Thillart and A. Murk. 2009b. PCBs and the energy cost of migration in the European eel (*Anguilla anguilla* L.). Aquat. Toxicol. 92: 213–220.

Versonnen, B.J., G. Goemans, C. Belpaire and C.R. Janssen. 2004. Vitellogenin content in European eel (*Anguilla anguilla*) in Flanders, Belgium. Environ. Pollut. 128: 363–371.

von Westernhagen, H., P. Cameron, V. Dethlefsen and D. Janssen. 2006. Chlorinated hydrocarbons in North Sea whiting (*Merlangius merlangus* L.), and effects on reproduction. I. Tissue burden and hatching success. Helgolander Meeresun. 43: 45–60.

White, P., B. McHugh, R. Poole, E. McGovern, J. White, P. Behan, B. Foley and A. Covaci. 2014. Application of congener based multi-matrix profiling techniques to identify potential PCDD/F sources in environmental samples from the Burrishoole Catchment in the West of Ireland. Environ. Pollut. 184: 449–456.

13

American Eel (*Anguilla rostrata*) Stock Status in Canada and the United States

Brian M. Jessop[1], and *Laura M. Lee[2]*

Introduction

The 2003 Québec Declaration of Concern (Dekker et al. 2003) raised alarm about the declining stock abundance of the temperate-zone anguillid eels—*Anguilla anguilla, A. japonica,* and *A. rostrata.* Since then, much new research has been conducted on eel biology and stock status and many concerned nations have implemented new stock management measures. The stock status of the American eel (*A. rostrata*) has since been reviewed in the following reports: (a) Committee on the Status of Endangered Wildlife in Canada (COSEWIC 2006, 2012), (b) U.S. Fish and Wildlife Service (USFWS 2007; Shepard 2015), (c) Canadian Science Advisory Secretariat (DFO 2010, 2014), (d) Atlantic States Marine Fisheries Commission (ASMFC 2012), and (e) International Union for the Conservation of Nature (IUCN) (Jacoby et al. 2014). With the exception of COSEWIC (2012), which raised the at-risk status for the American eel from

[1] Department of Fisheries and Oceans, Bedford Institute of Oceanography, P.O. Box 1006, Dartmouth, NS, B2Y 4A2, Canada (retired).
[2] North Carolina Division of Marine Fisheries, 3441 Arendell St., P.O. Box 769, Morehead City, NC 28557, USA.
E-mail: laura.lee@ncdenr.gov
* Corresponding author: welljess@ns.sympatico.ca

"special concern" to "threatened", and Jacoby et al. (2014) which assessed the status as "endangered", all other reports have concluded that the stock status is "of special concern" or "depleted" or not "threatened or endangered" or have made no specific assessment. The IUCN (Jacoby et al. 2014) report notes that the status is on the cusp of the "vulnerable" and "endangered" categories. The province of Ontario has designated the eel as "endangered" within the province (COSEWIC 2012).

The Canadian Ministers of the Departments of the Environment and of Fisheries and Oceans have, to date, not accepted either the COSEWIC (2006) "of special concern" or COSEWIC (2012) "threatened" designation for the American eel but did initiate an Eel Recovery Assessment review (DFO 2014; Cairns et al. 2014; Chaput et al. 2014a,b; Pratt et al. 2014; Young and Koops 2014). These DFO reports assessed current trends in eel abundance indices relative to short- and medium-term abundance targets with the ultimate goal of long-term eel stock sustainability, described stock status threats and limitations, threat mitigation options, and the results of population modelling. The Canadian status review process for the American eel under the Species at Risk Act (SARA) is ongoing. Briefly, the process of listing, or not, an aquatic species (not currently on Schedule 1 of listed species) begins with a COSEWIC assessment that is then submitted to the Minister of Fisheries and Oceans for review. This Minister advises the Minister of the Environment who then publishes a public response statement. The Department of Fisheries and Oceans (DFO) then begins an information gathering process that includes public consultation after which the DFO advises the Minister on whether to list the species by adding it to Schedule 1 of SARA, to refer the matter back to COSEWIC for reanalysis, or to recommend not to list. The Minister of Fisheries and Oceans advises the Minister of Environment accordingly and this Minister advises the Governor in Council which then puts the recommendation into law as appropriate. Listed species are then subject to SARA protection and recovery measures while those not listed may be subject to DFO protective or other management measures (Environment Canada 2003).

The DFO is responsible for the management of the American eel in Canada except in Ontario and Quebec where provincial agencies are responsible. In 2004, an integrated eel management plan was proposed by these agencies (Canadian Eel Working Group 2009) and in 2006 the first draft was developed but it has not yet been formally accepted although each agency has implemented actions consistent with the plan (DFO 2010).

In the United States, the U.S. Fish and Wildlife Service (USFWS) and the National Marine Fisheries Service (NMFS) have joint responsibility for various aspects of the management of the American eel because of its catadromous life history while the Atlantic States Marine Fisheries Commission (ASMFC) coordinates the conservation and management of interstate Atlantic coastal species, including the American eel. In 2004, a petition to list the American eel as endangered under the federal Endangered Species Act (ESA) resulted in the 2007 review (USFWS 2007) that declared the species "not threatened or endangered". In 2010, the USFWS and NMFS was petitioned by the Council for Endangered Species Act Reliability to again review the status of the American eel and to list it as endangered under the federal Endangered Species Act (ESA). This led to a 90-day finding that found sufficient grounds to initiate a review of the American eel status in the U.S. to determine if a listing is warranted (USFWS 2011). The U.S. Fish and Wildlife Service has found that listing the American eel as threatened or endangered within all or a significant portion of its range is presently unwarranted (USFWS 2015; Shepard 2015).

The 2014 Québec Declaration reported a hopeful, and regionally variable, increase in eel recruitment in recent years, progress in implementing new fishery management plans, and called for the maintenance or increase of current protection levels and their expansion to as yet unprotected areas (Dekker and Casselman 2014). The following discussion provides a synthesis of the current stock status of the American eel in Canada and the United States based upon the available reports, particularly the DFO (2014) reports and the ASMFC (2012) report. The release of Shepard (2015) was too late to be included in this review but its findings are consistent with it.

Species ecology and geographic distribution

The American eel (*Anguilla rostrata*) is facultatively catadromous and, during its continental life, can occupy fresh, estuarine, or near-shore coastal marine waters (Fig. 1). Juveniles may migrate periodically between freshwater and estuarine or marine waters (Jessop et al. 2008). Spawning occurs in the Sargasso Sea and larval eels (leptocephali) are primarily passively distributed north-eastward along the Atlantic coast of North America via the Gulf Stream, with smaller numbers believed to be transported south-westward into the

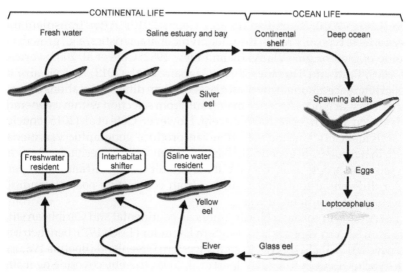

Fig. 1. General life cycle of the American eel. (With permission from D. Cairns.)

Caribbean Sea and the Gulf of Mexico, before metamorphosing into glass eels and moving shoreward into coastal and fresh waters where juvenile growth occurs (Helfman et al. 1987; Miller et al. 2014). American eels exhibit great individual variability in, and latitudinal clines in, biological characteristics such as length, weight, age, and growth rate, depending upon the geographic location (Helfman et al. 1987; Jessop 2010; DFO 2014). Eels from more northern regions and regions most distant from the spawning area tend to be exclusively female, where females silver at sizes and ages greater than do males. Female juvenile American eels, but not male juveniles, initiate sexual maturation and the spawning migration at ages and sizes that increase northward from the area of first continental arrival of elvers which occurs between northern Florida and North Carolina (Wang and Tzeng 1998; Jessop 2010). Whether increasing ages and sizes occur at the southern extremes is unknown but they are unlikely if the temperature-size rule applies (Jessop 2010). The sex distribution, size, and age composition of eels in the Gulf of Mexico, Caribbean Islands, Central America, and northern South America is poorly known.

The American eel is semelparous, as are all anguillid eels, and dies after spawning. The species is panmictic, with reproduction occurring randomly with respect to parental geographic origin and with no evidence of important spatial or temporal differentiation in neutral genetic markers, which have no adaptive function (Côté et al. 2013). The role of epigenetic effects remains to be answered (Bossdorf et al. 2008). Divergence in phenotypic traits, despite the lack of neutral genetic differentiation in marine fishes, has been interpreted as evidence for the adaptive significance of population structure (Hutchings et al. 2007). Consequently, the species (population) has been considered a single designatable unit (DU) for management purposes within the Canadian distribution (COSEWIC 2012). However, phenotypic traits such as size and age at migration, growth rate, and sex ratio vary widely across their geographic range (Jessop 2010; Velez-Espino and Koops 2010). After transplantation, American eels typically retain the phenotypic characteristics, e.g., growth rate, sex ratio of their origin (Vladykov and Liew 1982; Côté et al. 2009; Verreault et al. 2009; Pratt and Threader 2011). Gagnaire et al. (2012) hypothesize that genetic difference among American glass eels from different locations along the Atlantic coast of North America may result from selection within a generation that is erased at each reproductive event. However, Côté et al. (2015) conclude that systematically varying selection can produce geographic variations in growth that result from genetically-based differences. The mechanism by which such selection occurs is uncertain but may result from non-random dispersal and/or differential mortality resulting from contrasting coastal conditions upon entry into continental waters.

The American eel has a North American continental and Caribbean island range from eastern Venezuela to southern Labrador (Tesch 1973; Benchetrit and McCleave 2015). It also occurs in south-western Greenland (Boëtius 1985) and some hybrids occur in Iceland (Albert et al. 2006), thereby covering over 50° of latitude in its geographic range. At the extremes of the distribution range, the entire migration loop may exceed 10,000 km. The Canadian and U.S. Atlantic coastal and inland (excluding the Gulf of Mexico) eel habitats were divided into seven (7) RPA zones (Fig. 2) for review purposes to accommodate the observed geographic distributions in American eel size, age, and sex ratio, types and magnitudes of threats, and potential contributions to the spawning stock (DFO 2014). Reliable data are scarce on the occurrence and extent of commercial fisheries and landings throughout the American eel range in the Caribbean Islands, Central America, and northern South America, although small-scale glass eel fisheries have occurred or been recently developed in

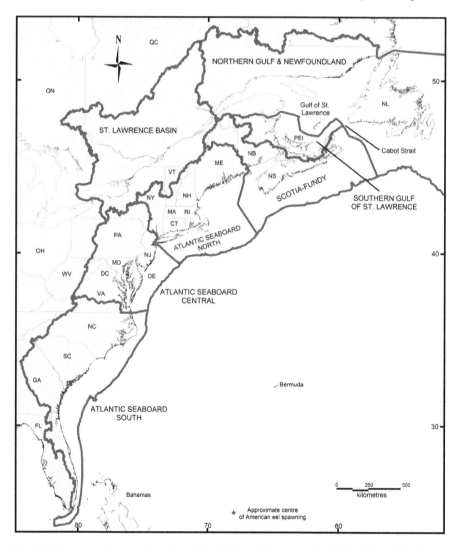

Fig. 2. RPA zones of eastern Canada and the U.S. used to summarize biological and life history characteristics of the American eel. (With permission from D. Cairns.)

Cuba, Haiti, and the Dominican Republic (Cairns et al. 2014; Benchetrit and McCleave 2015). These areas will not be considered further.

Trends in commercial landings and abundance indices

The available Canadian eel abundance indicators can be divided into fishery-dependent, e.g., reported commercial fishery landings and trap CPUE indices and fishery-independent indices, e.g., electrofishing and fishway counts. Canadian (Fig. 3A,B) and United States (Fig. 3C) commercial fishery landings have varied widely over time and across regions, e.g., RPA zones

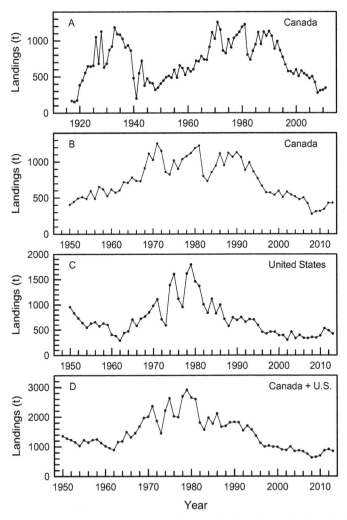

Fig. 3. Reported landings (t) of American eels in Canada (1917–2013), and in Canada, the United States, and Canada-U.S. combined (1950–2013).

differ regionally in the life stage fished, e.g., elvers in the Bay of Fundy area of Nova Scotia and New Brunswick and the Atlantic coastal region of Nova Scotia, yellow and silver eels in the Maritime provinces and primarily silver eels in the St. Lawrence River basin. Commercial landings may not be a reliable indicator of stock abundance and can be influenced by factors such as market conditions and completeness of statistics collection (ICES 2013). Since 1950, the pattern in Canadian and U.S. landings has differed in timing, with Canadian landings rising to a high about 1970 and fluctuating until the early 1990s before declining to recent lows (Fig. 3B) while American landings rose to a high in the mid-1970s and early 1980s before declining (Fig. 3C). The combined Canadian and U.S. landings follow a pattern similar to that for the U.S. (Fig. 3D). Recent landing levels have also been seen in the past (Fig. 3A) and the decline in Canadian landings after about 2005 is likely the consequence of reductions in fishing effort due to management restrictions (DFO 2010). The majority of abundance indices are from fresh water. From the 21 informative Canadian abundance indices available, eight composite indices (a generalized linear model (GLM) composite of several indicators varying in life stage, habitat, and RPA zone) and one single index were examined (Table 1) (Cairns et al. 2014; DFO 2014). Analyses were made for time intervals corresponding roughly to one generation (16 years), two generations (32 years), or over the available time series. The overall conclusion was of a general decline in

Table 1. American eel abundance indices evaluated for Canadian Recovery Potential Assessment (from Table 2, DFO (2014)).

Life stage	Habitat	Region	Index type, description (number of indicators), and time series
Recruitment	Fresh water	Scotia-Fundy	Composite, elver counts (2), 1990–2012
		St. Lawrence Basin	Single, Moses-Saunders eel ladder index (1), 1975–2012
			Composite, Quebec eel ladder counts (2), 1998–2011
Standing stock	Fresh water	St. Lawrence Basin	Composite, Lake Ontario survey indices (2), 1972–2012
		Northern Gulf and Newfoundland	Composite, Fence counts (3), 1971–2011
		Southern Gulf of St. Lawrence	Composite, electrofishing (2), 1952–2012
		Scotia-Fundy	Composite, electrofishing (3), 1985–2012
	Estuary/marine	Southern Gulf of St. Lawrence	Composite, commercial CPUE (2), 1996–2012
Spawner production	Fresh water	St. Lawrence Basin	Composite, trapnet catches (4), 1971–2012

American eel abundance in Canada over the past 32 years (two generations) or more, although two recruitment and one fishery indicators have increased over the past 16 years.

The Moses-Saunders Dam index to the upper St. Lawrence River and Lake Ontario is the longest juvenile eel recruitment series (up to 38 years to 2012) and has shown the most severe decline (about 99% at its lowest point), although the recent trend has increased (Fig. 4A). A composite recruitment index to two fishways in the middle St. Lawrence River area with shorter time series (14 years) showed a significant positive trend. The elver recruitment index to the East River, Chester on the Atlantic coast of Nova Scotia has, over the 19-year period from 1996–2014, recently become significantly positive (Fig. 4C).

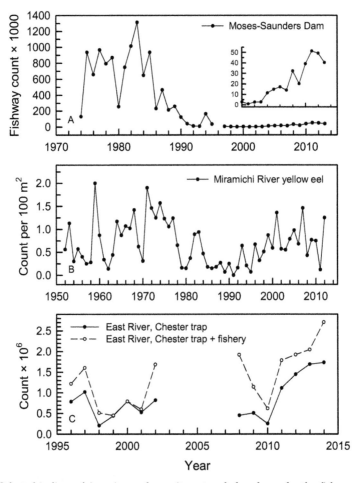

Fig. 4. Selected indices of American eel recruitment and abundance for the fishway count at Moses-Saunders Dam on the upper St. Lawrence River, Ontario, the Miramichi River, New Brunswick (Gulf of St. Lawrence), and the East River, Chester, Nova Scotia (Atlantic coast) elver recruitment index.

Standing stock composite indices, other than the Lake Ontario index that is directly affected by recruitment through the Moses-Saunders Dam and has severely declined (100%), showed a variety of trends depending upon the time period examined (Cairns et al. 2014). A non-statistically significant decline was found in the northern Gulf and Newfoundland area over a 41-year period. No significant change occurred for the southern Gulf of St. Lawrence area (Miramichi River) freshwater juvenile eel abundance index over a 61-year period but the significant increase over the most recent 32-year period tapers to non-significance over the past 16 years (Fig. 4B). The standing stock index for the Bay of Fundy and Atlantic coast of Nova Scotia, based on electrofishing surveys, declined slightly (3%) but significantly over the 25 years between 1985 and 2009 but more steeply (39%) over the most recent 16 years. The southern Gulf of St. Lawrence index (Prince Edward Island, Gulf New Brunswick and Gulf Nova Scotia) was the only index of standing stock to increase in trend (8% annually) over the 17-year period from 1996 to 2012. A composite index of four trapnet indices (three fishery-dependent and one fishery-independent) of silver eel catches from the St. Lawrence River basin showed a significant annual decline of 1.9% over the period 1971–2012 (total decline of 41%) but a non-significant decline over the most recent 16-year period. Mark-recapture studies of silver eel escapement from the St. Lawrence River basin in 1996–1997 gave estimated escapements of 488,000–397,000 eels and in 2010–2011 estimates of 155,000–160,000 eels, for a decline of 64%. These escapements are affected by the decline in recruitment and standing stock in the upper St. Lawrence River and Lake Ontario.

The trends in the composite indices were analyzed using the mean annual predicted values from the GLM and do not consider the uncertainty in those values, which tends to increase the associated uncertainty. No overall composite index for eastern Canada was developed nor could stock contribution weights be assigned to regions (Cairns et al. 2014). A high contribution to the total spawning stock has been proposed for eels from the St. Lawrence River basin based on their high fecundity and female sex ratio, the large historic habitat availability, and the large attraction flow from the river (COSEWIC 2012), but no substantive evidence for this position exists because of uncertainties in relative spawner production by region, fecundity-size-age trade-offs that vary with latitude, larval distribution patterns (random or regional), and a mechanism whereby river discharge could influence larval/glass eel detrainment from the Gulf Stream, particularly where most of the flow of the St. Lawrence River turns sharply south-westward as the Nova Scotia Current as it exits the Gulf of St. Lawrence (Jessop 1998). Given the scarcity of available indices, all (but particularly indices not directed primarily at eels) were assumed to be representative and proportional to the life history stage and RPA zone and consistent in sampling methods over the time series.

In the United States, elver fisheries presently occur only in South Carolina and Maine, which has the major elver fishery. Yellow eel fisheries using primarily baited pots and traps occur in all Atlantic coastal states and some

Gulf of Mexico states and are the dominant fishery. Small silver eel fisheries occur in New York and Maine (ASMFC 2012). The ASMFC fishery management plan for the American eel requires states to conduct an annual glass eel recruitment survey (ASMFC 2000). There were 20 sites that collected at least ten years of data as of 2013. From these 20 YOY (young-of-the-year) abundance indices, five GLM composite indices (ASMFC 2012) were developed—one for each of the five regions along the U.S. east coast (Table 2). A Mann-Kendall test was applied to test for the presence of a significant trend in the indices (Gilbert 1987). The null hypothesis is that the time series is independent and identically distributed—there is no significant trend across time. The test allows for missing values and can account for tied values if present.

Table 2. Summary of the young-of-the-year (YOY) recruitment indices used to develop the United States regional indices of Fig. 5.

Region	State/Jurisdiction	Site	Gear	Start Year
Gulf of Maine	Maine	West Harbor Pond	Irish Elver Ramp	2001
	New Hampshire	Lamprey River	Irish Elver Trap	2001
	Massachusetts	Jones River	Sheldon Elver Trap	2001
	Massachusetts	Parker River	Sheldon Elver Trap	2004
Southern New England	Rhode Island	Gilbert Stuart Dam	Irish Elver Ramp	2000
	Rhode Island	Hamilton Fish Ladder	Irish Elver Ramp	2004
	New York	Carman's River	Fyke Net	2000
Delaware Bay/ Mid-Atlantic Coastal Bays	New Jersey	Patcong Creek	Fyke Net	2000
	Delaware	Millsboro Dam	Fyke Net	2000
	Maryland	Turville Creek	Irish Elver Ramp	2000
Chesapeake Bay	Potomac River	Clarks Millpond	Irish Elver Ramp	2000
	Potomac River	Gardys Millpond	Irish Elver Ramp	2000
	Virginia	Brackens Pond	Irish Elver Ramp	2000
	Virginia	Kamps Millpond	Irish Elver Ramp	2000
	Virginia	Warehams Pond	Irish Elver Ramp	2003
	Virginia	Wormley Creek	Irish Elver Ramp	2001
South Atlantic	South Carolina	Goose Creek	Fyke Net	2000
	Georgia	Altamaha Canal	Fyke Net	2001
	Georgia	Hudson Creek	Fyke Net	2003
	Florida	Guana River Dam	Dip Net	2001

Trends among the U.S. YOY regional recruitment indices were highly variable, with wide 95% confidence intervals, and were inconsistent among regions during the relatively short time series (Fig. 5). The Mann-Kendall test did not detect any significant increasing or decreasing trends in YOY indices

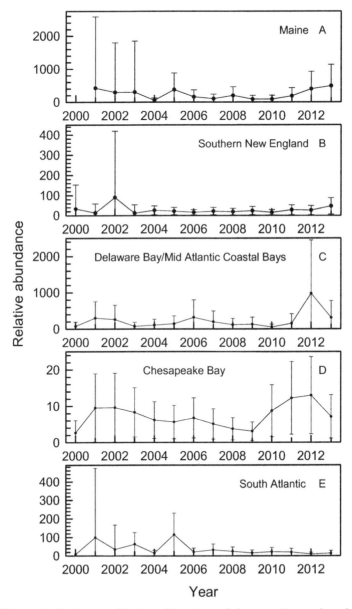

Fig. 5. GLM-standardized regional indices of American eel elver recruitment along the Atlantic coast of the United States (2000–2013). The error bars are 95% confidence intervals that have been truncated at zero. The number of indices in each regional index are shown in Table 2.

over the time series. Three of the five indices—Gulf of Maine, Delaware Bays/ Mid-Atlantic Coastal Bays, and Chesapeake Bay—exhibited their highest values in either 2012 or 2013. The Gulf of Maine and Southern New England indices demonstrate some evidence of increased recruitment in the most recent years. The highest values for the Southern New England and South Atlantic indices occurred in the early part of the time series. One important caveat in interpreting these indices is that the individual indices come from sites with varying sampling gears, timing, distance from obstructions (if an obstruction is present), distance from tidal influence, channel widths, and so on. Such differences are difficult to account for in the standardization process and had to be largely ignored; however, it is assumed that these indices would provide an indication of population-wide recruitment failure if it were to occur. The pattern in the yellow eel regional composite index depends on the time frame examined—no overall trend over the entire 44 years, an initial significant decline (Mann-Kendall test) over the last 30 years (Fig. 6), that levelled off after about 1989, and limited variability with a slightly increasing trend over the most recent 20 years (ASMFC 2012). A number of individual indices declined significantly over their time periods as determined by three trend analysis methods (Mann-Kendall, Manly, and ARIMA) and were judged to be cause for concern. The overall conclusion of the ASMFC (2012) report was "of declining, or at least, neutral abundance of American eels in the U.S. in recent decades". Management efforts to reduce American eel mortality were recommended, as were efforts to improve our understanding of the distribution and occurrence of eels, and of fishery catch and effort data. The benefits of a coast-wide stock assessment that includes both Canadian and

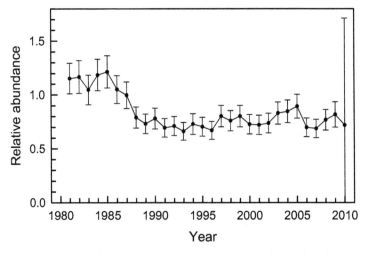

Fig. 6. GLM-standardized index of yellow-phase American eel abundance along the Atlantic coast of the United States (1981–2010; 30-year index). The error bars are 95% confidence intervals that have been truncated at zero. The wide 2010 95% CI results from n = 3 indices in the composite index as compared with n = 7 in all other years except 2009 with n = 6.

U.S. data and inter-jurisdictional cooperation in monitoring, assessing, and managing the species have been discussed since at least 1997 (Peterson 1997), reflecting the changed conditions since the 1970s when most delegates at the 1980 North American Eel Conference thought that international cooperation was unnecessary (OMNR 1982).

Habitat requirements and availability

The American eel has historically occupied, with varying density, all accessible rivers draining into the Atlantic Ocean and the Gulf of Mexico between Venezuela and the middle of Labrador, although their presence in the upper reaches and tributaries of the Mississippi River appears to be limited or non-existent (D. Cairns, DFO, personal communication). Cairns et al. (2014, Table 2.1.1) estimated the potential freshwater watershed habitat available in Canada and the U.S. Atlantic coastal RPA zones, assuming Niagara Falls as a historic barrier, to be 1,766,726 km^2. This includes watershed areas upstream of natural and artificial barriers so the area colonisable by eels is less. No comprehensive estimate of freshwater habitat potentially accessible (but not necessarily so) to eels is available for Canada but the U.S. was estimated to have 103,501 km^2, 92,538 km^2 of which is along the U.S. Atlantic seaboard, the difference being due to the U.S. watersheds draining into the St. Lawrence River system (Cairns et al. 2012, Table 4.4.1). The continental freshwater area presently accessible to eels is unknown but Pratt et al. (2014) provide a review of losses in available habitat. Eel distribution is currently substantially reduced due to artificial barriers to upstream migration (USFWS 2007; COSEWIC 2012), with as much as 84% of stream length inaccessible to eels in the eastern U.S., particularly in the region between Maine and Connecticut (Busch et al. 1998). The estuarine and coastal habitat suitable for yellow eels was estimated at 8,910 km^2 for Canada and 14,360 km^2 for the eastern seaboard of the U.S., totalling 23,270 km^2. Commercial eel fisheries occur in many estuarine and coastal areas and research surveys commonly find eels in such areas that are unfished. The absence of a commercial fishery does not necessarily indicate the absence of eels because an area may lack a tradition of eel fishing or be subject to unfavourable fishery regulations. Cairns et al. (2012) estimated that eel fisheries occurred in about 6.4% of the approximately 9,000 km^2 area of sheltered waters along the Atlantic coast and the Gulf of St. Lawrence and St. Lawrence River estuary. The average eel catch was about 30 kg/km^2/year for those fisheries. Assessments of American eel stock status have typically focused on freshwater habitat production (ASMFC 2012; COSEWIC 2012; IUCN 2014) because estimates of estuarine and coastal habitat availability and of eel density in these habitats were unavailable, thereby potentially overestimating the degree of stock decline by ignoring estuarine and coastal habitat production.

Threats and limiting factors

Threats to the American eel population include directed fisheries, physical obstructions such as hydro-electric dams and road culverts that limit or delay habitat access, habitat alterations, water quality, parasites and diseases, and ecosystem changes (ASMFC 2012; COSEWIC 2012; Chaput et al. 2014a). Threats can be quantified in their level of concern by evaluating the severity, geographic extent, temporal frequency, and certainty of threat causation. Changes in oceanic environmental conditions are also an important influence on eel population abundance but may be regarded less as a threat than as a limitation because they are beyond immediate human influence.

The potential negative effects of fisheries on eel stocks, particularly on the production of spawners, are well known but few estimates of exploitation rates are available and those may not be widely applicable in other areas. The mean exploitation rate of a dip-net fishery for elvers in a small Atlantic coastal river (East River, Chester) was estimated at 40.0% (range 12–59%) over 12 years between 1996 and 2014 (Jessop, unpublished data), the yellow eel fishery in the southern Gulf of St. Lawrence was estimated to harvest 7–8% of the standing stock biomass (Chaput et al. 2014a), and the silver eel fishery in the lower St. Lawrence River harvested 19–24% of the silver eels migrating from further upstream during a two-year period (Caron et al. 2003). Low annual exploitation rates by yellow eel fisheries in fresh water can substantially reduce silver eel production because of the age (mean 19 years) at which they mature in northern waters. The links between elver fishery exploitation and abundance at later stages, particularly spawners, are poorly understood but depend substantially on the relative effects of fishery and natural mortality rates at the elver stage, including whether, and to what degree, density-dependent mortality effects occur.

Large hydro-electric dams, particularly the Beauharnois Dam (completed early in the 1940s) and the Moses-Saunders Dam (completed in 1959) on the upper St. Lawrence River, have seriously affected the upstream and downstream passage of American eels into and from the large watersheds of Lake Ontario and the Ottawa River (Pratt et al. 2014). Until the construction of a fishway in 1974 and new, more efficient fishways in 2002 and 2004, upstream passage at the Beauharnois Dam occurred via the nearby ship navigation lock (Verdon and Desrochers 2003; Pratt et al. 2014). Between 1933 and 1965, a series of smaller dams were built in the area of the Beauharnois Dam and it is likely that eel passage has been blocked or impeded since the early 1940s (Verdon and Desrochers 2003). No upstream passage was available at the Moses-Saunders Dam between 1958 and 1974 when a fishway was installed (McGrath et al. 2003), and another fishway was built in 2006 (Pratt et al. 2014), but some passage likely occurred at the nearby ship locks. Upstream fishway passage is only moderately efficient and no downstream passage is provided at either dam. These dams have substantially influenced the decline in upstream eel abundance and may have reduced silver eel production by over 800,000 eels, a number comparable to the number of silver eels caught during the commercial catch peak during the 1920s and 1930s (Verdon and Desrochers 2003; Verreault et al. 2004; MacGregor et al. 2010). In the Canadian sector of the St. Lawrence River watershed, about 87% of the historically available habitat remains accessible but access is restricted or delayed to about 12,140 km^2 of habitat (COSEWIC 2012). Almost 5,000 dams have been identified in the St. Lawrence River watershed, of which about 86% were assessed as impassable for the upstream passage of eels and 86% were assessed as causing substantial delay to downstream passage (Tremblay et al. 2011). In the province of New Brunswick, the Mactaquac Dam on the Saint John River, which discharges into the Bay of Fundy and is the largest river in the province, has a trap-and-truck fish passage facility unsuited to eel passage that essentially blocks upstream passage of eels to about 17,000 km^2 of watershed between the dam and Grand Falls, which is a historic natural barrier (Pratt et al. 2014). Eels are widely distributed in the almost 30,000 km^2 of watershed downstream of the Mactaquac Dam. The downstream movement of eels through the hydro-electric turbines results in variable mortality rates depending upon factors such as blade impact, pressure changes, and shear forces, resulting in turbine mortalities of 18%–26% at each dam (Chaput et al. 2014a). The importance of obstructed fish passage and habitat loss relative to effects on spawning success and recruitment from oceanic environmental conditions is uncertain but several studies have found correlations between the North Atlantic Oscillation (NAO) index and recruitment indices (Cairns et al. 2014). Casselman (2003) attributed the severe decline in the number of eels recruiting to the upper St. Lawrence River to a major recruitment failure of the eel population on the assumption that density-dependent effects influence the number of eels moving upstream.

American eels may be afflicted by a variety of endemic diseases and parasites but the introduced swim bladder nematode *Anguillicola crassus*, which is native to the Japanese eel, has spread throughout much of the American eel distribution in eastern North America (Nilo and Fortin 2001). The pathological effects of this parasite are presumed to be similar to those for the European eel and may include various types of internal organ lesions that may affect both instream mortality and the potential ability for successful migration and spawning (Chaput et al. 2014a). Chaput et al. (2014a) also discuss a variety of other threats and limiting factors, including negative changes to water quality (including pollutants) and quantity, habitat alterations such as sedimentation, shoreline erosion, dredging, and effects of introduced aquatic organisms, etc. They also classify the level of threat concern and severity, describe the geographic scale and distribution of threats, their temporal extent (occurrence and frequency), and the certainty that such threats can be linked to a change in eel production. Physical obstructions (habitat loss) and commercial eel fisheries were assessed to be of medium or high threat across all regions, with other threats such as turbine mortality, the presence or threat of introduction of invasive species, and water quality and contaminants varying regionally from low to medium.

Population modelling

American eel population dynamics have been modelled to assess the sensitivity to perturbations of various model assumptions and the effects of anthropogenic mortality under those assumptions by Young and Koops (2014). A key model assumption is the nature of larval distribution along the Atlantic coast of North America. The hypothesis of random distribution of larvae of different parental origins or of abundance, for which there is no evidence, is often associated (COSEWIC 2006; ASMFC 2012; Gagnaire et al. 2012) with the hypothesis of panmixia, which is well supported (Côté et al. 2013). Evidence supportive of non-random dispersal includes consistent geographic variation in the abundance of larval and glass eels/elvers (McCleave 1993; Jessop 1998; Dutil et al. 2009), including hybrid distribution such as to Iceland (Als et al. 2011; Pujolar et al. 2014), and the absence of a mechanism to explain the timing of larval metamorphosis hypothesis, which varies with latitude and distance from the spawning area (Jessop 2010). European and American eel hybrids occur almost exclusively in Iceland, which suggests hybridization in a specific area of the Sargasso Sea spawning area and subsequent non-random larval dispersal (Als et al. 2011; Pujolar et al. 2014). However, Als et al. (2011) conclude that non-random advection of hybrid eel larvae from a specific area in the Sargasso Sea is unlikely and that another distribution mechanism must act, such as a migration behaviour and larval development (metamorphosis) that is intermediate between that of the two parental species. No explanation exists for how such an intermediate duration might be governed or what triggers metamorphosis. Kleckner and McCleave (1985) discuss mechanisms of larval transport out of the Gulf Stream and into continental shelf slope waters, with metamorphosis probably occurring or being completed after transition to the continental shelf, which may occur via directional swimming (Miller et al. 2014). What triggers this detrainment from the Gulf Stream and subsequent metamorphosis is unknown although Kleckner and McCleave (1985) suggest that metamorphosis is initiated as larvae become developmentally competent or respond to an environmental cue.

Hormones, such as thyroid, melatonin, and cortisol among many others, may influence metamorphosis (Otake 2003), migration, and settling in glass eels, and maturation and migration in silvering eels (Edeline et al. 2004; Oliveira and Sánchez-Vásquez 2010; Reiter et al. 2010; Rousseau et al. 2014). Hormones are also genetically mediated, integrating both genetic and environmental influences as well as regulating gene transcription, neuronal, and metabolic activities (Dufty et al. 2002; Edeline et al. 2009). Glass eel/elver recruitment occurs later in the year at northern latitudes than at southern latitudes and the converse is true for the timing of silver eel escapement (ASMFC 2012). Daylight hours and water temperatures increase seasonally and with increasing latitude (within latitudinal limits) during the spring larval migration period, they are high during the juvenile growing season, and decrease as the spawning season approaches (Jessop 2010). It can be hypothesized that the timing of larval metamorphosis is controlled by an internal biological clock in which the pineal organ, eye retinas, and melatonin production may be the central components (Oliveira and Sánchez-Vásquez 2010; Reiter et al. 2010). The diel clock mechanism affects a variety of other gene expression regulatory networks. Many biological clock components are maternally inherited and present at a very early embryonic stage. Gonadal development in female eels begins one or more years after continental arrival as an elver (Davey and Jellyman 2005) and environmental imprinting may occur at that time. Consequently, the time and size at larval metamorphosis may be linked to the distance to the spawning area of the female parent and associated increased age and size at adult maturation via a phenotypically expressed genetic mechanism that has developed due to regional genetic diversity not detected by the analysis of neutral genetic markers typically examined by studies concluding genetic panmixia in anguillid eels (Jessop 2010). Thus, the older, larger eels observed in northern waters would produce larvae that metamorphose at an older age and larger size thereby reaching the northern areas of their parentage (and perhaps with a degree of regional fitness), with analogous results over the geographic range of the eel. Divergence in phenotypic traits, despite the lack of neutral genetic differentiation, has been interpreted as evidence for the adaptive significance of population structure (Hutchings et al. 2007). Spatially varying selection may also produce genetic differences in eels from different locations (Côté et al. 2014, 2015). One implication of this hypothesis is a slow rate of stock recovery for areas of seriously depleted abundance, particularly at the extremes of the geographic range, and the necessity for regional as well as continental management.

Young and Koops (2014) used a matrix model with the population divided into seven zones of similar life-history composition, a two sex structure, and four larval distribution scenarios. The larval distribution hypotheses were: (a) full maternal effects, with larval distribution fully controlled by female contribution to the spawning stock, (b) full water attraction, with larvae attracted to each zone proportional to the zonal water discharge to the ocean, (c) hybrid maternal and water attraction effects, with 95% of larvae distributed according to female maternal effects, and (d) hybrid maternal and straying, with 95% of larvae distributed according to maternal effects and 5% straying in equal proportion to adjacent zones.

Although the model was not calibrated to observed data and is limited by the validity of its assumptions, it offers useful insight into the dynamics associated with different hypotheses. The model was very sensitive to assumptions such as leptocephalus distribution and mortality rate, elver mortality rate, and relative eel abundance in different zones. Model results varied greatly by zone within and among mortality factors and under different larval distribution hypotheses. For example, the different larval distribution hypotheses produced large variation in zone population growth results even when maternal effects were strong. Low levels of straying had large effects on population dynamics and health. Under random distribution, the entire population was affected by decline or growth in a zone, which acted as a sink or source. Most importantly, transient dynamics and population change momentum caused initial population growth where rates would otherwise suggest population stability or long-term decline. Initial population decline also occurred where rates suggested growth. Consequently, possible transient dynamics and population growth momentum should be considered in future modelling studies.

Threat mitigation

Feasible methods for the mitigation of identifiable threats to the eel population were assessed by Chaput et al. (2014b), of which the major threats are habitat loss and fishery exploitation. Since the late 1990s and early 2000s, reductions to fishing mortality have been accomplished by license buy-backs and fishery closures in Ontario and Quebec. The Quebec license buy-backs were partly funded by Hydro-Quebec as mitigation for turbine mortality at hydro-electric dams and have reduced the fishery catch of yellow and silver eels in the St. Lawrence River system by about 53%. Amongst other actions, DFO's Gulf Region has increased minimum size limits and reduced the fishing season, the Maritimes Region (Atlantic and Bay of Fundy drainages of New Brunswick and Nova Scotia) has increased minimum size limits, introduced a minimum size trap escape mechanism, frozen license numbers, and reduced elver fishery quotas, and the Newfoundland and Labrador Region has increased the minimum retention size and reduced the number of licenses issues, the number of gear units fished, and the fishing season (DFO 2010). Reference points for allowable fishery mortality or biomass have not been established. The fisheries management mitigation measures have to date reduced the total eel fishery mortality by more than 50% relative to 1997–2002 and has reduced landed weights by 27% in the Maritimes Region although the increase in landings in the southern Gulf of St. Lawrence has been attributed to a regional increase in abundance (Chaput et al. 2014b).

No improvements to upstream fish passage for eels at major hydro-electric dams have occurred since the 1970s with the exception of those to the Beauharnois and Moses-Saunders Dams on the St. Lawrence River between 2002 and 2006. No major dams presently have downstream passage management plans; where turbines occur, turbine passage is the norm although two small hydro-electric dams in Quebec have bypass mechanisms and turbine shut-downs are used at several small dams in Virginia and Maine (Therrien and Ewaschuk 2010). Options for mitigating the effects of habitat loss due to obstruction and due to turbine mortality during downstream passage such as removal of barriers and improved fish passage are examined by Chaput et al. (2014b). Another option to mitigate habitat loss due to obstruction is enhancement by elver stocking such as occurred in Lake Ontario between 2006 and 2010 (Pratt and Threader 2011) and in the Richelieu River tributary of the St. Lawrence River (Verreault et al. 2010). The growth rate of elvers introduced into Lake Ontario and the Richelieu River from the Maritimes was high and the eels matured at sizes and sex ratios characteristic of their origin rather than that of the eels that had naturally migrated to the region, perhaps due to genetic differences (Côté et al. 2014). The stocking program has ceased following the detection of the swim bladder nematode *A. crassus* in eels introduced into Lake Ontario and recognition of potential genetic differences between transplanted and local origin eels.

In the U.S., the Federal Power Act requires non-federal hydro-electric projects which meet certain criteria to obtain a license from the Federal Energy Regulatory Commission (FERC) every 30–50 years. The license regulates what facilities can be built and what types of mitigation measures must be implemented to protect environmental resources. The licensing requirements have resulted in the improvement of upstream and downstream passage facilities at some existing locations including the requirement for eel ways for upstream passage and night-time shut downs for downstream passage as conditions of fishway prescriptions at a number of projects. In some extreme cases, the FERC may even find that the environmental benefits of dam removal outweigh the costs of keeping the dam.

Recovery targets

Canadian eel stock recovery targets have been defined for abundance and distribution based on short-term (one generation or about 16 years), medium-term (three generations or about 50 years), and long-term (> 50 years) objectives (Cairns et al. 2014). Recovery was defined as meeting the recovery targets. The short-term distribution target seeks to maintain the current geographic distribution and to increase distribution to productive areas so as to increase escapement. This target has been met in all Canadian RPA zones. The medium-term distribution objective is to increase eel distribution to accessible areas that will generate productivity equivalent to what has been lost over the past three generations.

The short-term recovery target for abundance is to stop any decline in abundance indices (recruitment, standing stock, spawner production) and reduce mortality from all sources by 50% relative to the 1997–2002 average (DFO 2010; Chaput et al. 2014b). Over the medium and long terms, the goal is to rebuild regional and national abundance to the levels of abundance indices found in the mid-1980s (defined as 1981–1989). For the four Canadian RPA zones, of 7 recruitment and standing stock indices, only the index of freshwater standing stock in the southern Gulf of St. Lawrence and the recruitment indices for the St. Lawrence River basin and Scotia-Fundy RPA zones presently exceed their short-term recovery targets. Recruitment increases have yet to produce improvements in standing stock indices.

Conclusion

The American eel poses a challenge for management and conservation due to its unique life history. The threat status designations proposed or applied throughout its range suggest general concern for the species. Although the species can occupy a wide variety of habitats, fisheries, physical obstructions, habitat alterations, water quality, parasites and diseases, and ecosystem changes pose threats to this species almost everywhere it occurs. These concerns have led to new research on the species, implementation of new management measures, threat mitigation, and establishment of recovery targets. Continuation of population modelling efforts, based on improved data collection, will be essential towards further understanding the dynamics of this species. These measures, along with evidence of increases in recent recruitment, may pave the road to recovery for the American eel.

Acknowledgements

We thank D. Cairns for providing several graphics and Canadian eel data and for reviewing the manuscript, Y. Carey for recent East River, Chester elver index data, and J. Casselman for the Moses-Saunders Dam eel index data. We thank Wilson Laney, Sheila Eyler, Mark Cantrell, and Larry Miller for information on the FERC licensing process. We thank all of the state agencies and corporations that contributed to the ASMFC stock abundance indices and to the ASMFC for sharing some of that data.

Keywords: American eel, *Anguilla rostrata*, stock status, review, fishery management

References

Albert, V., B. Jónsson and L. Bernatchez. 2006. Natural hybrids in Atlantic eels (*Anguilla anguilla, A. rostrata*): evidence for successful reproduction and fluctuating abundance in space and time. Molecular Ecology 15: 1903–1916.

Als, T.D., M.M. Hansen, G.E. Maes, M. Castonguay, L. Riemann, K. Aarestrup, P. Munk, H. Sparholt, R. Hanel and L. Bernatchez. 2011. All roads lead to home: panmixia of European eel in the Sargasso Sea. Molecular Ecology 11. doi: 10.1111/j.1365-294X.2011.05011.x.

ASMFC (Atlantic States Marine Fisheries Commission). 2000. Interstate fishery management plan for American eel (*Anguilla rostrata*). ASMFC, Fishery Management Report No. 36, Washington, D.C.

ASMFC (Atlantic States Marine Fisheries Commission). 2012. American eel benchmark stock assessment. Atlantic States Marine Fisheries Commission, Arlington, VA.

Benchetrit, J. and J.D. McCleave. 2015. Current and historical distribution of the American eel *Anguilla rostrata* in the countries and territories of the Wider Caribbean. ICES J. Mar. Sci. doi: 10.1093/icesjms/fsv064.

Boëtius, J. 1985. Greenland eels, *Anguilla rostrata* LeSueur. Dana 4: 41–48.

Bossdorf, O., C.L. Richards and M. Pigliucci. 2008. Epigenetics for ecologists. Ecology Letters 11: 106–115.

Busch, W.-D.N., S.J. Lary, C.M. Castiglione and R. McDonald. 1998. Distribution and availability of Atlantic coast freshwater habitats for American eel (*Anguilla rostrata*). United States Fish and Wildlife Service Administrative Report 98-2, Amherst, New York.

Cairns, D.K., J.-D. Dutil, S. Proulx, J.D. Mailhiot, M.-C. Bédard, A. Kervella, L.G. Godfrey, E.M. O'Brien, S.C. Daley, E. Fournier, J.P.N. Tomie and S.C. Courtenay. 2012. An atlas and classification of aquatic habitat on the east coast of Canada, with an evaluation of usage by the American eel. Can. Tech. Rep. Fish. Aquat. Sci. 2986.

Cairns, D.K., G. Chaput, L.A. Poirier, T.S. Avery, M. Castonguay, A. Mathers, R.G. Bradford, T. Pratt, G. Verreault, K. Clarke, G. Veinnot and L. Bernatchez. 2014. Recovery Potential Assessment for the American Eel (*Anguilla rostrata*) for eastern Canada: life history, distribution, status indicators, and demographic parameters. DFO Can. Sci. Adv. Sec. Res. Doc. 2013/134.

Canadian Eel Working Group. 2009. American Eel Management Plan, Draft: February 26, 2009. Fisheries and Oceans Canada, Ontario Ministry of Natural Resources, and Ministère des Ressources naturelles et de la Faune du Québec.

Caron, F., G. Verreault and E. Rochard. 2003. Estimation of the Population size, exploitation rate, and escapement of silver-phase American Eels in the St. Lawrence watershed. Amer. Fish. Soc. Symp. 33: 235–242.

Casselman, J.M. 2003. Dynamics of resources of the American eel, *Anguilla rostrata*: declining abundance in the 1990s. pp. 255–274. *In*: K. Aida, K. Tsukamoto and K. Yamauchi (eds.). Eel Biology. Springer-Verlag, Tokyo.

Chaput, G., T.C. Pratt, D.K. Cairns, K.D. Clarke, R.G. Bradford, A. Mathers and G. Verreault. 2014a. Recovery Potential Assessment for the American Eel (*Anguilla rostrata*) for eastern Canada: description and quantification of threats. DFO Can. Sci. Advis. Sec. Res. Doc. 2013/135.

Chaput, G., D.K. Cairns, S. Bastien-Daigle, C. LeBlanc, L. Robichaud, J. Turple and C. Girard. 2014b. Recovery Potential Assessment for the American Eel (*Anguilla rostrata*) for eastern Canada: mitigation options. DFO Can. Sci. Advis. Sec. Res. Doc. 2013/133.

COSEWIC. 2006. COSEWIC assessment and status report on the American eel *Anguilla rostrata* in Canada. Committee on the Status of Endangered Wildlife in Canada. Ottawa.

COSEWIC. 2012. COSEWIC assessment and status report on the American eel *Anguilla rostrata* in Canada. Committee on the Status of Endangered Wildlife in Canada. Ottawa.

Côté, C.L., M. Castonguay, G. Verreault and L. Bernatchez. 2009. Differential effects of origin and salinity rearing conditions on growth of glass eels of the American eel *Anguilla rostrata*: implications for stocking programmes. J. Fish Biol. 74: 1934–1948.

Côté, C.L., P.-A. Gagnaire, V. Bourret, G. Verreault, M. Castonguay and L. Bernatchez. 2013. Population genetics of the American eel (*Anguilla rostrata*): FST = 0 and North Atlantic Oscillation effects on demographic fluctuations of a panmictic species. Molecular Ecol. 22: 1763–1776.

Côté, C.L., M. Castonguay, M.S. Kalujnaia, G. Cramb and L. Bernatchez. 2014. In absence of local adaptation, plasticity and spatially varying selection rule: a view from genomic reaction norms in a panmictic species (*Anguilla rostrata*). BMC Genomics 15: 403.

Côté, C.L., S.A. Pavey, J.A. Stacey, T.C. Pratt, M. Castonguay, C. Aude and L. Bernatchez. 2015. Growth, female size, and sex ratio variability in American eel of different origins in both controlled conditions and the wild: implications for stocking programs. Trans. Amer. Fish Soc. 144: 246–257.

Davey, A.J.H. and D.J. Jellyman. 2005. Sex determination in freshwater eels and management options for manipulation of sex. Rev. Fish Biol. Fisheries 15(1-2): 37–52.

Dekker, W. (coordinator) and J.M. Casselman (coordinator). 2014. Are eels climbing back up the slippery slope? The 2003 Quebec Declaration of Concern about eel declines—11 years later. Fisheries 39(12): 613–614.

Dekker, W., J.M. Casselman, D.K. Cairns, K. Tsukamoto, D. Jellyman and H. Lickers. 2003. Worldwide decline of eel resources necessitates immediate action. Québec Declaration of Concern. Fisheries 28(12): 28–30.

DFO. 2010. Status of American eel and progress on achieving management goals. DFO Can. Sci. Advis. Sec. Sci. Advis. Rep. 2010/062.

DFO. 2014. Recovery potential assessment of American Eel (*Anguilla rostrata*) in eastern Canada. DFO Can. Sci. Advis. Sec. Sci. Advis. Rep. 2013/078.

Dufty, Jr. A.M., J. Clobert and A.P. Møller. 2002. Hormones, developmental plasticity and adaptation. Trends Ecol. Evol. 17(4): 190–196.

Dutil, J.-D., P. Dumont, D.K. Cairns, P.S. Galbraith, G. Verreault, M. Castonguay and S. Proulx. 2009. *Anguilla rostrata* glass eel migration and recruitment in the estuary and Gulf of St. Lawrence. J. Fish Biol. 74: 1970–1984.

Edeline, E., S. Dufour, C. Briand, D. Fatin and P. Elie. 2004. Thyroid status is related to migratory behavior in *Anguilla anguilla* glass eels. Mar. Ecol. Prog. Ser. 282: 261–270.

Edeline, E., S. Dufour and P. Elie. 2009. Proximate and ultimate control of eel continental dispersal. pp. 433–461. *In*: G. van den Thillart, S. Dufour and J. Rankin (eds.). Spawning Migration of the European Eel. Reproduction Index, a Useful Tool for Conservation Management. Fish and Fisheries Series 30. Springer, Wageningen, The Netherlands.

Environment Canada. 2003. Species at Risk Act, a Guide. October 2003. Government of Canada, Ottawa.

Gagnaire, P.-A., E. Normandeau, C. Côté, M.M. Hansen and L. Bernatchez. 2012. The genetic consequences of spatially varying selection in the panmictic American eel (*Anguilla rostrata*). Genetics 190: 725–736.

Helfman, G.S., D.E. Facey, L.S. Hales, Jr. and E.L. Bozeman, Jr. 1987. Reproductive ecology of the American eel. Amer. Fish. Soc. Symp. 1: 42–56.

Hutchings, J.A., D.P. Swain, S. Rowe, J.D. Eddington, V. Puvanendran and J.A. Brown. 2007. Genetic variation in life-history reaction norms in a marine fish. Proc. Royal Soc. London Series B: Biol. Sci. 274: 1693–1699.

Jacoby, D., J. Casselman, M. DeLucia, G.A. Hammerson and M. Gollock. 2014. *Anguilla rostrata*. The IUCN Red List of Threatened Species. Version 2014.3. <http://www.iucnredlist.org/>. Downloaded on 16 November 2014.

Jessop, B.M. 1998. The management of, and fishery for, American eel elvers in the Maritime Provinces, Canada. Bull. Fr. Pêche Pisic. 349: 103–116.

Jessop, B.M. 2003. The run size and biological characteristics of American eel elvers in the East River, Chester, Nova Scotia, 2000. Can. Tech. Rep. Fish. Aquat. Sci. No. 2444.

Jessop, B.M. 2010. Geographic effects of American eel (*Anguilla rostrata*) life history characteristics and strategies. Can. J. Fish. Aquat. Sci. 67: 326–346.

Jessop, B.M., D. Cairns, I. Thibault and W.-N. Tzeng. 2008. Life history of American eel *Anguilla rostrata*: new insights from otolith microchemistry. Aquat. Biol. 1: 205–216.

Kleckner, R.C. and J.D. McCleave. 1985. Spatial and temporal distribution of American eel larvae in relation to North Atlantic Ocean current systems. Dana 4: 67–92.

MacGregor, R., J. Casselman, L. Greig, W.A. Allen, L. McDermott and T. Haxton. 2010. DRAFT Recovery Strategy for the American Eel (*Anguilla rostrata*) in Ontario. Ontario Recovery Strategy Series. Prepared for Ontario Ministry of Natural Resources, Peterborough, Ontario.

McCleave, J.D. 1993. Physical and behavioural control on the oceanic distribution and migration of leptocephali. J. Fish Biol. 43(Suppl. A): 243–273.

Miller, M.J., P. Munk, M. Castonguay, R. Hanel and J.D. McCleave. 2014. A century of research on the larval distribution of the Atlantic eels: a re-examination of the data. Biol. Rev. DOI 10.1111/brv.12144.

Nilo, P. and R. Fortin. 2001. Synthèse des connaissances et établissement d'une programmation de recherche sur l'anguille d'Amérique (*Anguilla rostrata*). Université du Québec à Montréal, Département des Sciences biologiques pour la Société de la faune et des parcs du Québec, Direction de la recherche sur la faune. Québec.

Oliveira, C. and F.J. Sánchez-Vásquez. 2010. Reproduction rhythms in fish. pp. 185–215. *In*: E. Kulczykowska, P. Wlodzimierz and B.G. Kapoor (eds.). Biological Clock in Fish. Science Publishers, Enfield, NH, USA.

OMNR (Ontario Ministry of Natural Resources). 1982. Proceedings of the 1980 North American Eel Conference. Ont. Fish. Tech. Rep. Ser. No. 4.

Peterson, R.H. (ed.). 1997. The American eel in eastern Canada: stock status and management strategies. Proceedings of Eel Management Workshop, January 13–14, 1997, Quebec City, QC. Can. Tech. Rep. Aquat. Sci. 2196.

Pratt, T.C. and R.W. Threader. 2011. Preliminary evaluation of a large-scale American eel conservation stocking experiment. N. Amer. J. Fish. Manage. 31: 619–628.

Pratt, T.C., R.G. Bradford, D.K. Cairns, M. Castonguay, G. Chaput, K.D. Clarke and A. Mathers. 2014. Recovery Potential Assessment for the American Eel (*Anguilla rostrata*) in eastern Canada: functional description of habitat. DFO Can. Sci. Advis. Sec. Res. Doc. 2013/132.

Pujolar, J.M., M.W. Jacobsen, T.D. Als, J. Frydenberg, E. Magnussen, B. Jónsson, X. Jiang, L. Cheng, D. Bekkevold, G.E. Maes, L. Bernatchez and M.M. Hansen. 2014. Assessing patterns of hybridization between North Atlantic eels using diagnostic single nucleotide polymorphisms. Heredity 112: 627–637.

Reiter, R.J., D.-X. Tan and L.C. Manchester. 2010. Melatonin in fish: circadian rhythm and functions. pp. 71–91. *In*: E. Kulczykowska, P. Wlodzimierz and B.G. Kapoor (eds.). Biological Clock in Fish. Science Publishers, Enfield, NH, USA.

Rousseau, K., A.-G. Lafont, J. Pasquier, G. Maugars, C. Jolly, M.-E. Sébert, S. Aroua, C. Pasqualini and S. Dufour. 2014. Advances in eel reproduction physiology and endocrinology. pp. 1–43. *In*: F. Trischitta, Y. Takei and P. Sébert (eds.). Eel Physiology. CRC Press, Boca Raton, FL, USA.

Shepard, S.L. 2015. American eel biological species report. U.S. Fish and Wildlife Service, Hadley, Massachusetts.

Tesch, F.-W. 1973. The eel: biology and management of anguillid eels. Chapman and Hall, London.

Therrien, J. and M. Ewaschuk. 2010. Best management practices guide for American eel and waterpower in Ontario. Ontario Waterpower Association, Peterborough.

Tremblay, V., C. Cossette, J.-D. Dutil, G. Verreault and P. Dumont. 2011. Assessment of upstream and downstream passability for eel at dams. Can. Tech. Rep. Fish. Aquat. Sci. 2912.

USFWS (U. S. Fish and Wildlife Service). 2007. Endangered and threatened wildlife and plants; 12-month finding on a petition to list the American eel as threatened or endangered. Federal Register 72(22): 4967–4997.

USFWS (U. S. Fish and Wildlife Service). 2011. Endangered and threatened wildlife and plants; 90-day finding on a petition to list the American eel as threatened or endangered. Federal Register 76(189): 60431–60444.

USFWS (U. S. Fish and Wildlife Service). 2015. Endangered and threatened wildlife and plants; 12-mont findings on petitions to list 19 species as endangered or threatened. Federal Register 80(195): 60834–60850.

Velez-Espino, L.A. and M.A. Koops. 2010. A synthesis of the ecological processes influencing variation in life history and movement patterns of American eel: towards a global assessment. Rev. Fish Biol. Fisheries 20: 163–186.

Verdon, R. and D. Desrochers. 2003. Upstream migratory movements of American eel *Anguilla rostrata* between Beauharnois and Moses-Saunders power dams on the St. Lawrence River. Amer. Fish. Soc. Symp. 33: 139–151.

Verreault, G., P. Dumont and Y. Mailhot. 2004. Habitat losses and anthropogenic barriers as a cause of population decline for American eel (*Anguilla rostrata*) in the St. Lawrence watershed, Canada. ICES CM 2004/S:04. 2004 ICES Annual Science Conference held September 22–25, Vigo, Spain. Preliminary report.

Verreault, G., W. Dargere and R. Tardif. 2009. American eel movements, growth, and sex ratio following translocation. Amer. Fish. Soc. Symp. 58: 129–136.

Verreault, G., P. Dumont, J. Dussureault and R. Tardif. 2010. First record of migrating silver American eels (*Anguilla rostrata*) in the St. Lawrence Estuary originating from a stocking program. J. Great Lakes Res. 36: 794–797.

Vladykov, V.D. and P.K.L. Liew. 1982. Sex of adult American eels (*Anguilla rostrata*) collected as elvers in two different streams along the eastern shore of Canada, and raised in the same freshwater pond in Ontario. pp. 88–93. *In*: K.H. Loftus (ed.). Proceedings of the 1980 North American Eel Conference, Ontario Ministry of Natural Resources. Ontario Fish. Tech. Rep. No. 4.

Wang, C.-H. and W.-N. Tzeng. 1998. Interpretation of geographic variation in size of American eel, *Anguilla rostrata*, elvers on the Atlantic coast of North America using their life history and otolith ageing. Mar. Ecol. Prog. Ser. 168: 35–43.

Young, J.A.M. and M.A. Koops. 2014. Recovery potential assessment for the American eel (*Anguilla rostrata*) for eastern Canada: recovery potential assessment population modelling. DFO Can. Sci. Advis. Sec. Res. Doc. 2013/131.

Management and Fisheries of Australasian Eels (*Anguilla australis, Anguilla dieffenbachii, Anguilla reinhardtii*)

Donald J. Jellyman

Introduction

Five species of *Anguilla* occur in Australia and New Zealand (hereafter called Australasia), *A. australis, A. dieffenbachii, A. reinhardtii, A. obscura,* and *A. bicolor,* but of these the latter two species occur in only limited areas of northeast Australia and do not constitute significant fisheries (Jellyman 2003). Accordingly, the following discussion is confined to *A. australis, A. dieffenbachii,* and *A. reinhardtii.* Out of these three species, *A. dieffenbachii* is endemic to New Zealand, *A. australis* occurs throughout New Zealand and along the southern part of the east coast of Australia, while *A. reinhardtii,* a tropical species, predominates in the central and northern regions of the east coast of Australia and is also found in Tasmania and in small numbers in the upper North Island of New Zealand. Spawning grounds are unknown for all these species, although results from pop-up tags (Jellyman and Tsukamoto 2010) indicate that *A. dieffenbachii* might spawn in the South Fiji basin. Likewise, results from a Lagrangian simulation model of near-surface flows indicate

National Institute of Water and Atmosphere, PO Box 8602, Christchurch 8053, New Zealand.
  Email: don.jellyman@niwa.co.nz

possible overlapping spawning grounds of *A. australis, A. dieffenbachii, A. reinhardtii* in the same general area (Jellyman and Bowen 2009).

Although separated by only 1500 km, Australia and New Zealand exhibit very different landforms and climate types, with Australia being extensive (the world's 6th largest country), arid (the world's driest inhabited continent), and having a diverse freshwater fish fauna (256 species) of which three-quarters are found only in Australia. In contrast, New Zealand is only 3.5% the size of Australia, with a moist and cool temperate climate, and a sparse freshwater fish fauna (38 species) of which 89% are endemic.

Importance of eels

The New Zealand fish fauna emerged in the absence of specialised predatory species, and freshwater eels, being proficient scavengers fulfill this role; partly for this reason, but also because they are the largest native fish species, eels are the apex predators and dominate the biomass of fish, typically accounting for > 70%–80% of the total fish biomass (Rowe et al. 1999). Their widespread availability, and the fact that they were easily caught and preserved, meant that the eels were a staple food item for New Zealand's indigenous Māori people (McDowall 2011; Jellyman 2014) who developed a sophisticated understanding of the various life history stages and associated means of capture. Capture by traditional means is still practiced in some regions today (Fig. 1).

While settlement of New Zealand by Māori dates back 700 years, occupation of Australia by Aboriginal people goes back perhaps 60,000 years. Aboriginals also developed a considerable knowledge of fish migrations, especially in terms of how they were influenced by the wet-dry seasons and their life history cycles. Aboriginals created what is probably the world's first aquaculture venture when 8000 years ago they flooded a series of natural wetlands and swamps at Lake Condah (Victoria) and linked these with a series of channels, weirs and dams in order to culture and catch eels. The production of eels from this 100 km^2 area is thought to have been sufficient to sustain up to 10,000 people (Builth 2002). Aboriginal capture methods included hooks and lures, "bobbing" (earthworms threaded on a string), traps and cages, and natural poisons from plants that were used to stupefy or kill eels (Pease and Carpenter 2004). In both countries, but especially New Zealand, eels feature prominently in the art, traditions and stories of the respective indigenous people.

Current status

There are different opinions about the taxonomic status of *A. australis*. Small but consistent morphological differences between western (Australian) and eastern (New Zealand) stocks (Schmidt 1928; Ege 1939; Watanabe et al. 2006) have been interpreted as indicating separate spawning areas, although results from

Fig. 1. Traditional means of capturing silver eels, Wairewa, New Zealand. (Top): blind channels dug into the gravel bar separating the lake from the sea. (Bottom): gaffing eels at night when they enter the trenches (photos courtesy Jack Jacobs, Wairewa, New Zealand).

allozyme and mtDNA analyses are equivocal, and have been interpreted as showing genetic homogeneity (Dijkstra and Jellyman 1999; Aoyama et al. 2000; Smith et al. 2001) and therefore panmixia. A more recent interpretation based on genetic variability of microsatellites is spawning segregation of eastern and western stocks, but with some genetic interchange between them (Shen and Tzeng 2007). Resolution of this issue is of importance in understanding whether silver eels from Australia might contribute towards recruitment of glass eels to New Zealand and/or vice versa.

Eel management in Australia is principally at a state level, whereas in New Zealand the management is national, although catch allowances are determined by region. The status of eels varies by state in Australia, with *A. australis* considered threatened in northern New South Wales (Baker 2010), but there is no equivalent concern for *A. reinhardtii*. However, most states regard their eel fisheries as fully exploited, and any future expansion in production must be through aquaculture (Baker 2010).

In New Zealand, there are no concerns at present about the status of *A. australis* (the Ministry for Primary Industries 2014), and a review of this species' conservation status by the Department of Conservation (Goodman et al. 2014) considered that it was not threatened. The same review (Goodman et al. 2014) classified *A. dieffenbachii* as "At Risk" and declining. Due to such concerns, a more comprehensive review of the species was undertaken (Parliamentary Commissioner for the Environment 2013), which made a number of recommendations designed to reduce the harvest of this species and to clarify its status. A subsequent expert review panel commissioned by the fishery managers (the Ministry for Primary Industries), concluded that *A. dieffenbachii* biomass was substantially reduced relative to pristine biomass, but since the late 2000s there has been a slowing and perhaps a halting of declines in regional catch trends (Haro et al. 2013). However, the review panel also noted limitations with current monitoring data and made recommendations for improvement.

Species ecology and geographic distribution

Like all *Anguilla* species, Australasian eels are marine spawners, and have a relatively long larval life ranging from 7 to 10 months (Arai et al. 1999; Marui et al. 2001). Glass eel recruitment to freshwater occurs in late winter-spring (Jellyman 2003); the exception is *A. reinhardtii*, which being a tropical species recruits year-round in the tropical part of its range (Beumer and Sloane 1990), but seasonally (March–May) in the temperate part (Sloane 1984a,b). Eels occupy a wide range of freshwater habitats including inshore marine, estuarine, lakes, streams and rivers. *A. australis* is essentially a still-water species (McDowall 1990), and commonly inhabits lowland streams and rivers, and coastal lakes and lagoons. In contrast, *A. dieffenbachii* principally inhabits flowing water habitats, and penetrates well inland in order to colonise high country rivers and lakes. *A. reinhardtii* mainly inhabits flowing rather than still waters, and is found in a variety of habitats ranging from river mouths to headwater streams (Pusey et al. 2004).

Juvenile eels, elvers, undertake substantial summer upstream migrations (Jellyman 1977; Sloane 1984a,b) until they reach ~ 30 cm and become more sedentary. Habitat requirements change with increasing size (Jellyman et al. 2003), but for all species and sizes, access to daytime refuge cover is essential.

Trends in commercial landings and abundance indices

About 80% of the commercial eel catch in Australia comes from Victoria, and *A. australis* comprises up to 95% of the catch in this state (Baker 2010). Annual catch in Victoria ranges from 125–450 t, with an average of ~ 280 t, with approximately half of this production resulting from translocation of juvenile eels to lakes and dams (Leahy et al. 2007). Many of these juvenile eels are sourced from the tailraces of hydro dams in Tasmania, and their release into Victorian waterways is carried out according to a series of guidelines and protocols (McKinnon 2006). The overall trend in reduced harvest in the 2000's (Fig. 2) is largely a reflection of persistent drought conditions that affected the harvest of both wild and transplanted eels (McKinnon 2006). Both glass eels and adult eels are harvested, but overseas export of glass eels is not allowed. The adult eel fishery is regarded as fully developed (Baker 2010), and much of the wild eel fishery targets silver eels of both species (McKinnon 2006). Management of the fishery is based on limited entry, gear restrictions, and allocation of waters to individual fishers. Almost half of the major rivers are closed to eel fishing, for reasons of stock conservation but also because of bycatch issues for air-breathing species like platypus and turtles.

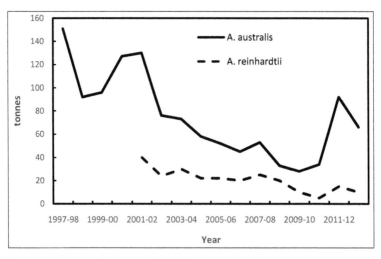

Fig. 2. Victorian commercial eel catch, 1997–2013.

Annual harvest in New South Wales is about 85 t (Department of the Environment 2013), which is twice that of Queensland and Tasmania; New South Wales has the largest fishery for *A. reinhardtii* in Australia (Pease and

Walford 2004). While in its infancy, the eel culture industry shows promise, with *A. reinhardtii* being the main species cultured in New South Wales with eels > 650 g receiving premium prices (McCormack 2014). Up to 300 kg of glass eels can be harvested in this state, although much culture is based on the growing of 200–300 g eels. In contrast, the most farmed eel in Queensland is *A. australis*, partly because it is similar in size to the local freshwater eel species in Japans (*A. japonica*) main market. Almost all wild-caught eels are exported live to Asia where good markets exist for large *A. reinhardtii*. The wild eel fishery in Queensland is managed via gear restrictions, minimum size (30 cm), and permanent closure of all natural waters meaning that fishing can only be carried out in impounded waters. Management in Tasmania is via closure of ~ 50% of natural waterways, limited licences, and protection of glass eels or silver eels (Department of the Environment 2004). Eels are uncommon in South Australia where issues like mouth closure of the Murray River are a major issue for migratory fish (Baker 2010).

The commercial eel fishery in New Zealand has a longer history, commencing in the early 1960's and growing rapidly to reach > 2000 t within a decade (Jellyman 2007). To cap this escalating catch, a series of national regulations were invoked including minimum size (currently 220 g by regulation, but effectively 300 g by voluntary industry agreement), restricted licences, development of regional management plans that included recognition of customary fisheries by Māori, and allocation of catch quota based on previous fishing history. An annual meeting of stakeholders (fishery and habitat managers, Māori, and representatives from the eel industry) makes recommendations to the Ministry for Primary Industries about management of the fishery, including any adjustments for commercial quota. Capture of glass eels is not allowed except for aquaculture trials, and capture of silver eels is only allowed in one lake. Trends in the total annual catch (Fig. 3) show a marked decline since the mid 1990's when increased restrictions were placed on the fishery.

In the absence of glass eel harvest, neither country has a time series of recruitment strength. The Waikato River in the North Island of New Zealand almost certainly receives the largest recruitment of any river in Australasia, and a comparison of trial glass eel catches in this river at an interval of 30 years indicated significant reduction in recruitment over this period (Jellyman et al. 2009). As a surrogate for national data on recruitment, New Zealand monitors trends in the abundance of the annual upstream migration of juvenile eels that are caught each summer below hydro dams and are transferred upstream. Analysis of these data are complicated by seasonally and annually varying hydro discharges, changes in fishing gear and effort, and variable cohort strength of eels. The most robust and longest-term data for two North Island sites (Fig. 4) show generally similar trends in the abundance of both *A. australis* and *A. dieffenbachii*, but considerable variability between years; numbers of *A. australis* always exceed *A. dieffenbachii*, but both species show episodic strong recruitment years.

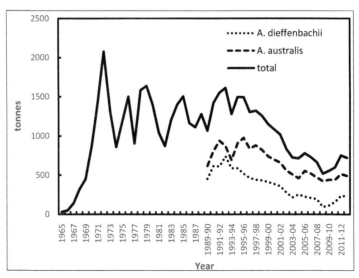

Fig. 3. New Zealand commercial eel catch, 1965–2012 (note that the reporting year changed in 1989).

Habitat modifications and threats

Along with differing size, Australia and New Zealand also contrast in terms of rainfall and river flows. New Zealand has abundant rainfall (although typically higher on west coasts than east coasts), while apart from coastal regions, much of Australia has limited rainfall. Associated habitat issues are changes to natural flow regimes (e.g., timing, magnitude, frequency and duration of flows; Pusey et al. 2004), and there are concerns that reduced net outflow might result in fewer glass eels being attracted to river mouths (Baker 2010). Other flow-related issues include water quality problems, river mouth closure, and up- and downstream access at weirs and dams. Much of the annual variability in commercial harvest is due to droughts, which may also result in fish kills due to elevated water temperatures, pH and salinity (Leahy et al. 2007) (Fig. 5). For example, in a lake in western Victoria, an estimated 50,000 eels died over two seasons due to drought (Leahy et al. 2007), while up to 100 t were estimated to have died in a stock-enhanced lake west of Melbourne in 1999 (McKinnon 2006).

A seasonal feature of some Australian lowland waterways is the so-called blackwater event. If flooding occurs after an extensive dry period when there has been a buildup of large amounts of organic material like leaf litter, much of this material can be washed into waterways where the consequent increased bacterial activity leads to a rapid depletion of dissolved oxygen (e.g., King et al. 2012). Native fish and crustaceans are particularly vulnerable to such events, although eels have a greater tolerance than other native species (McKinnon 2006). Management of water is a major environmental issue

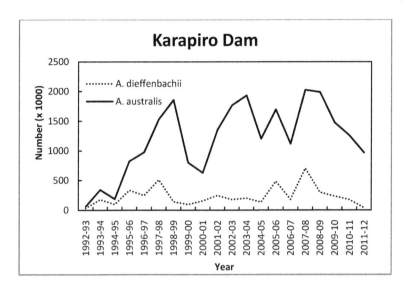

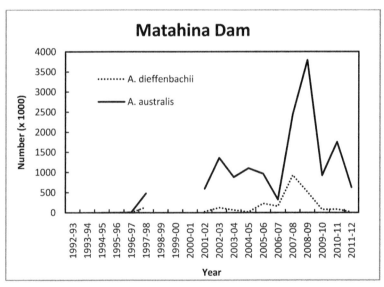

Fig. 4. Number of juvenile eels transferred each year at two hydro dams, North Island, New Zealand.

throughout Australia and excessive use of surface and groundwater has led to inadequate river flows, with consequent negative impacts on biodiversity, salinization of soils, and increased accumulation of contaminants. All such effects have a negative impact on eel habitats, as do the installation of water storage dams and weirs that affect eel movements.

New Zealand has lost 85% of its wetlands since European settlement around 200 years ago (the Ministry for the Environment 1997), representing

Fig. 5. Fish kill-dead eels, Lake Bolac, Victoria, Australia (photo courtesy Paul Leahy, Environment Protection Authority Victoria, Australia).

a huge loss of habitat especially for *A. australis*. Associated with this has been bankside channelization and installation of pumping stations that seldom have screens to exclude eels. Excessive growth of macrophytes in lowland waterways is variously controlled by mechanical dredging and cutting, and biocides, all of which result in major short-term loss or modification of habitat, especially for *A. australis*. Introduced willows (*Salix* spp.) provide important cover for eels, especially in many South Island waterways, but these trees can also impede flood flows and cause undesirable changes to river channels, and are often poisoned or removed by habitat managers. The biomass of larger eels has been directly related to the amount of suitable daytime cover (Burnet 1952). Practices like channelization and macrophyte removal reduce available cover, but an experiment in Australia that added logs and debris to a waterway resulted in an increase in species richness and abundance, although changes for eels (*A. reinhardtii*) were minimal (Brooks et al. 2004).

An estimated 10% of the total catchment area of the North Island and 22% of the South Island of New Zealand is affected by hydro development with associated problems for recruitment of juvenile eels and escapement of silver eels (Jellyman 2009). While all major hydro stations now have facilities for the capture and upstream transfer of juvenile eels, provision of downstream passage for silver eels is an ongoing problem (e.g., Boubée and Jellyman 2009). Small diameter bypasses (Boubée and Williams 2006), spillway openings (Watene and Boubée 2005) and catch-and-transfer of silver eels (Boubée et al. 2008) can be effective on a site-specific basis, but there is no universal solution to the problem of passage provision of silver eels; most migrations are associated with freshets and floods (Todd 1981; Boubée et al. 2001), when high flows and discoloured water make capture or diversion of silver eels difficult.

Establishment of reserve areas is a major component of eel conservation and management in both countries. Jellyman (1993) estimated that 36% of the total area of New Zealand lakes but < 1% of rivers were in some form of reserve, although most of the area of lakes was at high elevation within the National Parks, where habitat was limited and access sometimes compromised by downstream hydro development. More recently, Graynoth et al. (2008a) estimated that 49% of the total tonnage of *A. dieffenbachii* was either in reserves or in waterways that are rarely fished. As *A. australis* do not penetrate that far inland, they have been less impacted by river channelization and hydro development. Ironically, conversion of native forest to grassland habitat would have benefitted *A. australis* as biomass in pastoral catchments is significantly greater than in forested catchments (Rowe et al. 1999).

Growth rates of all species are relatively slow, and generation times are long. Mean annual growth rates from a river in Tasmania (calculated from data in Sloane 1984a,b) were 13.5 mm/year for *A. australis* and 21.5 mm/year for *A. reinhardtii*. Further north, in New South Wales, Walsh et al. (2006) recorded more rapid growth for *A. reinhardtii* of 30–60 mm/year. Growth rates in the North Island of New Zealand exceed those in the South Island—for example, the mean annual increment for *A. australis* has been estimated as 41 mm/year in the North Island and 33 mm/year in the South Island, with rates for *A. dieffenbachii* being 34 mm/year and 23 mm/year for North and South Islands respectively (Jellyman 2009). As a consequence of slow growth, the silver eels are relatively old at migration, with the range in the ages of female *A. reinhardtii* in Australia being 10–30 years (Walsh et al. 2004); the recorded range in the mean age of female eels in New Zealand is 13–41 for *A. australis* and 23–42 for *A. dieffenbachii* (Jellyman 2009).

Fyke nets are the method of capture used in the commercial fishery of both countries (Fig. 6). Bycatch of air-breathing animals like platypus and turtles is

Fig. 6. A fyke net full of eels, South Island, New Zealand—an uncommon sight these days (photo courtesy Vic Thompson, Mossburn Enterprises, New Zealand).

a matter of concern in Australia, and fyke nets must have a bycatch reduction device (BRD) fitted to minimise impacts on these non-target species; nets cannot be set fully submerged so that an air space is provided in the codend (Grant et al. 2004). In New Zealand, *A. australis* are particularly active during times of high water when they actively forage on terrestrial organisms in newly-inundated areas (Jellyman 1989), and this behaviour is exploited by fishers who often fish for *A. australis* around the margins of flooded lakes using unbaited nets to intercept foraging eels (Fig. 7). In contrast, *A. dieffenbachii* is fished using baited nets set along the edges of rivers; capture efficiency can be high—for instance, in a New Zealand river, capture efficiency (total catch on a single night/estimated total population) ranged from 55%–89% (Jellyman and Graynoth 2005). Obviously such efficient means of fishing can bring about significant changes in eel populations, and most commercial eel fishing is done on a rotational basis, with rotations varying between one and three years. Commercial catch quota in New Zealand is set for four fish stock areas in the North Island and six in the South Island, with additional quota set for customary and recreational catch—the allocation for customary catch is generally about 20% of the total allowable commercial catch.

Population modelling

No population modelling has been carried out in Australia. In New Zealand, various studies have been undertaken in order to review exploitation rates (Hoyle and Jellyman 2002), the natural mortality of juvenile eels (Graynoth et al. 2008b), and the status of longfin eels (Graynoth et al. 2008a; Fu et al.

Fig. 7. Retrieving large fyke net, Te Waihora/Lake Ellesmere, New Zealand.

2012). These studies have mainly been driven by ongoing concerns about the ability of long-lived fish like *A. dieffenbachii* to cope with an efficient harvesting regime. Trends in CPUE are used as a major monitoring tool in New Zealand. These trends vary by region, year and species, with major influences being the climate and the experience of fishers. In general, *A. australis* in the North Island show increasing CPUE, while *A. dieffenbachii* show initial declines and then a more constant level over the past decade. For the South Island, *A. australis* tends to show increased CPUE over recent years, while *A. dieffenbachii* show initial declines from the early 1990's and reach more stable situations post-2000 (Fig. 8).

Stock enhancement

Transplants of juvenile eels are widely carried out in both countries. Approximately one-third of the Australian commercial catch comes from transplants of juvenile eels to shallow lakes (Baker 2010). In New Zealand, transplants are made each year from below hydro dams, although inter-

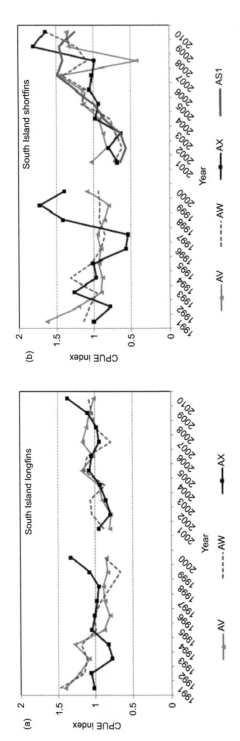

Fig. 8. Trends in South Island CPUE indices for *A. dieffenbachii* (panel a) and *A. australis* (panel b) for key management areas: Westland (AX), Otago (AV), Southland (AW), and Te Waihora/Lake Ellesmere (AS1). CPUE indices are the year effect from a GLM model of the relationship between daily catch and environmental variables. Reprinted from the Ministry for Primary Industries (2014).

catchment transfers are seldom allowed. Nationally, about 10 million juvenile eels are transferred, and this has resulted in the creation of viable commercial eel fisheries in hydro lakes, where upstream access has been compromised by dams. Successful trial transfers of sub-commercial sized *A. dieffenbachii* have been made to a high country lake, and resulting growth significantly exceeded that prior to transfer (Beentjes and Jellyman 2003).

Future prospects

The Australian states have generally adopted conservative means of eel management, and the fisheries are regarded as fully developed, with any future expansion anticipated to be from aquaculture. There are markets for both Australian species, although the market for *A. reinhardtii* requires a much larger eel. The wild eel fishery is largely dependent upon transplants of juvenile eels into shallow lakes, but projections of increased drought along the east coast of Australia raise concerns about the longterm viability of such ventures. Concerns in New Zealand are mainly associated with habitat protection, provision of up- and downstream access in rivers, and obtaining a better understanding of the status of the endemic *A. dieffenbachii*. Given the serious declines in the Northern hemisphere eel species (e.g., Dekker et al. 2003), there seems little likelihood of either country allowing the offshore sale of glass eels. Future emphasis in Australia is likely to focus on reducing mortality and improved passage of eels at dams and weirs, further transplants to suitable waterbodies while minimising fish kill events, promotion of aquaculture, and establishing biological reference points to trigger management actions (Baker 2010). In New Zealand, future emphasis will be on the development of regional eel management plans, improving the quality of the data used to monitor the fisheries, especially trends in recruitment, and conservative management especially of *A. dieffenbachii*.

Acknowledgements

Thanks to Paul Leahy, Environment Protection Authority Victoria, Australia, for permission to use Fig. 5, and to Jack Jacobs (Wairewa runanga) and Vic Thompson (Mossburn Enterprises) for permission to use Figs. 1 and 6 respectively.

Keywords: commercial fishery, customary fishery, Māori, Aboriginal, stock status, conservation, habitat modification, hydro, wetlands, enhancement

References

Aoyama, J., S. Watanabe, M. Nishida and K. Tsukamoto. 2000. Discrimination of catadromous eels of the genus *Anguilla* using polymerase chain reaction—restriction fragment length polymorphism analysis of the mitochondrial 16S ribosomal RNA domain. Trans. Am. Fish. Soc. 129: 873–878.

Arai, T., T. Otake, D.J. Jellyman and K. Tsukamoto. 1999. Differences in the early life history of the Australasian shortfinned eel *Anguilla australis* from Australia and New Zealand, as revealed by otolith microstructure and microchemistry. Mar. Biol. 135: 381–389.

Baker, J.L. 2010. Anguillidae. *In*: J.L. Baker (ed.). Marine Species of Conservation Concern in South Australia. Vol. 1. Bony and Cartilaginous Fishes. Reef Watch, Conservation Council of SA. http://www.conservationsa.org.au/files/marine/samcc/Anguillidae_FINAL_Nov_2012. pdf.

Beentjes, M.P. and D.J. Jellyman. 2003. Enhanced growth of longfin eels, *Anguilla dieffenbachii* transplanted into Lake Hawea, a high country lake in South Island, New Zealand. N.Z. J. Mar. Freshw. Res. 37: 1–11.

Beumer, J. and R. Sloane. 1990. Distribution and abundance of glass-eels *Anguilla* spp. in east Australian waters. Int. Rev. gesamten Hydrobiol. 75: 721–736.

Boubée, J. and D. Jellyman. 2009. Facilitating the upstream and downstream passage of indigenous fish at large dams. pp. 57–63. *In*: E. McSaveney (ed.). Dams—Operating in a Regulated Environment. IPENZ Proc. Tech. Groups 35/1.

Boubée, J., D. Jellyman and C. Sinclair. 2008. Eel protection measures within the Manapouri hydro-electric power scheme, South Island, New Zealand. Hydrobiol. 609: 71–82.

Boubée, J.A., C.P. Mitchell, B.L. Chisnall, D. West, E.J. Bowman and A. Haro. 2001. Factors regulating the downstream migration of mature eels (*Anguilla* spp.) at Aniwhenua Dam, Bay of Plenty, New Zealand. N.Z. J. Mar. Freshw. Res. 35: 121–134.

Boubée, J.A.T. and E.K. Williams. 2006. Downstream passage of silver eels at a small hydroelectric facility. Fish. Manag. Ecol. 13: 165–176.

Brooks, A.P., P.C. Gehrke, J.D. Jansen and T.B. Abbe. 2004. Experimental reintroduction of woody debris on the Williams River, NSW: geomorphic and ecological responses. River Res. Applic. 20: 513–536.

Builth, H. 2002. The Archaeology and Sociology of Gunditjmara: A Landscape Analysis from Southwest Victoria. Unpublished Ph.D. thesis, Flinders University, Adelaide.

Burnet, A.M.R. 1952. Studies on the ecology of the New Zealand longfinned eel, *Anguilla dieffenbachii* Gray. Australia. J. Mar. Freshw. Res. 3: 32–63.

Dekker, W., J.M. Casselman, D.K. Cairns, K. Tsukamoto, D.J. Jellyman and H. Lickers. 2003. Worldwide decline of eel resources necessitates immediate action. Quebec declaration of concern. Fisheries 28: 28–30.

Department of the Environment. 2004. Recommendations to the Inland Fisheries Service on the ecologically sustainable management of the Tasmanian freshwater eel fishery. Internal report. 2p.

Department of the Environment. 2013. New South Wales managed fisheries: estuary general fishery—supporting document for reassessment. Australian Government, Department of the Environment. http://www.environment.gov.au/system/files/pages/fd8874ac-0fd5-4338-99ec-c9236e93ac82/files/application-2013-supporting-document.pdf.

Dijkstra, L.H. and D.J. Jellyman. 1999. Is the subspecies classification of the freshwater eels *Anguilla australis australis* Richardson and *A. a schmidtii* Phillipps still valid? Mar. Freshw. Res. 50: 261–263.

Ege, V. 1939. A revision of the genus *Anguilla* Shaw. A systematic, phylogenetic and geographical study. Dana Rep. 3: 1–256.

Fu, D., M.P. Beentjes and A. Dunn. 2012. Further investigations into the feasibility of assessment models for New Zealand longfin eels (*Anguilla dieffenbachii*). Final Research Report for the Ministry of Fisheries Project EEL200702: 1–77.

Goodman, J.M., N.R. Dunn, P.J. Ravenscroft, R.M. Allibone, J.A.T. Boubée, B.O. David, M. Griffith, N. Ling, R.A. Hitchmough and J.R. Rolfe. 2014. Conservation status of New Zealand freshwater fish, 2013. New Zealand Threat Classification Series 7. Department of Conservation, Wellington. 12p.

Grant, T.R., M.B. Lowry, B. Pease, T.R. Walford and K. Graham. 2004. Reducing the by-catch of platypuses (*Ornithorhynchus anatinus*) in commercial and recreational fishing gear in New South Wales. Proc. Linnean Soc. N.S.W. 125: 259–272.

Graynoth, E., D.J. Jellyman and M.L. Bonnett. 2008a. Spawning escapement of female longfin eels. N.Z. Fisheries Assess. Rep. 2008/7: 1–57.

Graynoth, E., R.I.C.C. Francis and D.J. Jellyman. 2008b. Factors influencing juvenile eel (*Anguilla* spp.) survival in lowland New Zealand streams. N.Z. J. Mar. Freshw. Res. 42: 153–172.

Haro, A., W. Dekker and N. Bentley. 2013. Independent review of the information available for monitoring trends and assessing the status of New Zealand freshwater eels. Report to New Zealand Ministry for Primary Industries. 19p.

Hoyle, S.D. and D.J. Jellyman. 2002. Longfin eels need reserves: modelling the effects of commercial harvest on stocks of New Zealand eels. Mar. Freshw. Res. 53: 887–895.

Jellyman, D.J. 1977. Summer upstream migration of juvenile freshwater eels in New Zealand. N.Z. J. Mar. Freshw. Res. 11: 61–71.

Jellyman, D.J. 1989. Diet of two species of freshwater eel (*Anguilla* spp.) in Lake Pounui, New Zealand. N.Z. J. Mar. Freshw. Res. 23: 1–10.

Jellyman, D.J. 1993. A review of the fishery for freshwater eels in New Zealand. New Zealand Freshwater Research Report 10. Christchurch, NIWA. 51p.

Jellyman, D.J. 2003. The distribution and biology of the South Pacific species of *Anguilla*. pp. 275–292. In: K. Aida, K. Tsukamoto and K. Yamauchi (eds.). Eel Biology. Springer, Tokyo.

Jellyman, D.J. 2007. Status of New Zealand freshwater eel stocks and management initiatives. ICES J. Mar. Sci. 64: 1379–1386.

Jellyman, D.J. 2009. Forty years on—the impact of commercial fishing on stocks of New Zealand freshwater eels (*Anguilla* spp.). Am. Fish. Soc. Symp. 58: 37–56.

Jellyman, D.J. 2014. Freshwater eels and people in New Zealand—a love/hate relationship. pp. 143–153. In: K. Tsukamoto and M. Kuroki (eds.). Eels and People. Springer-Verlag, Tokyo.

Jellyman, D.J. and E. Graynoth. 2005. The use of fyke nets as a quantitative capture technique for freshwater eels (*Anguilla* spp.) in rivers. Fish. Manage. Ecol. 12: 237–247.

Jellyman, D.J. and M. Bowen. 2009. Modelling larval migration routes and spawning areas of Anguillid eels of New Zealand and Australia. Am. Fish. Soc. Symp. 69: 255–274.

Jellyman, D.J. and K. Tsukamoto. 2010. Vertical migrations may control maturation in migrating female *Anguilla dieffenbachii*. Mar. Ecol. Prog. Ser. 404: 241–247.

Jellyman, D.J., M.L. Bonnett, J.R.E. Sykes and P. Johnstone. 2003. Contrasting use of daytime habitat by two species of freshwater eel (*Anguilla* spp.) in New Zealand rivers. Am. Fish. Soc. Symp. 33: 63–78.

Jellyman, D.J., D.J. Booker and E. Watene. 2009. Recruitment of glass eels (*Anguilla* spp.) in the Waikato River, New Zealand. Evidence of declining migrations? J. Fish. Biol. 74: 2014–2033.

King, A.J., Z. Tonkin and J. Lieshcke. 2012. Short-term effects of a prolonged blackwater event on aquatic fauna in the Murray River, Australia: considerations for future events. Mar. Freshw. Res. 63: 576–586.

Leahy, P., A. Leonard and R. Johnson. 2007. Findings of Western Victorian lakes eel death investigation 2004–06. Scientific Report, Publication 1114. EPA, Victoria, Australia. 36p. http://beachreport.epa.vic.gov.au/~/media/Publications/1114.pdf.

Marui, M., T. Arai, M.J. Miller, D.J. Jellyman and K. Tsukamoto. 2001. Comparison between the early life history between New Zealand temperate eels and Pacific tropical eels revealed by otolith microstructure and microchemistry. Mar. Ecol. Prog. Ser. 213: 273–284.

McCormack, R. 2014. Eels. New South Wales Aquaculture Association Inc. http://nswaqua.com.au/index.php?s=eels.

McDowall, R.M. 1990. New Zealand Freshwater Fishes: A Natural History and Guide, Heinemann-Reed, Auckland.

McDowall, R.M. 2011. Ikawai. Freshwater fishes in Māori culture and economy, Canterbury University Press, Christchurch.

McKinnon, L.J. 2006. A Review of Eel Biology: Knowledge and Gaps. Report to EPA Victoria, Audentes Investments Pty. Ltd.

Ministry for the Environment. 1997. The state of New Zealand's Environment 1997. The Ministry for the Environment. Paged in sections.

Ministry for Primary Industries. 2014. Fisheries Assessment Plenary, November 2014: stock assessments and stock status. Compiled by the Fisheries Science Group, Ministry for Primary Industries, Wellington, New Zealand. 618p.

Parliamentary Commissioner for the Environment. 2013. On a pathway to extinction? An investigation into the status and management of the longfin eel. Report, Parliamentary Commissioner for the Environment. 95p.

Pease, B. and H. Carpenter. 2004. Cultural significance of the indigenous fishery. pp. 108–110. *In*: B.C. Pease (ed.). Description of the Biology and An Assessment of the Fishery for Adult Longfinned Eels in NSW. NSW Department of Primary Industries, Fisheries Final Report Series No. 69.

Pease, B. and T. Walford. 2004. Commercial fisheries. pp. 94–107. *In*: B.C. Pease (ed.). Description of the Biology and An Assessment of the Fishery for Adult Longfinned Eels in NSW. NSW Department of Primary Industries, Fisheries Final Report Series No. 69.

Pusey, B., M. Kennard and A. Arthington. 2004. Freshwater Fishes of North-Eastern Australia. CSIRO Publishing, Melbourne.

Rowe, D.K., B.L. Chisnall, T.L. Dean and J. Richardson. 1999. Effects of land use on native fish communities in east coast streams of the North Island of New Zealand. N.Z. J. Mar. Freshw. Res. 33: 141–151.

Schmidt, J. 1928. The fresh-water eels of Australia with some remarks on the shortfinned species of *Anguilla*. Rec. Australia. Mus. 5: 179–201.

Shen, K.N. and W.-N. Tzeng. 2007. Genetic differentiation among populations of the shortfinned eel *Anguilla australis* from East Australia and New Zealand. J. Fish Biol. 70 Suppl. B: 177–190.

Sloane, R.D. 1984a. Upstream migration by young pigmented freshwater eels (*Anguilla australis australis* Richardson) in Tasmania. Aust. J. Mar. Freshw. Res. 35: 61–73.

Sloane, R.D. 1984b. Distribution, abundance, growth and food of freshwater eels (*Anguilla* spp.) in the Douglas River, Tasmania. Aust. J. Mar. Freshw. Res. 35: 325–337.

Smith, P.J., P.G. Benson, C. Stanger, B.L. Chisnall and D.J. Jellyman. 2001. Genetic structure of New Zealand eels *Anguilla dieffenbachii* and *A. australis* with allozyme markers. Ecol. Freshw. Fish 10: 132–137.

Todd, P.R. 1981. Timing and periodicity of migrating New Zealand freshwater eels (*Anguilla* spp.). N.Z. J. Mar. Freshw. Res. 15: 225–235.

Walsh, C.T., B.C. Pease and D.J. Booth. 2004. Variation in the sex ratio, size and age of longfinned eels within and among coastal catchments of south-eastern Australia. J. Fish Biol. 64: 1297–1312.

Walsh, C.T., B.C. Pease, S.D. Hoyle and D.J. Booth. 2006. Variability in growth of longfinned eels among coastal catchments of south-eastern Australia. J. Fish Biol. 68: 1693–1706.

Watanabe, S., J. Aoyama and K. Tsukamoto. 2006. Confirmation of morphological differences between *Anguilla australis australis* and *A. australis schmidtii*. N.Z. J. Mar. Freshw. Res. 40: 325–331.

Watene, E.M. and J.A.T. Boubée. 2005. Selective opening of hydroelectric dam spillway gates for downstream migrant eels in New Zealand. Fish. Manage. Ecol. 12: 69–75.

15

Fisheries, Stocks Decline and Conservation of Anguillid Eel

Wann-Nian Tzeng[1,2]

Introduction

There are 19 species of freshwater eels (Genus *Anguilla*) in the world, which are all catadromous fishes, spawning in the open ocean but growing in continental freshwaters, estuaries, and lagoons (Tesch 2003). Freshwater eels are divided into temperate and tropical species. The temperate eels, *A. anguilla*, *A. rostrata*, and *A. japonica*, in the northern hemisphere have declined to less than 10% of their population level in the 1970s in the past 3 decades. The reason for their decline is not clear but might be due to global climate change, overfishing, and habitat degradation, etc. (e.g., Knights 2003, Tzeng et al. 2012, Chen et al. 2014). Because of their population decline, the catch of temperate glass eels is not enough to meet the demand for aquaculture and thus the tropical glass eels have been used to supplement the temperate glass eels, particularly the Japanese eel, for aquaculture. To avoid the over-exploitation of tropical glass eels for aquaculture, both the Philippine and Indonesian governments have prohibited the export of tropical glass eels to Asian countries for aquaculture. For sustainable eel aquaculture industry and eel resources conservation, we need to understand the fishery status, the population decline, and why the eel couldn't recover its population size by self-regulation? The eel has been proposed to be able to adjust its sex ratio to maximize population level

[1] Institute of Fisheries Science, National Taiwan University, Taipei 10617, Taiwan.
[2] Department of Environmental Biology and Fisheries Science, National Taiwan Ocean University, Keelung 20224, Taiwan.
  E-mail: wnt@ntu.edu.tw

(Colombo and Rossi 1978). Over-exploitation of the temperate eel resources may have reduced the eel population to a level too low to recover (Quinn and Deriso 1999). This chapter mainly reviews the biology and fishery of the Japanese eel in Taiwan to examine possible conservation methods for the eel.

Evolution and geographical distribution

The anguillid eel is a catadromous fish, spawning in the open ocean but growing in continental freshwaters, estuaries, and lagoons (Tesch 2003). There are 19 species/subspecies of anguillid eels (Watanabe et al. 2009). The existing species *Anguilla borneensis* found in Indonesian waters was the progenitor of the 19 current eel species (Minegishi et al. 2008). All of the eel species originated in the Indo-Pacific Ocean (Fig. 1). *A. anguilla* and *A. rostrata* evolved from an ancient Indo-pacific species that migrated through the equatorial corridor known as the "Tethys Sea" approximately 100 million years ago (Aoyama et al. 2001).

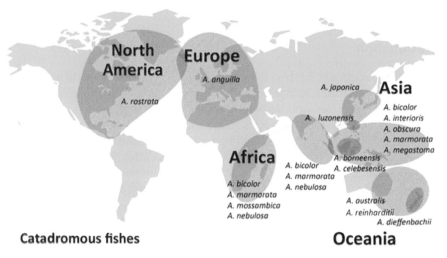

Fig. 1. Geographical distribution of the 19 species and subspecies of anguillid eels (Photo updated from Dr. Mari Kuroki).

The temperature preference differs among eel species. Thus the 19 eel species are divided into temperate and tropical species. *A. anguilla, A. rostrata, A. japonica, A. australis australis, A. australis schmidtii* and *A. dieffenbachii* are temperate species. The remaining 13 species/subspecies are tropical species (Fig. 1). *A. marmorata* is the most abundant tropical species with five sub-populations widely-distributed in the tropical Indian and Pacific Oceans

(Minegishi et al. 2008). The subspecies *A. bicolor pacifica* and *A. bicolor bicolor* are genetically different and occur in the Pacific Ocean and Indian Ocean, respectively. Two other nominal subspecies *A. australis australis* and *A. australis schmidtii* are distributed in the east Australian and New Zealand waters, and are probably two sub-populations that have not evolved to the subspecies level (Shen and Tzeng 2007).

The geographical distribution of the eel species depends on their spawning grounds, pelagic larval duration (PLD), larval transportation by oceanic currents, and other factors (Han et al. 2012; Kimura et al. 1994; Leander et al. 2013; Tsukamoto 1992, 2009; Wang and Tzeng 2000). The PLD plays an important role in determining the dispersal latitude and range of the eel population. For example, *A. japonica* and *A. marmorata* are two sympatric species in the western North Pacific that have their leptocephalus larvae transported by the North Equatorial Current from the same spawning ground in the waters west of the Mariana Islands (Tsukamoto 1992) and then transported by the Kuroshio Current along the Philippines coasts. However, the geographic distribution differs between these two species. *A. japonica* has a longer PLD and thus delays metamorphosis from leptocephalus to glass eel stage such that few *A. japonica* glass eels recruit to the Philippines and most *A. japonica* glass eels recruit to Taiwan, mainland China, Korea, and Japan (Chen and Tzeng 1996; Leander et al. 2013). In contrast, *A. marmorata* has a shorter PLD and is able to metamorphose from leptocephalus to glass eel earlier and then recruit to the Philippines. Similarly, the difference in PLD also occurs in *A. anguilla* and *A. rostrata* that leads to *A. anguilla* recruiting to Europe while *A. rostrata* recruits to central and north America although these two species spawn in the Sargasso Sea in the Atlantic Ocean (Wang and Tzeng 2000).

Five species are found in Taiwan (Tzeng 1982, 2014; Tzeng and Tabeta 1983; Leander et al. 2012). The Japanese eel *A. japonica* is the most abundant and important aquaculture species, followed by the giant mottled eel *A. marmorata* the three tropical species *A. celebesensis*, *A. luzonensis*, and *A. bicolor pacifica* are rare (Tzeng et al. 1995). The riverine habitat use differs among species; the temperate *A. japonica* prefers estuarine waters while the tropical *A. marmorata* prefers deeper pools in the upper reaches of rivers (Shiao et al. 2003).

Environmentally sex-determined hypothesis: a key life history strategy of eel

The eel doesn't have a sex chromosome. The sex of the eel is environmentally determined. The biological minimum size (length at first maturity or silvering) differs between males and females. The females tend to maximize their size while males minimize their age (Fig. 2). The life history strategy of the eel has evolved in order to use their food resources and space effectively. The eel tends to become female when the population level is lower and becomes male when the population level is higher. This unique life history strategy enables rapid recovery of the eel population when food resources and space are limited.

Fig. 2. Sex dimorphism in size of the American eel at silvering (Photo from Dr. Guy Verreault).

The sex of the eel is undifferentiated until the yellow eel stage after recruitment to the river. When the biomass of the eel is much lower than the carrying capacity of the ecosystem, the eel tends to become female and to become male when its population biomass approaches the upper limit of the carrying capacity of the environment (Fig. 3). This life history strategy adjusts to ecosystem carrying capacity by adjusting the sex ratio because the size at maturity is smaller in males than in females, thus permitting more males than females to occupy a given habitat. When the population biomass increases, the food resources and space available may not be enough to provide for the demands of the females. Thus, becoming male saves energy, improves survival, and reduces intra-specific competition.

In contrast, if population biomass decreases, the food resources and space are plentiful to supply the demands of the larger-sized female, enabling more fecundity and greater reproductive capacity so as to rapidly increase the eel population. The age-minimization life history strategy of males and size-maximization of females effectively uses the food resources and space in the growth-phase habitat to maximize the reproductive potential of the eel.

The sex ratio of the eel population is labile. An investigation on the sex ratios of the Japanese eel in the Kao-Ping River of Taiwan found that the wild Japanese eel population in the river was skewed to female during the period from 1998–June 2004, however, the sex ratio became male-dominated between July and October 2004 (Fig. 4). The sex ratio reversal of the Japanese

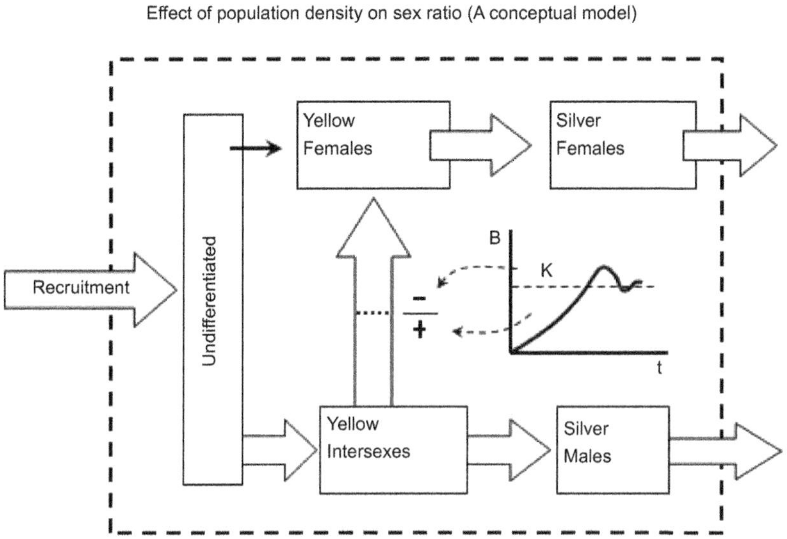

Effect of population density on sex ratio (A conceptual model)

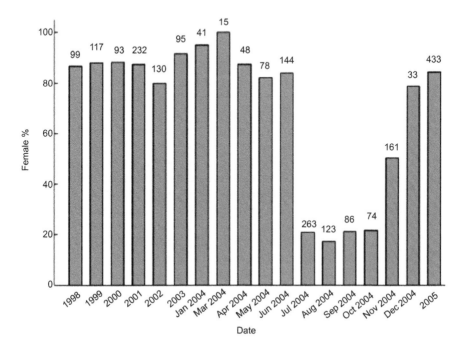

Fig. 3. Environmentally sex-determined hypothesis for anguillid eels. The sex of the eel is undifferentiated until recruitment to the continental growth habitat. B*t*: biomass; K: carrying capacity of the ecosystem. If B*t* is less than K, the sex tends to be female and male if B*t* is larger than K (modified from Colombo and Rossi 1978).

Fig. 4. Temporal changes in the sex ratio of Japanese eels in the Kao-Ping River of Taiwan during the period from 1998–2005 (Chu et al. 2006). Numerals in the figure indicate sample size.

eel from female-dominated to male-dominated might have resulted from the escapement of male-dominated eels from aquaculture ponds during typhoon Mindulle in July of 2004 (Table 1). However, the male-dominated population was maintained for only 4–5 months (Fig. 4), probably due to dispersion of the cultured eels in the river and/or due to overfishing by fishermen. The ratio of wild male eels was lower before and after the introduction of cultured male eels, indicating that the eels tend to differentiate into males in the intensively cultured ponds, but differentiated into females in the Kao-Ping River where population density was low. In other words, the sex ratio (female/male + female %) of eels was inversely related to population density (Han et al. 2010).

Table 1. Comparison of sex ratio between wild and cultured Japanese eel (Chu et al. 2006).

	Sample size (*n*)	Number of eels		Female percentage (%)
		Male	Female	
W1	1023	137	886	86.6
W2	96	22	74	77.1
WC	577	467	110	19.1
C	34	28	6	18.0

W1: wild yellow eel before July 2004; W2: wild yellow eel between July–November 2004; WC: yellow eel with blue-gray back between July–November 2004; C: cultured eel in 2004.

Plasticity of habitat-use

The Sr/Ca ratios in the otoliths of eels have been validated as reliable biological indicators to reconstruct the past environmental history of the eel migrating between freshwater and seawater (Tzeng 1996). The investigation of Sr/Ca ratios in otoliths of the Japanese eel in the Kao-Ping River estuary in southwest Taiwan (Fig. 5) indicated that the migratory environmental history of the eel in the continental life stage can be classified into three types: Type 1: Seawater contingent, Type 2: Freshwater contingent, and Type 3: Estuarine contingent (Fig. 6). The majority of the Japanese eels recruited to the continental growth habitat in Taiwan were of the estuarine contingent (75%), 22% were of the fresh water contingent, and 3% were of the seawater contingent, which can skip the fresh water life stage and complete the life cycle in seawater (Lin et al. 2012). This migration pattern has been called semi-catadromy (Tzeng et al. 2000b) or facultative catadromy (Tsukamoto and Arai 2001) (Fig. 7). A similar facultative migratory pattern has been found in American and European eels in different habitats (Tzeng et al. 1997; Jessop et al. 2002, 2008; Daverat et al. 2006; Thibault et al. 2007; Panfili et al. 2012; Capoccioni et al. 2014).

The habitat use of Japanese eels is also sex-dependent (Han and Tzeng 2007). The growth rate of the eel differs between sexes and among contingents (Tzeng et al. 2000a, 2003). Fishery management should consider the sex-dependent habitat use of the eel.

Fig. 5. Sampling sites (Red arrows) of Japanese eel in the Kao-Ping River estuary, southwestern Taiwan.

Fisheries and fishing gears

The eel is harvested at all life stages after recruitment to the continental growth habitat. Based on life stage, the eel fisheries can be divided into three categories, i.e., glass eel, yellow eel, and silver eel fisheries (Fig. 8). The behavior and habitat of the eel differs among life stages. Thus, the design of fishing gears and fishing methods differ among fisheries.

Glass eel fisheries

Glass eels are harvested in the estuaries, surf zones, and coastal waters for aquaculture during their upstream migration. Eels are nocturnal and are always fished at night during a flood tide. Thus, the fishing activity and daily catch per unit effort reaches the maximum around the new moon period (Tzeng 1985). Glass eels are caught by different fishing gears, depending upon the country and the fishing site (Fig. 9). The fishing gear has evolved from a small-scale hand net to a large-scale boat-driven net. Fyke nets are commonly used in Europe, for example in Ireland, and are set against the flood tide to catch glass eels migrating upstream in the estuary. Dip nets are used in the USA and UK. Triangle nets are used in China and Taiwan. The push-nets moved

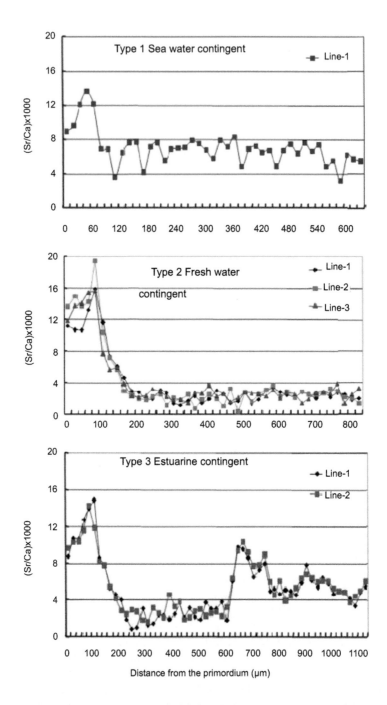

Fig. 6. Migratory environmental history as indicated by the temporal changes of the Sr/Ca ratio in the otoliths of Japanese eels. Type 1: Seawater contingent; Type 2: Freshwater contingent; Type 3: Estuarine contingent. Lines 1–2 are replicate measurements (Tzeng et al. 2003).

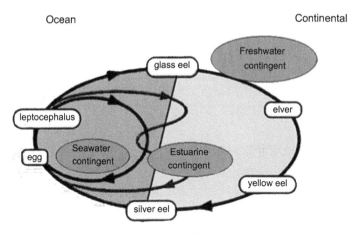

Fig. 7. The facultative migration pattern of anguillid eels (from Prof. Katsumi Tsukamoto).

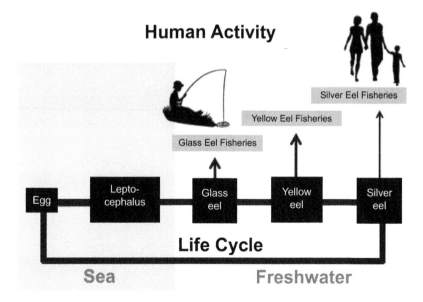

Fig. 8. Eels are harvested at the glass eel, yellow eel, and silver eel stages (Photo updated from Dr. Mari Kuroki).

by boat are used in France, Taiwan and Japan. Set nets are popularly used in China and Taiwan. The set net is a passive fishing gear used to catch glass eels migrating upstream in the estuary during the night flood tide. Because many set nets are often set together at the river mouth, boat passage may be disturbed.

Due to the high demand for glass eels for aquaculture, the glass eel has been over-exploited, particularly the Japanese and European glass eels. For example the annual mean catch of glass eels of the Japanese eel in Taiwan

Fig. 9. contd....

Fig. 9. contd.

Fig. 9. contd....

Fig. 9. contd.

Fig. 9. contd....

Fig. 9. contd.

Fig. 9. Fishing gears for glass eels in different countries. (a) Fyke nets used in Ireland (Photo from Prof. Kieran McCarthy), (b) Hand net with steel frame in Taiwan (Photo from Prof. Wann-Nian Tzeng), (c) Triangle nets in China and Taiwan (Photo from Prof. Shuo-zeng Dou), (d) Set nets in China and Taiwan (Photo from Prof. Shuo-zeng Dou), (e) Dip net in UK (Photo from Dr. David Righton), (f and g) Large-scale nets moved by boat in France and Taiwan (Photos from Dr. Eric Feunteun and Dr. Yu-San Han), (h) Glass eel fishing boats with light operated in the night in Japan (Photo from Prof. Shuo-zeng Dou).

has been less than 10 tonnes in the past 3 decades; however, the annual mean demand for glass eels for restocking (aquaculture) has exceeded 20 tonnes (Fig. 10). The high demand for glass eels for aquaculture has led to their over-exploitation in Taiwan, where approximately 50%–75% of the glass eels migrating up stream are fished in the estuary (Tzeng 1984). The over-exploitation of glass eels occurs not only in Taiwan but also in other Asian countries and is a major problem for eel conservation.

Yellow eel fisheries

Yellow eels are harvested in the rivers, estuaries, and lagoons during their growth phase where they search for food organisms such as shrimp and other small creatures. The fishing gears used to harvest the yellow eel are very diverse (Fig. 11), ranging from small-scale traditional eel tube and eel pots to large-scale modern nets. The eel tube is made of bamboo and popularly used in the tropical Taiwan where abounds with bamboo. Eel pots are widely used in Europe and New Zealand, and fyke nets are widely used in Europe (Tesch 2003). Snake nets are used in Taiwan. Four-armed scoop nets are used in Japan.

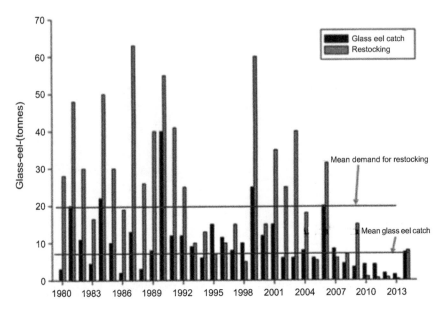

Fig. 10. The mean catch and demand of glass eels for restocking in Taiwan (updated from Dr. Yu-San Han).

The snake net in Taiwan was originally used for catching shrimps. Eels caught in the snake net are by catch when eels chase shrimps for food. Almost no professional fishers and fisheries for yellow eel occur in Taiwan because there are few eels in the rivers. Electric fishing gear was used in Taiwan in the past, but it has been prohibited many years ago.

Silver eel fisheries

Silver eels are harvested during their downstream spawning migration in autumn. The fishing gear used to harvest the silver eel varies widely among countries (Fig. 12). A classical eel bucket and wicker baskets that can be lowered into a flowing river to catch silver eels were used in the nineteenth century UK. An eel weir (*pd-tuna*) was used to catch migrating silver eels in New Zealand. Traps were set in the gaps along the wooden fence. The fishing gear used to catch migrating silver eels (mostly for *A. marmorata*) in Taiwan is very similar to that used in New Zealand, which was set at a small dam to catch the downstream migrating *A. marmorata* during floods in the autumn rainy season. However, the silver eel fishery in Taiwan no longer exists after the eel population declined in the 1980s.

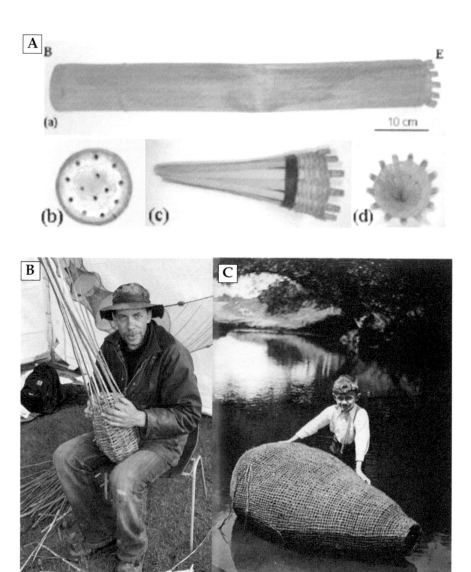

Fig. 11. contd....

Fig. 11. contd.

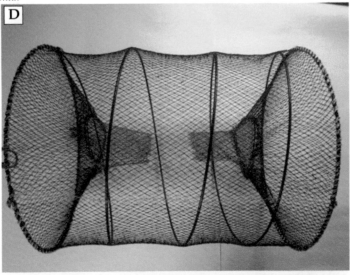

Fig. 11. contd....

Fig. 11. contd.

Fig. 11. contd....

Fig. 11. contd.

Fig. 11. Fishing gears for yellow eel in different countries. (A) Eel tube in Taiwan, (a) external view (B, bottom; E, entrance), (b) bottom view, holes for water flowing into the tube to diffuse the smell from the live bait "earthworm" to lure eels entering the tube from the funnel-like entrance, (c) and (d) side and top view of the entrance. (B) and (C) Eel pots used in UK and New Zealand (Photos from Dr. David Righton and Dr. Mari Kuroki), (D) Eel cage in Taiwan, (E) Fyke nets in UK (Photo from Dr. David Righton), (F) and (G) Snake nets in Taiwan, releasing (F) and recovery (G) of the net. (H) Electric fishing gear in Taiwan, and (I) Four-armed scoop nets in Japan (Photo from Dr. Mari Kuroki).

Aquaculture vs. capture fishery productions

The long-term trends in eel production are inversely correlated between aquaculture and the capture fishery (Fig. 13). The annual aquaculture production for both European eels and Japanese eels has rapidly increased annually since 1970. However, the annual production of the capture fishery for wild yellow and silver eels for both European and Japanese eels has decreased annually since 1970. Because the artificial propagation of glass eels has not yet reached commercial scale (Prof. Katsumi Tsukamoto, personal communication), the aquaculture industry completely relies on the wild-caught glass eels. The increasing demand for glass eels for aquaculture may lead to decreased glass eel recruitment to the river and subsequently a decreased eel population and production in the capture fishery.

Fig. 12. Fishing gears for silver eel in different countries. (a) A classical eel bucket and wicker baskets used in the nineteenth century UK (Photo from Dr. David Righton), (b) An eel weir used to catch migrating silver eels in New Zealand (Photo from Dr. Don Jellyman).

Decreased yellow and silver eel abundances in rivers will then cause a decreased escapement of spawners for reproduction which then leads to reduced recruitment of glass eels in the next year. The fishing of glass eels for

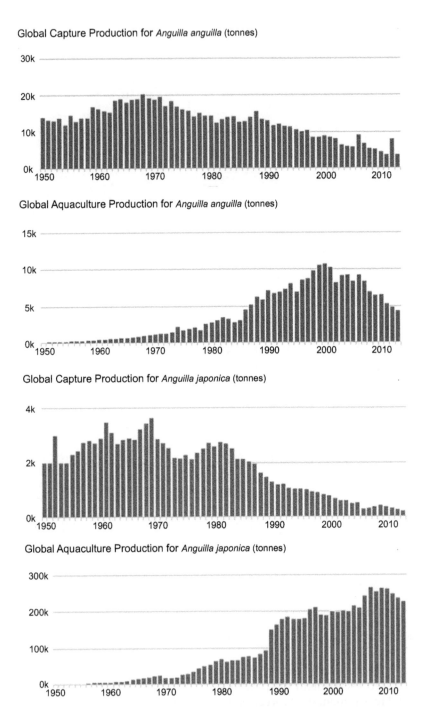

Fig. 13. The decreasing trend of capture fishery production versus the increasing trend of aquaculture production for European eels (*Anguilla anguilla*) and Japanese eels (*Anguilla japonica*) in Europe and Asia (data from FAO).

aquaculture in the estuary and the fishing of yellow eel and silver eel in the river is a chicken- and egg problem and should be controlled for the sustainable use of the eel resources.

Population decline of temperate eels

In the most recent 3 decades, the recruitment of the temperate-zone European eel *A. anguilla* and Japanese eel *A. japonica* has declined to less than 10% of their population level in the 1970s as has the recruitment of the American eel *A. rostrata* in the upper St. Lawrence River system, with lesser declines elsewhere (Fig. 14). The data have not been updated for European glass eel recruitment for the last few years because controls on their fisheries mean that the data are no longer comparable. The ICES/Working Group on Eel has used GLM approaches to estimate recent recruitment and notes recent increases (see http://www.ices.dk/sites/pub/Publication%20Reports/Expert%20 Group%20Report/acom/2013/WGEEL/wgeel_2013.pdf). Although the recruitment statistics of the temperate eels have been compiled in different ways, there is no doubt that the recruitment of these three temperate eels has dramatically declined in the most recent 3 decades. However, it is unknown if the tropical eels have also declined like the temperate eels because their catch data are unavailable.

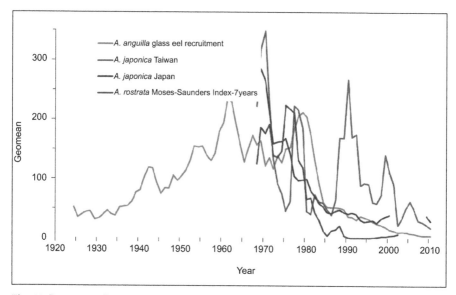

Fig. 14. Long-term fluctuations of the recruitment (3-year running average of indices as % of the 1979–1994 means, adjusted to the years of arrival) for the European eel, American eel, and Japanese glass eels from Japan and Taiwan. The Moses-Saunders index refers only to the upper St. Lawrence River and Lake Ontario and is not indicative of the degree of decline elsewhere in North America (Data partially from Prof. Brian Knights).

Regionally, the annual fluctuation of recruitment of Japanese eels has been greater in Taiwan than in Japan (Fig. 14). This may result from Taiwan being located at the marginal distribution of the Japanese eel population. In general, the fluctuation of a population is greater at the margin than at the center of the distribution of the population.

Possible causes of population decline

The causes of the decline of the eel populations are not clear but are believed to be due to some combination of global climate change (Tzeng et al. 2012), overfishing, habitat degradation, swim bladder parasite infection, pollution, and other factors. The recruitment of European eels has dramatically decreased since 1980 but stock abundance has declined less sharply, possibly because of increased density-dependent survival (Fig. 15). Spawning escapements may have declined proportionate to stock abundance, indicating that the recruitment might not completely depend on stock abundance or spawner escapement. The impact of global climate change and regime shifts such as the NAO (North Atlantic Oscillation) might also play an important role in the historical decline of recruitment (Knights 2011).

The recruitment of Japanese eels, as indicated by the catches of glass eels, has also dramatically decreased since the 1970s, but its stock, as represented by the catches of yellow and silver eels, has gradually decreased (Fig. 16). The declining pattern for the Japanese eel is very similar to that for the European

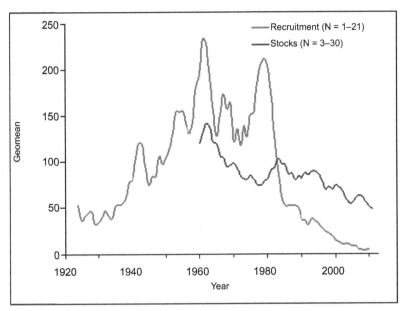

Fig. 15. Historical changes of the stock and recruitment of European eel (Prof. Brian Knights' unpublished data).

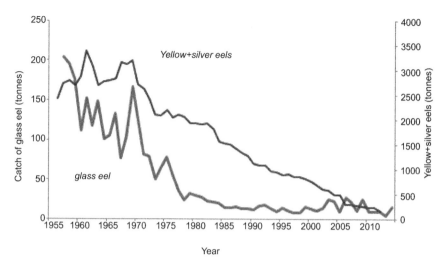

Fig. 16. Catches of glass eels, and yellow and silver Japanese eels in Japan (data from Prof. Katsumi Tsukamoto).

eel. The stable decrease of the stocks might be because the stock is composed of multiple age groups, but the recruitment is of single year classes and because continental survival has increased relative to the eel recruitment decline. If recruitment fails, it will immediately appear as a decreasing pattern.

The annual catches of the glass eels of the Japanese eel *Anguilla japonica* from Taiwan during the period from 1972–2014 (Fig. 17), were analyzed by their trend, and by autocorrelation and cross-correlation with the number of sunspots to understand whether the population of the Japanese eel is climbing back up the slippery slope from low stock abundance. The annual catch data of the Japanese glass eels was compiled from the monthly reports in the Taiwan Fisheries Yearbook (Fisheries Agency, Council of Agriculture, Executive Yuan, Republic of China), of the daily catches collected by the district Fisheries Association of Taiwan. The annual catch of Japanese glass eels was calculated from July to June of the following year, because the recruitment season of glass eels in Taiwan occurs mainly from October to April of the next year, and peaks between November and February (Tzeng 1985).

The catch of glass eels has decreased since the 1970s ($p < 0.01$), with the peak catches occurring in an 11-year period ($p = 0.02$), which corresponds to the maximal number of sunspots over the 11.2 year sunspot cycle. The glass eel catch (Fig. 17) and the number of sunspots were positively cross-correlated at time lags of –1, 0, and 1 year, respectively ($p = 0.027$, 0.006, and 0.01). The peak catch of glass eel has decreased from 40 tonnes in 1979, to 31.3 tonnes (1990), 18.3 tonnes (2001), and to 8.0 tonnes in 2014 (Fig. 17).

The decreasing trend of annual glass eel catches indicates that Japanese eel recruitment is still in decline. The 2014 increase in catch is due to a climate effect, not population recovery.

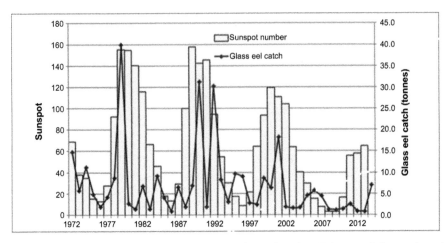

Fig. 17. Correlation between the annual Japanese glass eel catch in Taiwan and the number of sunspots.

Regarding the impact of habitat degradation on the eel, Chen et al. (2014) used chronological Landsat imagery to measure Japanese eel habitat loss from human activities in 16 rivers in East Asia, including Japan, Korea, Taiwan, and China. On average, 76.8% of the effective habitat area (Ae) was lost in these 16 rivers from the 1970s to the 2010. Taiwan and China had the highest percentages of Ae loss, with declines of 49.3% and 81.5%, respectively. Extensive habitat loss may play an important role, together with regional climate change and overfishing, in the decline of the Japanese eel in East Asia.

Eels in the IUCN Red List

The population status of the 19 anguillid eel species was assessed by the London Zoological Society (LZS) in July 2014 in London. The assessment results and the Red List of the eel species are available on the website of the International Union for Conservation of Nature (IUCN) (http://www.iucnredlist.org/). Eel species were assigned to different status categories of the IUCN (Fig. 18) as follows:

1. Four species were 'threatened' due to 'population' decline
   Vulnerable (VU)—*Anguilla borneensis*
   Endangered (EN)—*A. japonica*; *A. rostrata*
   Critically Endangered—*A. anguilla*

2. Five were found to be Near Threatened (NT)
   A. bengalensis
   A. bicolor
   A. celebesensis
   A. luzonensis (limited range)

3. Two were Least Concern (LC)
   A. marmorata
   A. mossambica
4. Three were Data Deficient (DD)
   A. interioris
   A. megastoma
   A. obscura
5. Five were Not Evaluated (NE) due to data not available.

In fact, the European eel was added to the global list of endangered species of CITES (Convention on International Trade in Endangered Species, Washington, 1973) Appendix II in 2007. A permit is now required from CITES for the export of live European eels and eel products from the European Union since 2010. *A. japonica* and *A. rostrata* will be reviewed to see if they should be added to the global list of endangered species of CITES in 2016. The other species are in the status of Near Threatened (NT), Least Concern (LC), Data Deficient (DD) and Not Evaluated (NE) where insufficient data were available. The assessment of the eel for conservation has not been completed and requires continuous evaluation.

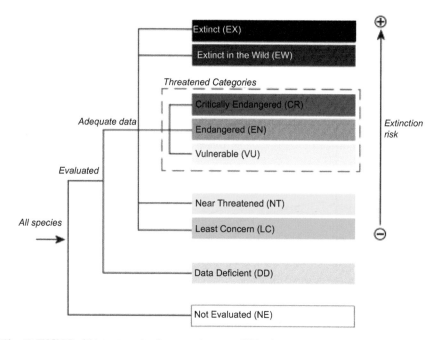

Fig. 18. IUCN Red List categories for assessing anguillid eels.

Stock assessment models

The reason(s) for the population declines of the above-mentioned three temperate eel species is not clear, but might be due to global climate change, habitat degradation, and overfishing, amongst other factors. It is not easy to slow down the pace of global climate change to support population recovery, but it is possible to ameliorate habitat degradation and to prevent overfishing by regulation of the fisheries.

Two models, the yield-per-recruit (YPR) and spawning stock biomass-per-recruit (SPR) models have been used to evaluate whether the yellow and silver eels of the local stock of Japanese eel (*Anguilla japonica*) in the Kao-Ping River in southwestern Taiwan were overfished (Lin et al. 2010). The current fishery status and uncertainty of biological reference points (e.g., F_{max}, $F_{0.1}$, $F_{40\%}$, and $F_{50\%}$) were assessed using Monte Carlo simulation. The probabilities that the current fishing mortality rates for yellow and silver eels (F_{cur}) were higher than F_{max} ranged from 15%–20%, and for being higher than $F_{0.1}$ ranged from 58%–66%. The probabilities of F_{cur} being higher than $F_{40\%}$ and $F_{50\%}$ ranged from 26%–93% and from 73%–99%, respectively. This indicated that the Japanese eel in the lower reaches of the Kao-Ping River appeared to be fully exploited in terms of YPR (Fig. 19). This local stock is also subject to a high risk of recruitment overfishing because the contribution of the SPR of the local stock to the entire Japanese eel stock is possibly below a minimal acceptable level (40%) (Fig. 20). Estimates of $F_{0.1}$ and F_{max}, and to a lesser extent those of $F_{40\%}$ and $F_{50\%}$, were relatively robust, but estimates of the maximal YPR and SPR at F_{cur} were highly sensitive to the uncertainty of the input parameters.

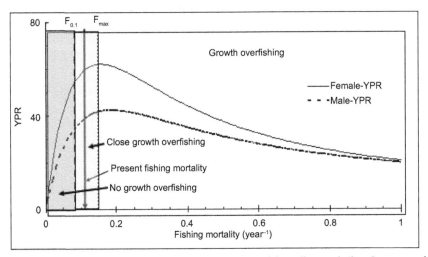

Fig. 19. Yield-per-recruit (YPR) model used in evaluation of the yellow and silver Japanese eels (*Anguilla japonica*) in the Kao-Ping River in southwestern Taiwan (Photo from Dr. Yu-Jia Lin).

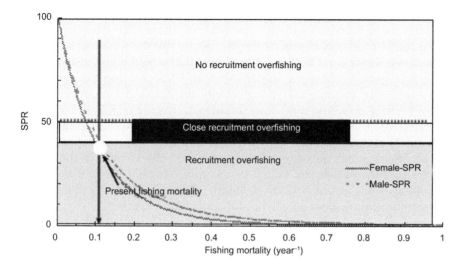

Fig. 20. Spawning stock biomass-per-recruit (SPR) model used in evaluation of the yellow and silver Japanese eels (*Anguilla japonica*) in the Kao-Ping River in southwestern Taiwan (Photo from Dr. Yu-Jia Lin).

Management approaches

Many management measures have been implemented to prevent the continuous decline of the temperate eel populations and to avoid the over-exploitation of new tropical glass eel fishery resources for aquaculture. Management measures include restrictions on fishing effort and season, improved upstream and downstream fish passage at obstructions, export restrictions, and restocking.

Management measures for the European eel in Europe

The European eel stock is at a historical low level and has been in constant decline at an alarming rate with a recruitment situated between 1% and 5% of that observed in the 1970s. The European eel has been included in Appendix II of the Convention on International Trade in Endangered Species (CITES) since 2007. For recovery of the European eel population, the European Union considered retaining 35% of the glass eels caught for release within EU territory and increasing the release to 60% until 2013 and letting at least 40% of silver eels escape from inland waters to the sea (Refer to the report from the commission to the council and European Parliament, Brussels, 21.10.2014 COM (2014) 640 final). The Eel Regulation obliges Member States with river basins in their national territory that constitute habitats for the eels, to establish and implement Eel Management Plans (EMPs). These EMPs should contain management measures to ensure the escapement to the sea of at least 40% of adult eels relative to the escapement levels that would have existed in the absence of human influences. In particular, an EMP may contain measures, such as reducing commercial fishing activity, restricting recreational fishing, restocking measures, structural measures to make rivers passable and improve river habitats, transportation of silver eels from inland waters, combatting predators, measures related to hydro-electric power turbines, aquaculture, as well as any other measure necessary for the achievement of the above mentioned 40% escapement target. In addition, Member States that permit the fishing of eels less than 12 cm in length (glass eels) have an obligation to reserve 60% of their glass eel catches for restocking purposes.

Management measures for the Japanese eel in Asia

Under the structure of the Asia-Pacific Economic Cooperation (APEC) agreement, Japan, Korea, China, and Taiwan established the non-government organization Alliance for Sustainable Eel Aquaculture (ASEA) in 2014 to increase collaboration in capture fishery and aquaculture, ecology and resource studies, and resource management, and to discuss effective laws for the implementation of eel resource conservation and management. These conservation measures include: (1) resource restoration, including prohibited fishing areas and seasons and effective restocking, (2) trade transparency to ensure open access to catch, production, aquaculture, and trade information, (3) promotion of trade normalization and prohibition of smuggling behavior, and (4) improved aquaculture control by reducing the Japanese glass eel aquaculture capacity. Accordingly the restocking of Japanese glass eel in 2015 would not be greater than 80% of that in 2014 (Table 2) and the restocking of exotic glass eels such as *A. Anguilla, A. rostrata, A. marmorata, A. bicolor pacifica,* and *A. mossambica* would not be greater than the level of the most recent 3 years (Table 3). The importation of the exotic glass eels for aquaculture might affect the conservation and management of those species if their importation is not controlled.

Table 2. The annual restocking (tonnes) of Japanese glass eels for aquaculture among Asian countries.

	2014	2015
Japan	27.9	21.6
Korea	13.9	11.1
Taiwan	12.5	10.0
China	45.0	36.0
Total	98.4	78.7

Table 3. The annual restocking (tonnes) of exotic glass eels for aquaculture among Asian countries.

	2012	2013	2014	2015
Japan	0.4	1.3	2.5	2.6
Korea	6.0	13.1	2.9	13.1
Taiwan	5.5	10.0	1.5	10.0
China	14.5	20.0	25.0	25.0
Total	26.43	44.4	31.4	50.7

The management approach for the Japanese eel differs among Asian countries. In Japan, the catch of juvenile eels is prohibited and a special license for catch is issued in order to capture seed for farming or research within a limited fishing period. Since 2013, the capture of silver eel has been prohibited or restricted in three of the principal glass-eel-fishing prefectures (Miyazaki, Kumamoto, and Kagoshima) with the aim of preserving spawning eels in these productive areas. Restocking rivers and lakes with young eels for stock enhancement has also been carried out for many years in Japan and studies on its effectiveness are currently being conducted.

In Taiwan, the Fishery Agency, Council of Agriculture stipulated that the Japanese glass eel fishing season be closed from March 1 to October 31 in the 13 prefectures of Taiwan except for Fa-Lien and Tai-Tung prefectures where few Japanese eel have been distributed since 2012, and that each prefecture select at least one river where adult eel fishing would be prohibited year-round. The I-Lan prefecture, which is the most productive area of the glass eel fishery, even prohibited whole rivers for fishing adult eels.

In addition, the release of eels to natural waters is considered an effective way to promote the recovery of eel populations. The Taiwan Fisheries Research Institute has implemented an eel release program since 1976 by releasing hormone-injected silver eels offshore and young eels in the river to promote the escapement of spawners. However, the release of young and adult eels seems not to reflect the catch of glass eels (Fig. 21). Because the Japanese eel is a panmictic population and exploited by Japan, Korea, China, and Taiwan, a release program needs international collaboration to efficiently promote the effects of the program.

The larval/oceanic phases of the *A. japonica* life history are by far the most well studied of all the anguillids. However, there are still significant gaps in our knowledge of this species that, coupled with their broad range and multiple life stages, makes conservation measures difficult to implement. In 2009 however, the Eel River Project was set up by the East Asia Eel Resource Consortium (EASEC) and was designed to sample and monitor glass eel recruitment year round across a number of key localities in Japan, mainland China, and Taiwan. Current resources appear to be focused on the aquaculture industry, although some recent studies have begun to address the stock status of the Japanese eel in Taiwan (e.g., Lin et al. 2010) and the ecology of reared eels returned to their natural habitat (e.g., Lin et al. 2012).

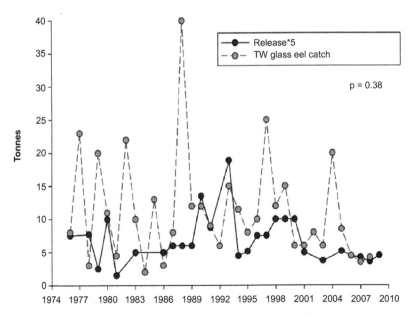

Fig. 21. Time series changes in the released amounts of young and silver eels and the glass eel catch in Taiwan. The correlation between release and catch was not significant (p = 0.38). Before 2003, hormone-injected silver eels were released offshore; after 2003, young eels were released in the river.

Management measures for the tropical eels in Indonesia and the Philippines

The anguillid eel in tropical areas is characterized by high species diversity that is believed to be due to the complicated current system and geomorphology of the archipelago countries, of which Indonesia and the Philippines are representative. There are 10 species and subspecies of tropical eels found in Indonesian waters including *A. marmorata, A. bicolor bicolor, A. borneensis, A. interioris, A. celebesensis, A. bicolor pacifica, A. obscura, A. megastoma, A. luzonensis* and one new species (Dr. Hagi Yulia Sugeha, personal communication), and 9 species reported in the Philippines, i.e., *A. luzonensis, A. celebesensis, A. malgumora, A. marmorata, A. japonica, A. bicolor pacifica, A. bicolor bicolor, A. bengalensis bengalensis* and *A. australis australis* (Dr. Apolinario Yambot, personal communication). The species composition varies by country and habitat. *A. marmorata* is dominant and *A. luzonensis* is an endemic species in the Philippines, but *A. celebesensis* is dominant in Indonesia (Sugeha 2010; Sugeha et al. 2001).

The tropical eels in Indonesia and the Philippines are now targeted as the next candidates for aquaculture to compensate for the shortage of Japanese glass eels in East Asia. To avoid the over-exploitation of tropical glass eels for export for aquaculture use, the Philippine government placed a ban on the export of glass eels (Refer to Fisheries Administrative No. 242, Series of 2012, Department of Agriculture, Republic of the Philippines) that prohibits anguillid eels less than 15 cm in total length from being exported from the Philippines. However, in consideration of the livelihood of fishermen and the country's economic development, the Indonesian Institute of Sciences (LIPI) recommended an eel regulation that was issued by the Ministry of Marine and Fisheries of the Government of Indonesia in 2009 to prohibit Indonesian tropical eels of less than 300 grams from export. This law was to encourage foreign eel aquaculture companies to culture the glass eel within Indonesia up to market size and then to process the eel for local markets and/or export and thus to avoid the illegal smuggling of tropical glass eels from Indonesia to East Asia for aquaculture. The species cultured in Indonesia is mostly *A. celebesensis, A. marmorata* and *A. bicolor bicolor.*

On the other hand, information on the current fisheries, population status, and recruitment level of the eel in Indonesia and the Philippines is lacking in respect to eel conservation except for a few studies on species composition, early life history, and recruitment behaviour of the glass eel (Arai et al. 2003; Briones et al. 2007; Jamandre et al. 2007; Leander et al. 2013; Sugeha 2010; Sugeha et al. 2001). To avoid over exploitation and population collapse, it is necessary to strengthen the studies for the conservation of tropical eels.

Conclusion and suggestions

The populations of temperate eels such as the Japanese eel, European eel, and American eel have declined over the past 3 decades. As with other animals, eel populations should have a compensation mechanism in the spawner-recruit relationship such that when the population level is lower, i.e., at a lower spawner abundance (S), the recruits per spawner (R/S) should increase as spawner abundance increases (Quinn and Deriso 1999). The European eel population has been managed by increasing the escapement of spawners for several years but its population level is still very low. It is necessary to continuously manage the fisheries of glass eels, yellow eels, and silver eels so as to increase the escapement and subsequent recruitment of the eel. If the spawner-recruit relationship becomes depensatory, the eel stocks may not recover after being fished to a very low abundance, even when fishing is stopped (Quinn and Deriso 1999). This will be very dangerous. We should avoid overfishing to prevent the eel stock from becoming critically low in abundance.

On the other hand, most of the growth habitat of the eel species has been lost, such as that of Japanese eel where 76.8% of the effective habitat area of the eel was lost during the 1970s–2010 in China (Chen et al. 2014). In addition, the recruitment of the eel is easily affected by global climate changes (Knights 2003; Tzeng et al. 2012). Accordingly, habitat restoration and protection and the potential effects of climate change may need to be integrated into the management planning for the anguillid eel resources.

Acknowledgements

I thank the Ministry of Science and Technology (former National Science Council) of the Republic of China for long-term financial support for the eel research, my previous students of National Taiwan University for assistance on the research, Brian Knights for providing unpublished data, Shuo-zeng Dou. Eric Feunteun, Yu-San Han, Don Jellyman, Mari Kuroki, Kieran McCarthy, David Righton, Katsumi Tsukamoto and Guy Verreault for providing photos and Brian M. Jessop for comments on the various drafts of the manuscript.

Keywords: Eel species, life history characteristics, fishing gear, aquaculture and wild fisheries, population decline, yield-per-recruit model and spawner-per-recruit model, management approaches

References

Aoyama, J., M. Nishida and K. Tsukamoto. 2001. Molecular phylogeny and evolution of the freshwater eel, genus *Anguilla*. Mol. Phylogenet. Evol. 20: 450–459.

Arai, T., M.J. Miller and K. Tsukamoto. 2003. Larval duration of the tropical eel *Anguilla celebesensis* from Indonesia and Philippines coasts. Mar. Ecol. Prog. Ser. 251: 255–261.

Briones, A.A., A.V. Yambot, J.-C. Shiao, Y. Iizuka and W.-N. Tzeng. 2007. Migratory pattern and habitat use of tropical eels *Anguilla* spp. (Teleostei: Anguilliformes: Anguillidae) in the Philippines, as revealed by otolith microchemistry. Raffles B Zool. 141–149.

Capoccioni, F., D.-Y. Lee, Y. Iizuka, W.-N. Tzeng and E. Ciccotti. 2014. Phenotypic plasticity in habitat use and growth of the European eel (*Anguilla anguilla*) in transitional waters in the Mediterranean area. Ecol. Freshw. Fish 23: 65–76.

Chen, J.-Z., S.-L. Huang and Y.-S. Han. 2014. Impact of long-term habitat loss on the Japanese eel *Anguilla japonica*. Estuarine Coast. Shelf Sci. 151: 361–369.

Cheng, P.-W. and W.-N. Tzeng. 1996. Timing of metamorphosis and estuarine arrival across the dispersal range of the Japanese eel *Anguilla japonica*. Mar. Ecol. Prog. Ser. 131: 87–96.

Chu, Y.-W., Y.-S. Han, C.-H. Wang, C.-F. You and W.-N. Tzeng. 2006. The sex-ratio reversal of the Japanese eel *Anguilla japonica* in the Kaoping River of Taiwan: The effect of cultured eels and its implication. Aquaculture 261(4): 1230–1238.

Colombo, G. and R. Rossi. 1978. Environmental influence on growth and sex ratio in different eels populations (*Anguilla anguilla* L.) of Adriatic coasts. pp. 313–320. *In*: D.S. McLusky and A.J. Berry (eds.). Physiology and Behaviour of Marine Organisms. Proceedings of the 12th European Symposium on Marine Biology, Stirling, Scotland, September 1977. Pergamon Press, Oxford and New York.

Daverat, F., K.E. Limburg, I. Thibault, J.-C. Shiao, J.J. Dodson, F. Caron, W.-N. Tzeng, Y. Iizuka and H. Wickström. 2006. Phenotypic plasticity of habitat use by three temperate eel species *Anguilla anguilla*, *A. japonica* and *A. rostrata*. Mar. Ecol. Prog. Ser. 308: 231–241.

Han, Y.-S. and W.-N. Tzeng. 2007. Sex-dependent habitat use by the Japanese eel *Anguilla japonica* in Taiwan. Mar. Ecol. Prog. Ser. 338: 193–198.

Han, Y.-S., C.-L. Hung, Y.-F. Liao and W.-N. Tzeng. 2010. Population genetic structure of the Japanese eel *Anguilla japonica*: panmixia at spatial and temporal scales. Mar. Ecol. Prog. Ser. 401: 221–232.

Han, Y.-S., H. Zhang, Y.-H. Tseng and M.-L. Shen. 2012. Larval Japanese eel (*Anguilla japonica*) as sub-surface current bio-tracers on the East Asia continental shelf. Fish. Oceanogr. 21: 281–290.

Jamandre, B.W.D., K.-N. Shen, A.V. Yambot and W.-N. Tzeng. 2007. Molecular phylogeny of Philippine freshwater eels *Anguilla* spp. (Actinopterygi: Anguilliformes: Anguillidae) inferred from mitochondrial DNA. Raffles B Zool. 51–59.

Jessop, B.M., J.-C. Shiao, Y. Iizuki and W.-N. Tzeng. 2002. Migratory behaviour and habitat use by American eels *Anguilla rostrata* as revealed by otolith microchemistry. Mar. Ecol. Prog. Ser. 233: 217–229.

Jessop, B.M., D.K. Cairns, I. Thibault and W.-N. Tzeng. 2008. Life history of American eel *Anguilla rostrata*: new insights from otolith microchemistry. Aquat. Biol. 1: 205–216.

Kimura, S., K. Tsukamoto and T. Sugimoto. 1994. A model for the larval migration of the Japanese eel: roles of the trade winds and salinity front. Mar. Biol. 119: 185–190.

Knights, B. 2003. A review of the possible impacts of long-term oceanic and climate changes and fishing mortality on recruitment of anguillid eels of the Northern Hemisphere. Sci. Total Environ. 310: 237–244.

Knights, B. 2011. Eel biology and the status of recruitment and stocks. pp. 22–40. *In*: D.A. Bunt and A.M. Don (eds.). Proc. Conf. Eel Management and the State of the Art. Inst. Fish. Mgmt.

Leander, N.J., K.-N. Shen, R.-T. Chen and W.-N. Tzeng. 2012. Species composition and seasonal occurrence of recruiting glass eels (*Anguilla* spp.) in the Hsiukuluan River, eastern Taiwan. Zool. Stud. 51(1): 59–71.

Leander, N.J., W.-N. Tzeng, N.-T. Yeh, K.-N. Shen, R.-T. Chen and Y.-S. Han. 2013. Effects of metamorphosis timing and the larval growth rate on the latitudinal distribution of sympatric fresh water eels, *Anguilla japonica* and *Anguilla marmorata*, in the western North Pacific 52(1): 30–45.

Lin, S.-H., Y. Iizuka and W.-N. Tzeng. 2012. Migration behavior and habitat use by juvenile Japanese eels *Anguilla japonica* in continental waters as indicated by mark-recapture experiments and otolith microchemistry. Zool. Stud. 51(4): 442–452.

Lin, Y.-J., Y.-J. Chang, C.-L. Sun and W.-N. Tzeng. 2010. Evaluation of the Japanese eel fishery in the lower reaches of the Kao-Ping River, southwestern Taiwan using a per-recruit analysis. Fish. Res. 106: 329–336.

Minegishi, Y., J. Aoyama and K. Tsukamoto. 2008. Multiple population structure of the giant mottled eel, *Anguilla marmorata*. Mol. Ecol. 17: 3109–3122.

Panfili, J., A.M. Darnaude, Y.-J. Lin, M. Chevalley, Y. Iizuka, W.-N. Tzeng and A.J. Crivelli. 2012. Habitat residence during continental life of the European eel *Anguilla anguilla* investigated using linear discriminant analysis applied to otolith Sr:Ca ratio. Aquat. Biol. 15: 175–185.

Quinn, T.J. II and R.B. Deriso. 1999. Quantitative Fish Dynamics. Oxford University Press, New York.

Shen, K.N. and W.-N. Tzeng. 2007. Genetic differentiation among populations of the shortfinned eel *Anguilla australis* from East Australia and New Zealand. J. Fish Biol. 70 (Supplement B): 177–190.

Shiao, J.-C., Y. Iizuka, C.-W. Chang and W.-N. Tzeng. 2003. Disparities in habitat use and migratory behavior between tropical eel *Anguilla marmorata* and temperate eel *A. japonica* in four Taiwanese rivers. Mar. Ecol. Prog. Ser. 261: 233–242.

Sugeha, H.Y. 2010. Recruitment mechanism of the tropical eels Genus *Anguilla* in the Poso estuary, central Sulawesi Island, Indonesia. Jurnal Perikanan (J. Fish. Sci.) XII(2): 86–100.

Sugeha, H.Y., T. Arai, M.J. Miller, D. Limbong and K. Tsukamoto. 2001. Inshore migration of the tropical eels *Anguilla* spp. recruiting to the Poigar River estuary on north Sulawesi Island. Mar. Ecol. Prog. Ser. 221: 233–243.

Teng, H.-Y., Y.-S. Lin and C.-S. Tzeng. 2009. A new *Anguilla* species and a reanalysis of the phylogeny of freshwater eels. Zool. Stud. 48: 808–822.

Tesch, F.-W. 2003. The eel (5th ed.). Blackwell Publishing Ltd., Oxford, UK. 408pp.

Thibault, I., J.J. Dodson, F. Caron, W.-N. Tzeng, Y. Iizuka and J.-C. Shiao. 2007. Facultative catadromy in American eels: testing the conditional strategy hypothesis. Mar. Ecol. Prog. Ser. 344: 219–229.

Tsukamoto, K. 1992. Discovery of the spawning area for Japanese eel. Nature 356: 789–791.

Tsukamoto, K. 2009. Oceanic migration and spawning of anguillid eels. J. Fish Biol. 74: 1833–1852.

Tsukamoto, K. and T. Arai. 2001. Facultative catadromy of the eel *Anguilla japonica* between freshwater and seawater habitats. Mar. Ecol. Prog. Ser. 220: 265–276.

Tzeng, W.-N. 1982. New record of the elver, *Anguilla celebesensis* Kaup, from Taiwan. Chinese Bioscience 19: 57–66 (in Chinese with English abstract).

Tzeng, W.-N. 1984. An estimate of the exploitation rate of *Anguilla japonica* elvers immigrating into the coastal waters off Shuang-Chi River, Taiwan. Bull. Inst. Zool., Academia Sinica 23(2): 173–180.

Tzeng, W.-N. 1985. Immigration timing and activity rhythms of the eel, *Anguilla japonica*, elvers in the estuary of northern Taiwan with emphasis on environmental influences. Bull. Jap. Soc. Fish. Oceanogr. No. 47/48: 11–28.

Tzeng, W.-N. 1996. Effects of salinity and ontogenetic movements on strontium: calcium ratios in the otoliths of the Japanese eel, *Anguilla japonica* Temminck and Schlegel. J. Exp. Mar. Biol. Ecol. 199: 111–122.

Tzeng, W.-N. 2014. Freshwater eels and humans in Taiwan. pp. 129–142. *In*: K. Tsukamoto and M. Kuroki (eds.). Humanity and the Sea: Eels and Humans. Springer, Japan.

Tzeng, W.-N. and O. Tabeta. 1983. First record of the short-finned eel *Anguilla bicolor pacifica* elvers form Taiwan. Bull. Jap. Soc. Sci. Fish. 49(1): 27–32.

Tzeng, W.-N., P.-W. Cheng and F.-Y. Lin. 1995. Relative abundance, sex ratio and population structure of the Japanese eel *Anguilla japonica* in the Tanshui River system of northern Taiwan. J. Fish Biol. 46: 183–201.

Tzeng, W.-N., K.P. Severin and H. Wickström. 1997. Use of otolith microchemistry to investigate the environmental history of European eel *Anguilla anguilla*. Mar. Ecol. Prog. Ser. 149: 73–81.

Tzeng, W.-N., H.-R. Lin, C.-H. Wang and S.-N. Xu. 2000a. Differences in size and growth rates of male and female migrating Japanese eels in Pearl River, China. J. Fish Biol. 57(5): 1245–1253.

Tzeng, W.-N., C.-H. Wang, H. Wickström and M. Reizenstein. 2000b. Occurrence of the semi-catadromous European eel *Anguilla anguilla* (L.) in the Baltic Sea. Mar. Biol. 137: 93–98.

Tzeng, W.-N., Y. Iizuka, J.-C. Shiao, Y. Yamada and H.P. Oka. 2003. Identification and growth rates comparison of divergent migratory contingents of Japanese eel (*Anguilla japonica*). Aquaculture 216: 77–86.

Tzeng, W.-N., Y.-H. Tseng, Y.-S. Han, C.-C. Hsu, C.-W. Chang, E.D. Lorenzo and C.-H. Hsieh. 2012. Evaluation of multi-scale climate effects on annual recruitment levels of the Japanese eel, *Anguilla japonica*, to Taiwan. PLoS ONE 7(2): e30805. doi: 10.1371/journal.pone.0030805.

Wang, C.-H. and W.-N. Tzeng. 2000. The timing of metamorphosis and growth rates of American and European eel leptocephali—a mechanism of larval segregative migration. Fish. Res. 46: 191–205.

Watanabe, S., J. Aoyama and K. Tsukamoto. 2009. A new species of freshwater eel *Anguilla luzonensis* (Teleostei: Anguillidae) from Luzon Island of the Philippines. Fish. Sci. 75: 387–392.

Index

temperate eel species 316
temperate eels 1, 11, 291, 292, 311, 316, 317, 321
temperate species 111–114, 117, 119, 130, 131
temperate-zone 311
temporal distributions 55
TEQ 239
terminology 108
Tethys Sea 292
Threat mitigation 252, 268, 270
threatened 277, 314, 315
Threats 252, 255, 264–266, 268, 270
tidal cycle 124
TL 4
toxicity 226, 227, 231, 232, 235, 236, 239, 245
trade normalization 318
trade transparency 318
transcriptional markers 36
transcriptome 37, 39, 45, 47, 48
transcriptomics 45, 47–49
Transfers and stocking 161
transportation 293, 318
Traps 304
Triangle nets 297, 303
tropical 1, 4, 6, 11, 12, 15, 16
tropical eels 1, 4, 6, 12, 89–96, 103–105, 207, 208, 212–214, 218, 311, 320, 321
tropical glass eel 291, 317, 320, 321
tropical species 111, 113, 115, 117, 119, 131
typhoon 296

U

UK 297, 303, 304, 308, 309
uncertainty 316

upper reaches 293
upstream migration 143–147, 149, 150, 164, 297
U.S. Fish and Wildlife Service (USFWS) 251, 252

V

vasculotoxicity 47
vertebrae counts 42
VOCs 228
Vulnerable (VU) 314, 315

W

water quality 264–266, 270, 280
water temperature 112, 148–151, 159
Western Europe 52
wicker baskets 304, 309
wild stocks 80
wild-caught glass eels 308
wooden fence 304

Y

yellow and silver Japanese eels 313, 316, 317
yellow eel 294, 296, 297, 299, 303, 304, 308, 311, 321
Yellow eel fisheries 299, 303
yellow eel stage 294
yield-per-recruit (YPR) 316, 322
Yield-per-recruit (YPR) model 316
yolk 214, 215
YOY (young-of-the-year) abundance indices 260

Printed and bound by CPI Group (UK) Ltd, Croydon, CR0 4YY

01/11/2024

01782623-0009